Lecture Notes in Mathematics

Volume 2351

This series reports on new developments in all areas of mathematics and their applications - quickly, informally and at a high level. Mathematical texts analysing new developments in modelling and numerical simulation are welcome. The type of material considered for publication includes:

1. Research monographs
2. Lectures on a new field or presentations of a new angle in a classical field
3. Summer schools and intensive courses on topics of current research.

Texts which are out of print but still in demand may also be considered if they fall within these categories. The timeliness of a manuscript is sometimes more important than its form, which may be preliminary or tentative. Please visit the LNM Editorial Policy (https://drive.google.com/file/d/1MOg4TbwOSokRnFJ3ZR3ciEeKs9hOnNX_/view?usp=sharing)

Titles from this series are indexed by Scopus, Web of Science, Mathematical Reviews, and zbMATH.

Pascal Remy

Asymptotic Expansions and Summability

Application to Partial Differential Equations

Springer

Pascal Remy
Laboratoire de Mathématiques de Versailles
Université de Versailles Saint-Quentin
Versailles, France

ISSN 0075-8434 ISSN 1617-9692 (electronic)
Lecture Notes in Mathematics
ISBN 978-3-031-59093-1 ISBN 978-3-031-59094-8 (eBook)
https://doi.org/10.1007/978-3-031-59094-8

Mathematics Subject Classification: 35C10, 35C20, 40B05, 35G05, 35G20

This Springer imprint is published by the registered company Springer Nature Switzerland AG
The registered company address is: Gewerbestrasse 11, 6330 Cham, Switzerland

To Marie, Guillaume and Agathe

Preface

This book attempts to extend the classical and well-known summability theory of formal power series in $\mathbb{C}[[t]]$ to formal power series in $\mathcal{O}(D_{\rho_1,\ldots,\rho_n})[[t]]$, the set of formal power series in t with analytic coefficients in a neighborhood $D_{\rho_1,\ldots,\rho_n}$ of the origin of $\mathbb{C}^n$ in order to apply it to formal solutions of partial differential equations, both linear and nonlinear. It is intended for different types of readers: on one hand, those who are familiar, to a certain degree, with the various topics discussed and who wish either to find a proof or precise references, or to deepen their knowledge; on the other hand, those who are new to this theory and who will be able to find here all the tools necessary to apprehend these various subjects.

This book is organized as follows. In a first part, we develop in detail all the elements necessary for the theory: Taylor expansions (Chap. 2), Gevrey formal power series (Chap. 3) and Gevrey asymptotics (Chap. 4). After defining the notion of the k-summability of a formal power series $\widetilde{u}(t,x) \in \mathcal{O}(D_{\rho_1,\ldots,\rho_n})[[t]]$ in a given direction (Chap. 5), we present two characterizations: the first one based on appropriate asymptotic conditions on some successive derivatives (Chap. 6), and the second one based on the classical approach by Borel and Laplace operators (Chap. 7). This second characterization is then generalized in Chap. 9 by considering for the Borel and Laplace operators, more general kernel functions than exponentials. These functions are studied in detail in Chap. 8.

With a view to applying this theory to formal solutions of partial differential equations, each theoretical result is illustrated by a simple example on which we introduce different theoretical tools needed in this kind of study (Nagumo norms, Newton polygon, fixed point method, combinatorial methods, etc.). This example is then completed by various general results, given without proof, but with appropriate references where the details can be found, thus providing a review of existing results.

In the last chapter (Chap. 10), we briefly present the recent theory on moment partial differential equations. These equations, introduced by W. Balser and M. Yoshino in 2010, are obtained by replacing the usual derivations by some moment derivations, the moment functions being related to the kernel functions presented in Chap. 8. Doing that, one defines a new and wide class of functional equations which includes, among them, many classical equations such as the partial differential equations, the fractional partial differential equations, the q-difference equations, etc.; hence, their great interest. We first establish several important properties of the moment derivations, including an integral representation which generalizes the Cauchy Formula, and we show that these moment derivations are compatible with the summability in the following sense: if a formal power series is k-summable in a certain direction, then its moment derivative is still k-summable in this direction. Then, we show how the methods used to study the summability of formal solutions of partial differential equations can be extended to formal solutions of moment partial differential equations.

Some complementary results, as well as on questions related to our topic as on theoretical aspects, are presented at the end of the book (Part IV).

This book thus offers a self-contained presentation of the theory of the k-summability in $\mathcal{O}(D_{\rho_1,\dots,\rho_n})[[t]]$, as well as a presentation of the various approaches that can be used in the study of formal solutions of partial differential equations and moment partial differential equations, at least from the point of view of the k-summability, that of multisummability being addressed in a future work. We hope it will achieve the goal we set when we wrote it.

Houilles, France
February 2023

Pascal Remy

About this Book

In this book, we extend the classical and well-known summability theory of formal power series in $\mathbb{C}[[t]]$ to formal power series with analytic coefficients at the origin of $\mathbb{C}^n$, in order to apply it to formal solutions of some partial differential equations, both linear and nonlinear.

The elements necessary for this theory (Taylor expansions, Gevrey formal power series and Gevrey asymptotics) are thoroughly developed. Then, we present three characterizations of the summability based on different approaches: the first one requires appropriate asymptotic conditions, the second one uses the classical approach by the Borel and Laplace operators, and the last one generalizes this classical approach by considering, for the Borel and Laplace operators, more general kernel functions than exponentials. For each theoretical result, we develop a simple example in order to introduce different theoretical tools needed in this kind of study (Nagumo norms, Newton polygon, fixed point method, combinatorial methods, etc.). This example is then completed by various general results, given without proof, but with appropriate references where the details can be found, thus providing a review of existing results.

In addition, we briefly present the recent theory on moment partial differential equations. These equations, introduced by W. Balser and M. Yoshino in 2010 by replacing the usual derivations by some moment derivations, the moment functions being related to some convenient kernel functions used for the summability, offer a new and wide class of functional equations which includes, among them, many classical equations such as the partial differential equations, the fractional partial differential equations, the q-difference equations, etc.; hence, their great interest. We first establish several important properties of the moment derivations, including an integral representation which generalizes the Cauchy Formula, and we show that these moment derivations are compatible with the summability in the following sense: if a formal power series is k-summable in a certain direction, then its moment derivative is still k-summable in this direction. Then, we show how the methods used to study the summability of formal solutions of partial differential equations can be extended to formal solutions of moment partial differential equations.

This book is therefore intended for different types of readers: on one hand, those who are familiar, to a certain degree, with the various topics discussed and who wish either to find a proof or precise references, or to deepen their knowledge; on the other hand, those who are new to this theory and who will be able to find here all the tools necessary to apprehend these various subjects.

Contents

Chapter 1
Introduction

Partial differential equations, also called *evolution equations*, are fundamental in many physical, chemical, biological, and ecological problems to represent, for instance, the motion of the isolated waves, localized in a small part of space in many fields such as optical fibers, neural physics, solid state physics, hydrodynamics, diffusion process, plasma physics and nonlinear optics (heat equation, Klein-Gordon equation, Euler-Lagrange equation, Burgers equation, Korteweg-de Vries equation, Boussinesq equation, etc.). When studying such equations, one of the major challenges is the determination of exact solutions, if any exists, and the precise analysis of their properties (dynamic, asymptotic behavior, etc.) in order to have a better understanding of the mechanism of the underlying physical phenomena and dynamic processes. To do that, one possible way is given by the *summation theory of divergent power series*.

But, what does this theory cover?

Initially applied within the framework of linear analytic ordinary differential equations with an irregular singular point at the origin 0 of the complex plane $\mathbb{C}$, the *summation theory of divergent power series* provides a powerful tool for constructing analytic solutions from the formal ones. Roughly speaking, denoting by $k_1 < \ldots < k_n$ the $n \geqslant 1$ positive slopes of the Newton polygon of such a differential equation and choosing a convenient direction θ, it allows to associate with each of the formal power series $\widetilde{f}(t) \in \mathbb{C}[[t]]$ appearing in the formal solutions a unique analytic function f, called *the sum of* $\widetilde{f}$ *in the direction* θ, whose $\widetilde{f}$ is the $1/k_1$-Gevrey asymptotic expansion on a convenient sector with vertex 0, bisected by θ and with opening larger than π/k_n.

Developed in the 1980s and 1990s, mainly by J.-P. Ramis, J. Martinet, Y. Sibuya, B. Malgrange, W. Balser, M. Loday-Richaud and G. Pourcin, and *a priori* independently of any type of equations, the summation theory is in fact separated into two distinct sub-theories: the *summability theory* and the *multisummability theory*. In the particular case of the previous differential equations, the first one

P. Remy, *Asymptotic Expansions and Summability*, Lecture Notes in Mathematics 2351, https://doi.org/10.1007/978-3-031-59094-8_1

applies when the Newton polygon of the given equation admits a unique positive slope, and the second one when it admits at least two positive slopes.

There exist many different approaches to characterize the summability and the multisummability of a formal power series $\widetilde{f}(t) \in \mathbb{C}[[t]]$ in a given direction θ: some require appropriate asymptotic conditions, some are based on the use of Borel and Laplace-type operators, whereas some others rely on cohomological arguments; but all are equivalent in the sense that they provide the same sum [65, Chapters 5 and 7] (see also [4, 63, 71, 72, 74, 75, 100, 102]). Thereby, gathered together, these various approaches provide a well stocked toolbox to handle summable and multisummable formal power series in $\mathbb{C}[[t]]$, both from a theoretical point of view and from an effective point of view with the implementation of algorithms of summation (see for instance [27, 29, 51, 93, 127–129]).

The summability and multisummability theories apply to the formal solutions of the nonlinear differential equations as well [16]. On the other hand, they fail in the case of the difference and q-difference equations, where not all the formal solutions are summable or multisummable in the classical sense. In these two cases, new and specific types of summation processes are needed, for instance those considered by B. L. J. Braaksma, B. F. Faber and G. K. Immink in the case of the difference equations [17, 46] (see also [18]), and those introduced by F. Marotte and C. Zhang in the case of the q-difference equations [73] (see also [30]).

In the case of the partial differential equations, the question of the summability and of the multisummability of the formal solutions is relatively recent and dates back some twenty years. One of the classical problem is the following.

Problem 1.1 Let us consider a formally well-posed non-Kowalewskian Cauchy problem in 1-dimensional time variable $t \in \mathbb{C}$ and n-dimensional spatial variable $x = (x_1, \ldots, x_n) \in \mathbb{C}^n$ of the form

$$\begin{cases} \partial_t^{\kappa} u - G(t, x, (\partial_t^i \partial_x^q u)_{(i,q)\in\Lambda}) = \widetilde{f}(t, x) \\ \partial_t^j u(t, x)_{|t=0} = \varphi_j(x),\ j = 0, \ldots, \kappa - 1 \end{cases} \tag{1.1}$$

where

- $\kappa \geqslant 1$ is a positive integer;
- Λ is a nonempty finite subset of $\{0, \ldots, \kappa - 1\} \times \mathbb{N}^n$ ($\mathbb{N}$ denotes the set of the nonnegative integers);
- ∂_x^q denotes the derivation $\partial_{x_1}^{q_1} \ldots \partial_{x_n}^{q_n}$ while $q = (q_1, \ldots, q_n) \in \mathbb{N}^n$;
- $G(t, x, (Z_{i,q})_{(i,q)\in\Lambda})$ is analytic in the domain $D_{\rho_0} \times D_{\rho_1,\ldots,\rho_n} \times \prod_{(i,q)\in\Lambda} D_{\rho_{i,q}}$, where D_ρ and $D_{\rho_1,\ldots,\rho_n}$ stand respectively for the open disc with center $0 \in \mathbb{C}$ and radius $\rho > 0$, and for the polydisc $D_{\rho_1} \times \ldots \times D_{\rho_n}$ centered at the origin $(0, \ldots, 0) \in \mathbb{C}^n$;

- the inhomogeneity $\widetilde{f}(t,x)$ is a formal power series in t with analytic coefficients in $D_{\rho_1,\dots,\rho_n}$ (we denote this by $\widetilde{f}(t,x) \in \mathcal{O}(D_{\rho_1,\dots,\rho_n})[[t]]$) which may be smooth, or not;[1]
- the initial conditions $\varphi_j(x)$ are analytic on $D_{\rho_1,\dots,\rho_n}$ (we denote this by $\varphi_j(x) \in \mathcal{O}(D_{\rho_1,\dots,\rho_n})$) for all $j = 0, \dots, \kappa - 1$.

and let us denote by $\widetilde{u}(t,x) \in \mathcal{O}(D_{\rho_1,\dots,\rho_n})[[t]]$ its unique formal power series solution in the time variable.
Question: how to characterize the summability and the multisummability of $\widetilde{u}(t,x)$?

Remark 1.2 The existence and the uniqueness of the formal solution $\widetilde{u}(t,x)$ is guaranteed by the fact that the Cauchy problem (1.1) is *formally well-posed*.[2] Furthermore, the fact that Problem (1.1) is also *non-Kowalewskian* implies that this formal solution does not converge in general. Hence, the natural question of its summation.

Considering t as the variable and x as a parameter, the notions of summability and multisummability of formal power series in $\mathcal{O}(D_{\rho_1,\dots,\rho_n})[[t]]$ are obtained by extending the classical notions of summability and multisummability of formal power series in $\mathbb{C}[[t]]$ to families parametrized by x in requiring similar conditions, which are however uniform with respect to x.

To our knowledge, Problem 1.1 as stated above has never been studied before. However, there are many partial results in the case where the function G, and thus Eq. (1.1), is of a particular type.

The first summability result was formulated in the 1999 article [67] by D. A. Lutz, M. Miyake and R. Schäfke in the case where Eq. (1.1) is the linear heat equation

$$\begin{cases} \partial_t u = \partial_x^2 u \\ u(0,x) = \varphi(x) \end{cases}, (t,x) \in \mathbb{C}^2. \tag{1.2}$$

Following this article, a lot of work has been done in the last two decades on the summability and multisummability of the formal solutions of Eq. (1.1) in the linear case, with more or less strong conditions on the function G (see for instance [7, 10, 12, 26, 44, 45, 69, 77, 79, 82, 88, 96, 105, 125, 134] and the references therein). More recently, some summability results have also been established in some particular cases of nonlinearity [68, 98, 107–110].

In the various works dealing with the summability of the formal solution $\widetilde{u}(t,x)$ of Eq. (1.1), which is what interests us exclusively in the present book, two approaches are essentially used, providing thus two complementary characterizations of the summability of $\widetilde{u}(t,x)$. The first one, based on a characterization of

[1] We denote $\widetilde{f}$ with a tilde to emphasize the possible divergence of the series $\widetilde{f}$.

[2] This notion, introduced by W. Balser and V. Kostov in [9], extends to formal power series solutions the classical notion of *well-posed problem* of J. Hadamard [37].

the summability in terms of appropriate asymptotic conditions on some successive derivatives, provides a characterization of the summability of $\widetilde{u}(t, x)$ both in terms of the inhomogeneity $\widetilde{f}(t, x)$ and of a finite number of its formal coefficients $\partial_x^n \widetilde{u}(t, x)_{|x=0}$; the second one, based on a characterization of the summability in terms of Borel-Laplace-type operators, provides a characterization of the summability of $\widetilde{u}(t, x)$ both in terms of the inhomogeneity $\widetilde{f}(t, x)$ and of the initial data $\varphi_j(x)$.

Although it is known that these different approaches are equivalent since the theory of summability in $\mathcal{O}(D_{\rho_1,\ldots,\rho_n})[[t]]$ "copies" the one in $\mathbb{C}[[t]]$, there is to our knowledge no book similar to [65] in which these approaches are all developed and their equivalence proved, except for the book [6] of W. Balser. However, since the presentation chosen in the latter uses Banach spaces, the reader must systematically reformulate the results obtained to make them usable in the case of formal power series in $\mathcal{O}(D_{\rho_1,\ldots,\rho_n})[[t]]$. Moreover, apart from the heat equation (1.2), there is no mention of applications to partial differential equations.

Part I
Asymptotic Expansions

All along this work, we consider functions of $n + 1$ complex variables t and $x = (x_1, \ldots, x_n)$, and their asymptotic expansions with respect to the variable t at the given point $t_0 = 0$ of the Riemann sphere, x being a parameter. Such asymptotics are studied on sectors with vertex 0, which are drawn either in the complex plane $\mathbb{C}$, precisely, $\mathbb{C}^* = \mathbb{C}\setminus\{0\}$ (the functions are then *single-valued* or *univalued*) or on the Riemann surface $\widetilde{\mathbb{C}}$ of the logarithm (the functions are *multivalued* or given in terms of polar coordinates). We denote by

- $\Sigma = \Sigma_{\alpha,\beta}(R)$ the open sector with vertex 0 consisting of all points $t \in \mathbb{C}$ satisfying the conditions $\alpha < \arg(t) < \beta$ and $0 < |t| < R$ (see Fig. 1);
- $\overline{\Sigma} = \overline{\Sigma}_{\alpha,\beta}(R)$ the closure of Σ in $\mathbb{C}^*$ or in $\widetilde{\mathbb{C}}$ (0 is always excluded) and we use the term *closed sector*.

Recall that a sector Σ' is a *proper subsector* of Σ, and we denote this by $\Sigma' \Subset \Sigma$, if its closure $\overline{\Sigma}'$ in $\mathbb{C}^*$ or in $\widetilde{\mathbb{C}}$ is included in Σ. Thus, the notation $\Sigma'_{\alpha',\beta'}(R') \Subset \Sigma_{\alpha,\beta}(R)$ means $\alpha < \alpha' < \beta' < \beta$ and $0 < R' < R$.

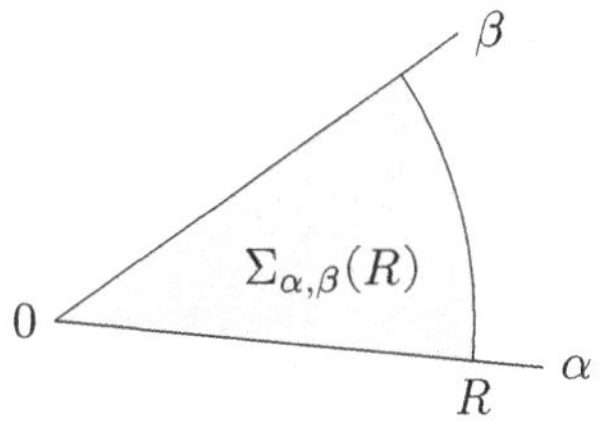

Fig. 1 A sector $\Sigma = \Sigma_{\alpha,\beta}(R)$

Chapter 2
Taylor Expansions

As in the case of formal power series in $\mathbb{C}[[t]]$ (see [65] for instance), the first step towards the summability of formal power series in $\mathcal{O}(D_{\rho_1,\dots,\rho_n})[[t]]$ is the *Poincaré asymptotic* which provides, if any exists, the *Taylor expansion* at $t=0$ of a holomorphic map $t \longmapsto u(t,x_0)$ on a given sector Σ, x_0 being an arbitrary fixed point on a convenient polydisc $D_{r,\dots,r}$.

In the sequel of this chapter, r denotes a positive real number and Σ an open sector with vertex 0 either in $\mathbb{C}^*$ or in $\widetilde{\mathbb{C}}$.

Definition 2.1 (Taylor Expansion) Let $u(t,x) \in \mathcal{O}(\Sigma \times D_{r,\dots,r})$ be a holomorphic function on $\Sigma \times D_{r,\dots,r}$ and

$$\widetilde{u}(t,x) = \sum_{j\geqslant 0} u_j(x)t^j \in \mathcal{O}(D_{r,\dots,r})[[t]]$$

a formal power series with analytic coefficients in $D_{r,\dots,r}$.

Given a fixed point $x_0 \in D_{r,\dots,r}$, the map $t \longmapsto u(t,x_0)$ is said to *admit* $\widetilde{u}(t,x_0)$ *as Taylor expansion at* 0 *on* Σ (or to *be asymptotic to* $\widetilde{u}(t,x_0)$ *at* 0 *on* Σ) if for all proper subsector $\Sigma' \Subset \Sigma$ of Σ and all $J \in \mathbb{N}^* = \mathbb{N}\backslash\{0\}$, there exists a positive constant $C_{x_0,\Sigma',J} > 0$ such that the following estimate holds for all $t \in \Sigma'$:

$$\left| u(t,x_0) - \sum_{j=0}^{J-1} u_j(x_0)t^j \right| \leqslant C_{x_0,\Sigma',J}\, |t|^J . \tag{2.1}$$

Observe that the constant $C_{x_0,\Sigma',J}$ may depend on x_0, Σ' and J, but no condition is required on the nature of this dependence. Observe also that the definition includes infinitely many estimates in each of which J is fixed. For instance, for $J=1$, they tell us that $t \longmapsto u(t,x_0)$ can be continuously continued at 0 on Σ with $u(0,x_0) = u_0(x_0)$; for $J=2$, they say that $t \longmapsto u(t,x_0)$ has a derivative at 0

P. Remy, *Asymptotic Expansions and Summability*, Lecture Notes
in Mathematics 2351, https://doi.org/10.1007/978-3-031-59094-8_2

on Σ, etc. However, all these estimates have nothing to do with the convergence or the divergence of the series as J goes to infinity.

The following lemma shows that the map $t \longmapsto u(t, x_0)$ admits at most one Taylor expansion at 0 on Σ.

Lemma 2.2 *Let $u(t, x) \in \mathcal{O}(\Sigma \times D_{r,\dots,r})$ and $\widetilde{u}(t, x), \widetilde{v}(t, x) \in \mathcal{O}(D_{r,\dots,r})[[t]]$:*

$$\widetilde{u}(t, x) = \sum_{j \geqslant 0} u_j(x)t^j, \quad \widetilde{v}(t, x) = \sum_{j \geqslant 0} v_j(x)t^j.$$

Assume that, for a fixed point $x_0 \in D_{r,\dots,r}$, the map $t \longmapsto u(t, x_0)$ admits both $\widetilde{u}(t, x_0)$ and $\widetilde{v}(t, x_0)$ as Taylor expansion at 0 on Σ.

Then, $u_j(x_0) = v_j(x_0)$ for all $j \geqslant 0$.

Proof Let us fix a proper subsector $\Sigma' \Subset \Sigma$ of Σ. From Definition 2.1, the following inequalities hold: for all $J \geqslant 1$, there exist positive constants $C_{x_0,\Sigma',J}, C'_{x_0,\Sigma',J} > 0$ such that, for all $t \in \Sigma'$,

$$\begin{aligned} \left| u(t, x_0) - \sum_{k=0}^{J-1} u_k(x_0)t^k \right| &\leqslant C_{x_0,\Sigma',J}\, |t|^J \\ \left| u(t, x_0) - \sum_{k=0}^{J-1} v_k(x_0)t^k \right| &\leqslant C'_{x_0,\Sigma',J}\, |t|^J \end{aligned} \tag{2.2}$$

Let us now prove the identities $u_j(x_0) = v_j(x_0)$ by induction on $j \geqslant 0$.

Applying first inequalities (2.2) with $J = 1$, we get

$$\begin{aligned} |u_0(x_0) - v_0(x_0)| &= |(u(t, x_0) - v_0(x_0)) - (u(t, x_0) - u_0(x_0))| \\ &\leqslant (C_{x_0,\Sigma',1} + C'_{x_0,\Sigma',1})\, |t| \end{aligned}$$

for all $t \in \Sigma'$; hence, $u_0(x_0) = v_0(x_0)$ by making t tend to 0.

Let us now suppose that identities $u_k(x_0) = v_k(x_0)$ hold for all $k = 0, \dots, j$ for a certain $j \geqslant 0$. Applying then inequalities (2.2) with $J = j + 2$, we get

$$\begin{aligned} \left|u_{j+1}(x_0) - v_{j+1}(x_0)\right| &= \left| \left(u(t, x_0) - \sum_{k=0}^{j+1} v_k(x_0)t^j \right) - \left(u(t, x_0) - \sum_{k=0}^{j+1} u_k(x_0)t^j \right) \right| \\ &\leqslant (C_{x_0,\Sigma',j+2} + C'_{x_0,\Sigma',j+2})\, |t|^{j+2} \end{aligned}$$

for all $t \in \Sigma'$; hence $u_{j+1}(x_0) = v_{j+1}(x_0)$ by making t tend again to 0.

This achieves the proof of Lemma 2.2. ■

Notation 2.3 We denote by

- $\mathcal{T}_r(\Sigma)$ the set of all the functions $u(t,x) \in \mathcal{O}(\Sigma \times D_{r,\dots,r})$ satisfying the following property: there exists a formal power series $\widetilde{u}(t,x) \in \mathcal{O}(D_{r,\dots,r})[[t]]$ such that, for any $x \in D_{r,\dots,r}$, the map $t \longmapsto u(t,x)$ admits $\widetilde{u}(t,x)$ as Taylor expansion at 0 on Σ;
- $T_{r,\Sigma} : \mathcal{T}_r(\Sigma) \longrightarrow \mathcal{O}(D_{r,\dots,r})[[t]]$ the map assigning to each $u \in \mathcal{T}_r(\Sigma)$ its *"Taylor expansion at* 0 *on* Σ", that is the formal power series in $\mathcal{O}(D_{r,\dots,r})[[t]]$ satisfying the property above.

Observe that the coefficient $u_j(x)$ of the Taylor expansion $\widetilde{u}(t,x)$ may be holomorphic on a common polydisc $D_{\rho_1,\dots,\rho_n}$ "greater" than $D_{r,\dots,r}$, that is satisfying $\min(\rho_1,\dots,\rho_n) \geqslant r$. Observe also that, due to the Analytic Continuation Principle, Lemma 2.2 implies the uniqueness of the Taylor expansion; hence, the fact that the map $T_{r,\Sigma}$ is well-defined.

Proposition 2.4 below specifies the algebraic structure of $\mathcal{T}_r(\Sigma)$.

Proposition 2.4

- *The set $\mathcal{T}_r(\Sigma)$ endowed with the usual algebraic operations and the usual derivation ∂_t is a $\mathbb{C}$-differential algebra. Moreover, it is stable under the anti-derivation ∂_t^{-1}.*
- *The map $T_{r,\Sigma} : \mathcal{T}_r(\Sigma) \longrightarrow \mathcal{O}(D_{r,\dots,r})[[t]]$ is a morphism of $\mathbb{C}$-differential algebras which commutes with ∂_t^{-1}.*

Proof Since $(\mathcal{O}(\Sigma \times D_{r,\dots,r}), +, \times, \partial_t)$ is a $\mathbb{C}$-differential algebra, it is sufficient to prove that $\mathcal{T}_r(\Sigma)$ is a $\mathbb{C}$-differential subalgebra of $\mathcal{O}(\Sigma \times D_{r,\dots,r})$ stable under the anti-derivation ∂_t^{-1}. To this end, let us consider $u, v \in \mathcal{T}_r(\Sigma)$ and

$$\widetilde{u}(t,x) = \sum_{j\geqslant 0} u_j(x)t^j, \quad \widetilde{v}(t,x) = \sum_{j\geqslant 0} v_j(x)t^j$$

the formal power series in $\mathcal{O}(D_{r,\dots,r})[[t]]$ such that $T_{r,\Sigma}(u) = \widetilde{u}$ and $T_{r,\Sigma}(v) = \widetilde{v}$.

Let us fix $x_0 \in D_{r,\dots,r}$ and let us consider, for all $J \geqslant 1$, the functions U_J and V_J defined on Σ by

$$U_J(t) = u(t,x_0) - \sum_{j=0}^{J-1} u_j(x_0)t^j \quad \text{and} \quad V_J(t) = v(t,x_0) - \sum_{j=0}^{J-1} v_j(x_0)t^j.$$

By assumption, we have the following conditions (see Definition 2.1): for all proper subsector $\Sigma' \Subset \Sigma$ of Σ, there exist two positive constants $C_{x_0,\Sigma',J}, C'_{x_0,\Sigma',J} > 0$ such that, for all $t \in \Sigma'$,

$$|U_J(t)| \leqslant C_{x_0,\Sigma',J}\, |t|^J \quad \text{and} \quad |V_J(t)| \leqslant C'_{x_0,\Sigma',J}\, |t|^J . \tag{2.3}$$

◁ *Stability under linear combination.* Let $\lambda, \mu \in \mathbb{C}$, $\Sigma' \Subset \Sigma$ and $J \geqslant 1$. Applying inequalities (2.3), it is clear that

$$\left|(\lambda u + \mu v)(t, x_0) - \sum_{j=0}^{J-1} (\lambda u_j(x_0) + \mu v_j(x_0)) t^j\right| = |\lambda U_J(t) + \mu V_J(t)|$$

$$\leqslant \Big(|\lambda|\, C_{x_0,\Sigma',J} + |\mu|\, C'_{x_0,\Sigma',J}\Big) |t|^J$$

for all $t \in \Sigma'$. Hence, the map $t \longmapsto (\lambda u + \mu v)(t, x_0)$ admits the formal series $(\lambda\widetilde{u} + \mu\widetilde{v})(t, x_0)$ as Taylor expansion at 0 on Σ.

◁ *Stability under multiplication.* Let $\Sigma' \Subset \Sigma$ and $J \geqslant 1$. For all $t \in \Sigma'$, we first have the identity

$$(uv)(t, x_0) - \sum_{j=0}^{J-1} \left(\sum_{k+\ell=j} u_k(x_0) v_\ell(x_0)\right) t^j = U_J(t) V_J(t) + U_J(t) \sum_{j=0}^{J-1} v_j(x_0) t^j$$

$$+ V_J(t) \sum_{j=0}^{J-1} u_j(x_0) t^j + t^J \sum_{j=J}^{2J-2} \left(\sum_{k+\ell=j} u_k(x_0) v_\ell(x_0)\right) t^{j-J}.$$

Applying then inequalities (2.3) and the fact that $|t| \leqslant R$, where $R > 0$ stands for the radius of Σ, we get

$$\left|(uv)(t, x_0) - \sum_{j=0}^{J-1} \left(\sum_{k+\ell=j} u_k(x_0) v_\ell(x_0)\right) t^j\right| \leqslant C''_{x_0,\Sigma',J}\, |t|^J$$

for all $t \in \Sigma'$, with a convenient positive constant $C''_{x_0,\Sigma',J} > 0$ independent of t. Consequently, the map $t \longmapsto (uv)(t, x_0)$ admits the formal series $(\widetilde{uv})(t, x_0)$ as Taylor expansion at 0 on Σ.

◁ *Stability under the derivation* ∂_t. Let $\Sigma' \Subset \Sigma$ and $J \geqslant 1$. We must prove that there exists a positive constant $A_{x_0,\Sigma',J} > 0$ such that

$$\left|\partial_t u(t, x_0) - \sum_{j=0}^{J-1} (j+1) u_{j+1}(x_0) t^j\right| \leqslant A_{x_0,\Sigma',J}\, |t|^J$$

for all $t \in \Sigma'$, that is

$$|\partial_t U_{J+1}(t)| \leqslant A_{x_0,\Sigma',J}\, |t|^J$$

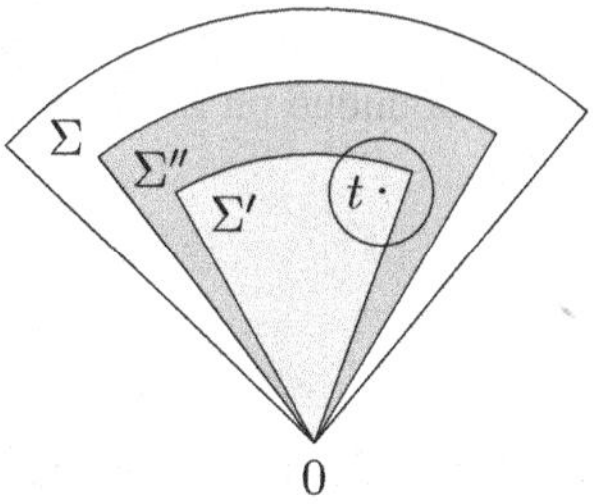

Fig. 2.1 The sectors $\Sigma' \Subset \Sigma'' \Subset \Sigma$

for all $t \in \Sigma'$. To do that, let us choose a sector Σ'' such that $\Sigma' \Subset \Sigma'' \Subset \Sigma$, and let δ be so small that, for all $t \in \Sigma'$, the closed disc $\overline{D}(t, |t|\,\delta)$ with center t and radius $|t|\,\delta$ be contained in Σ'' (see Fig. 2.1).

By assumption, we have $\left|U_{J+1}(t')\right| \leqslant C_{x_0,\Sigma'',J+1}\left|t'\right|^{J+1}$ for all $t' \in \overline{D}(t, |t|\,\delta)$ and all $t \in \Sigma'$. Applying then the Cauchy Integral Formula

$$\partial_t U_{J+1}(t) = \frac{1}{2i\pi}\int_{|t'-t|=|t|\delta} \frac{U_{J+1}(t')}{(t'-t)^2}dt',$$

we deduce, for all $t \in \Sigma'$, the inequalities

$$\begin{aligned}
|\partial_t U_{J+1}(t)| &\leqslant \frac{1}{2\pi}\max_{|t'-t|=|t|\delta}\left|U_{J+1}(t')\right|\frac{2\pi\,|t|\,\delta}{(|t|\,\delta)^2}\\
&\leqslant \frac{C_{x_0,\Sigma'',J+1}}{|t|\,\delta}(|t|\,(1+\delta))^{J+1}\\
&= A_{x_0,\Sigma',J}\,|t|^J \quad \text{with } A_{x_0,\Sigma',J} = \frac{C_{x_0,\Sigma'',J+1}(1+\delta)^{J+1}}{\delta}.
\end{aligned}$$

Thereby, the map $t \longmapsto \partial_t u(t, x_0)$ admits the formal series $\partial_t\widetilde{u}(t, x_0)$ as Taylor expansion at 0 on Σ.

◁ *Stability under the anti-derivation* ∂_t^{-1}. Let $\Sigma' \Subset \Sigma$ and $J \geqslant 1$. Writing $\partial_t^{-1}\widetilde{u}(t, x_0)$ in the form

$$\partial_t^{-1}\widetilde{u}(t, x_0) = \sum_{j\geqslant 0} u'_j t^j$$

with $u'_0 = 0$ and $u'_j = u_{j-1}(x_0)/j$ for all $j \geqslant 1$, we must prove there exists a positive constant $A_{x_0,\Sigma',J} > 0$ such that

$$\left|\partial_t^{-1}u(t, x_0) - \sum_{j=0}^{J-1} u'_j t^j\right| \leqslant A_{x_0,\Sigma',J}\,|t|^J \quad \text{for all } t \in \Sigma'.$$

The case $J = 1$ is obvious since the map $t \longmapsto u(t, x_0)$ is continuous on $\overline{\Sigma}' \cup \{0\}$; hence, bounded on Σ'. For $J \geqslant 2$, we have

$$\partial_t^{-1} u(t, x_0) - \sum_{j=0}^{J-1} u'_j t^j = \partial_t^{-1} U_{J-1}(t)$$

for all $t \in \Sigma'$; hence, using inequalities (2.3):

$$\left| \partial_t^{-1} u(t, x_0) - \sum_{j=0}^{J-1} u'_j t^j \right| \leqslant \frac{C_{x_0, \Sigma', J-1}}{J} |t|^J$$

for all $t \in \Sigma'$. Consequently, the map $t \longmapsto \partial_t^{-1} u(t, x_0)$ admits the formal series $\partial_t^{-1} \widetilde{u}(t, x_0)$ as Taylor expansion at 0 on Σ. ■

Remark 2.5 As we can see in the proof of Proposition 2.4, the technical condition "*for all* $\Sigma' \Subset \Sigma$" imposed in Definition 2.1 plays a fundamental role. Indeed, if the estimates were required for the sector Σ itself and not for any proper subsector $\Sigma' \Subset \Sigma$, then the Cauchy Integral Formula could not be applied and we could not assert any longer that the algebra $\mathcal{T}_r(\Sigma)$ is differential, which would be not appropriate to handle solutions of partial differential equations.

Remark 2.6 It is also worth to notice that the proofs of the stability under the derivation ∂_t and under the anti-derivation ∂_t^{-1} can not be extended to the derivations ∂_{x_ℓ} and the anti-derivations $\partial_{x_\ell}^{-1}$ since the conditions given in Definition 2.1 on the parameter x are too weak. In particular, the algebras $\mathcal{T}_r(\Sigma)$ are not stable under the derivations ∂_{x_ℓ} and under the anti-derivations $\partial_{x_\ell}^{-1}$.

The following proposition provides a necessary and sufficient condition for a holomorphic function $u(t, x) \in \mathcal{O}(\Sigma \times D_{r,\dots,r})$ to belong to $\mathcal{T}_r(\Sigma)$.

Proposition 2.7 *A function $u(t, x)$ belongs to $\mathcal{T}_r(\Sigma)$ if and only if $u(t, x)$ belongs to $\mathcal{O}(\Sigma \times D_{r,\dots,r})$ and a sequence $(u_\ell) \in (\mathcal{O}(D_{r,\dots,r}))^{\mathbb{N}}$ of holomorphic functions on $D_{r,\dots,r}$ exists such that*

$$\forall x \in D_{r,\dots,r}, \forall \ell \geqslant 0, \forall \Sigma' \Subset \Sigma : \quad \frac{1}{\ell!} \lim_{\substack{t \to 0 \\ t \in \Sigma'}} \partial_t^\ell u(t, x) = u_\ell(x).$$

Proof ◁*The* only if *part.* Let $u(t, x) \in \mathcal{T}_r(\Sigma)$ and

$$\widetilde{u}(t, x) = \sum_{j \geqslant 0} u_j(x) t^j \in \mathcal{O}(D_{r,\dots,r})[[t]]$$

such that $T_{r,\Sigma}(u) = \widetilde{u}$. From Proposition 2.4, we first derive for all $\ell \geqslant 1$ that $\partial_t^\ell u(t,x) \in \mathcal{T}_r(\Sigma)$ with $T_{r,\Sigma}(\partial_t^\ell u) = \partial_t^\ell \widetilde{u}$, where

$$\partial_t^\ell \widetilde{u}(t,x) = \sum_{j \geqslant 0} (j+\ell)\ldots(j+1) u_{j+\ell}(x) t^j.$$

The identities follow then by applying Definition 2.1 with $J = 1$.

◁*The* if *part.* Let us now consider $x \in D_{r,\ldots,r}$, $\Sigma' \Subset \Sigma$ and $J \geqslant 1$. The map $t \longmapsto u(t,x)$ admits the Taylor expansion with integral remainder

$$u(t,x) - \sum_{j=0}^{J-1} \partial_t^j u(t_0,x) \frac{(t-t_0)^j}{j!} = \int_{t_0}^{t} \partial_t^J u(t',x) \frac{(t-t')^{J-1}}{(J-1)!} dt'$$

for all $t, t_0 \in \Sigma'$. Observe that, a priori, we can not write such a formula for $t_0 = 0$ since 0 does not even belong to the definition set of u. However, by assumption, the limit of the left-hand side as t_0 tends to 0 in Σ' exists; hence, the limit of the right-hand side exists too and we can write

$$u(t,x) - \sum_{j=0}^{J-1} u_j(x) t^j = \int_0^t \partial_t^J u(t',x) \frac{(t-t')^{J-1}}{(J-1)!} dt'$$

for all $t \in \Sigma'$. Consequently,

$$\left| u(t,x) - \sum_{j=0}^{J-1} u_j(x) t^j \right| \leqslant \frac{|t|^J}{J!} \sup_{t' \in \Sigma'} \left| \partial_t^J u(t',x) \right| \leqslant C_{x,\Sigma',J} \, |t|^J$$

for all $t \in \Sigma'$, the constant $C_{x,\Sigma',J} = \frac{1}{J!} \sup_{t' \in \Sigma'} \left| \partial_t^J u(t',x) \right|$ being finite by assumption; hence the conclusion of the proof. ■

As an application of Proposition 2.7, we have the following two results. The first one tells us that several functions in $\mathcal{T}_r(\Sigma)$ may admit the same Taylor expansion at 0 on Σ; the second one provides us a sufficient condition for which a function $u(t,x) \in \mathcal{O}(\Sigma \times D_{r,\ldots,r})$ belongs to $\mathcal{T}_r(\Sigma')$ for any $\Sigma' \Subset \Sigma$. This last result will be very useful later for the study of the summability of the fomal power series solutions of the partial differential equations.

Corollary 2.8 *The linear map $T_{r,\Sigma}$ is not injective.*

Proof By means of a rotation, we can first assume that Σ is bisected by the direction $\arg(t) = 0$; hence,

$$\Sigma = \left\{ t; 0 < |t| < R \text{ and } |\arg(t)| < \frac{s\pi}{2} \right\}$$

for two convenient positive constants $R, s > 0$. Then, choosing the principal determination of the logarithm in $]-\pi, \pi]$, the function

$$u(t,x) = \exp\left(-t^{-1/s} + x_1 + \ldots + x_n\right) \in \mathcal{O}(\Sigma \times D_{r,\ldots,r})$$

satisfies the following property:

$$\forall x \in D_{r,\ldots,r}, \forall \ell \geqslant 0, \forall \Sigma' \Subset \Sigma : \quad \lim_{\substack{t \to 0 \\ t \in \Sigma'}} \partial_t^{\ell} u(t,x) = 0.$$

Hence, applying Proposition 2.7: $u \in \mathcal{T}_r(\Sigma)$ and $T_{r,\Sigma}(u) = 0$, which achieves the proof. ∎

Corollary 2.9 *Let $u(t,x) \in \mathcal{O}(\Sigma \times D_{r,\ldots,r})$.*

Let us assume that u and all its derivatives $\partial_t^{\ell} u$ for $\ell \geqslant 1$ are bounded on $\Sigma \times D_{r,\ldots,r}$.

Then, $u \in \mathcal{T}_r(\Sigma')$ for any proper subsector $\Sigma' \Subset \Sigma$.

Proof Let us fix a proper subsector $\Sigma' \Subset \Sigma$.

For all $x \in D_{r,\ldots,r}$ and all $\ell \geqslant 0$, the maps $t \longmapsto \partial_t^{\ell} u(t,x)$ are bounded on Σ. Therefore, the Removable Singularities Theorem applies and implies the existence of the limits $\lim_{\substack{t \to 0 \\ t \in \Sigma'}} \partial_t^{\ell} u(t,x)$ for all $x \in D_{r,\ldots,r}$ and all $\ell \geqslant 0$.

Let us denote by $u_\ell(x)$ these limits and let us prove that $u_\ell \in \mathcal{O}(D_{r,\ldots,r})$ for all $\ell \geqslant 0$. To do that, let us consider a sequence $(t_n) \in (\Sigma')^{\mathbb{N}}$ such that $t_n \to 0$, and, for a fixed $\ell \geqslant 0$, the sequence $(U_n^{(\ell)}) \in (\mathcal{O}(D_{r,\ldots,r}))^{\mathbb{N}}$ of holomorphic functions on $D_{r,\ldots,r}$ defined by $U_n^{(\ell)}(x) = \partial_t^{\ell} u(t_n, x)$. By assumption, there exists a positive constant $C_\ell > 0$ such that

$$\sup_{n \geqslant 0} \sup_{x \in D_{r,\ldots,r}} \left| U_n^{(\ell)}(x) \right| \leqslant C_\ell,$$

that is the sequence $(U_n^{(\ell)})$ is a bounded sequence of the space $\mathcal{O}(D_{r,\ldots,r})$. Applying then the Montel's Theorem, we deduce that $(U_n^{(\ell)})$ is locally compact, that is it admits a subsequence which converges to a holomorphic function $U^{(\ell)} \in \mathcal{O}(D_{r,\ldots,r})$. By unicity of the limit, we deduce that $U^{(\ell)} = u_\ell$; hence $u_\ell \in \mathcal{O}(D_{r,\ldots,r})$.

Corollary 2.9 follows then from Proposition 2.7. ∎

Chapter 3
Gevrey Formal Power Series

As it is already the case when working with ordinary differential equations, it appears soon that the Poincaré asymptotics is not sufficient to study the summability of formal power series solutions of partial differential equations. Indeed, the conditions required are too weak to provide functions that are solutions of the equation when the Taylor expansions themselves are solutions (recall that the algebras $\mathcal{T}_r(\Sigma)$ are not stable under the derivations ∂_{x_ℓ}, see Remark 2.6) or, better, to set a 1-to-1 correspondence between the functions solution and their Taylor expansion (which is not currently the case, see Corollary 2.8). This is due to the fact that the constants $C_{x_0,\Sigma',J}$ given in Definition 2.1 are not uniform with respect to x, and that their dependence in J is not known.

To get around these problems, it is necessary to strengthen Poincaré asymptotics into the so-called Gevrey[1] asymptotics. This is moreover motivated by the Maillet type theorems which provide growth conditions on the coefficients $u_j(x)$ of the formal power series solutions of partial differential equations, conditions which are uniform with respect to x (see for instance [38, 87, 111, 124, 135]).

In this section, we are first interested in the formal series in $\mathcal{O}(D_{\rho_1,\ldots,\rho_n})[[t]]$ which verify this type of conditions and which are classically referred as *Gevrey formal power series*.

[1] Maurice Gevrey, 1884–1957, was a French mathematician who introduced the now called "Gevrey classes" in 1918 after he defended his thesis *Sur les équations aux dérivées partielles du type parabolique* at Université de Paris in 1913.

P. Remy, *Asymptotic Expansions and Summability*, Lecture Notes in Mathematics 2351, https://doi.org/10.1007/978-3-031-59094-8_3

3.1 Algebras of Gevrey Formal Power Series

To define the notion of *Gevrey formal power series in* $\mathcal{O}(D_{\rho_1,\dots,\rho_n})[[t]]$, one extends the classical notion of *Gevrey formal power series in* $\mathbb{C}[[t]]$ to families parametrized by x in requiring similar conditions, the estimates being uniform with respect to x.

Definition 3.1 (s-Gevrey Formal Series) Let $s \geqslant 0$ be. A formal power series

$$\widetilde{u}(t,x) = \sum_{j \geqslant 0} u_j(x) t^j \in \mathcal{O}(D_{\rho_1,\dots,\rho_n})[[t]]$$

is said to be *Gevrey of order* s (in short, s*-Gevrey*) if there exist three positive constants $0 < r \leqslant \min(\rho_1, \dots, \rho_n)$, $C > 0$ and $K > 0$ such that the inequalities

$$\left|u_j(x)\right| \leqslant C K^j \Gamma(1+sj) \tag{3.1}$$

hold for all $x \in D_{r,\dots,r}$ and all $j \geqslant 0$.

In other words, Definition 3.1 means that $\widetilde{u}(t,x)$ is s-Gevrey in t, uniformly in x on a neighborhood of $x = (0, \dots, 0) \in \mathbb{C}^n$.

Remark 3.2 According to the Stirling Formula, the term $\Gamma(1+sj)$ may be replaced by $(j!)^s$ in inequality (3.1).

Notation 3.3 We denote by $\mathcal{O}(D_{\rho_1,\dots,\rho_n})[[t]]_s$ the set of all the formal series in $\mathcal{O}(D_{\rho_1,\dots,\rho_n})[[t]]$ which are s-Gevrey.

Observe that the set $\mathbb{C}\{t,x\}$ of germs of analytic functions at the origin of $\mathbb{C}^{n+1}$ coincides with the union $\bigcup_{\rho_1>0,\dots,\rho_n>0} \mathcal{O}(D_{\rho_1,\dots,\rho_n})[[t]]_0$. In particular, any element of $\mathcal{O}(D_{\rho_1,\dots,\rho_n})[[t]]_0$ is convergent and $\mathbb{C}\{t,x\} \cap \mathcal{O}(D_{\rho_1,\dots,\rho_n})[[t]] = \mathcal{O}(D_{\rho_1,\dots,\rho_n})[[t]]_0$. More precisely, if $\widetilde{u}(t,x) \in \mathcal{O}(D_{\rho_1,\dots,\rho_n})[[t]]_0$ with the condition

$$\left|u_j(x)\right| \leqslant C K^j$$

for all $x \in D_{r,\dots,r}$ and all $j \geqslant 0$ for some radius $0 < r \leqslant \min(\rho_1, \dots, \rho_n)$, then $\widetilde{u}(t,x)$ defines an analytic function on $D_{1/K} \times D_{r,\dots,r}$.

Observe also that, according to the classical property

$$\Gamma(a) \leqslant \Gamma(b) \quad \text{for all } a \geqslant 1 \text{ and all } b \geqslant \max(2,a), \tag{3.2}$$

which is a direct consequence of the increase of the Gamma function on $[2, +\infty[$ and of the relation $\Gamma(a) \leqslant 1 = \Gamma(2)$ for all $a \in [1,2]$, the relations

$$\Gamma(1+sj) \leqslant \Gamma(2+s'j) = (1+s'j)\Gamma(1+s'j) \leqslant e^{1+s'j}\Gamma(1+s'j)$$

hold for all $0 \leqslant s < s'$ and all $j \geqslant 0$, and, consequently, the sets $\mathcal{O}(D_{\rho_1,\ldots,\rho_n})[[t]]_s$ are filtered as follows:

$$\mathcal{O}(D_{\rho_1,\ldots,\rho_n})[[t]]_0 \subset \mathcal{O}(D_{\rho_1,\ldots,\rho_n})[[t]]_s \subset \mathcal{O}(D_{\rho_1,\ldots,\rho_n})[[t]]_{s'} \subset \mathcal{O}(D_{\rho_1,\ldots,\rho_n})[[t]]$$

for all s and s' satisfying $0 < s < s' < +\infty$. Moreover, one has also the inclusion

$$\mathcal{O}(D_{\rho_1,\ldots,\rho_n})[[t]]_s \subset \mathcal{O}(D_{\rho'_1,\ldots,\rho'_n})[[t]]_s$$

for all $s \geqslant 0$ and any radii $0 < \rho'_i \leqslant \rho_i$.

Following Proposition 3.4 specifies the algebraic structure of $\mathcal{O}(D_{\rho_1,\ldots,\rho_n})[[t]]_s$.

Proposition 3.4 *Let $s \geqslant 0$. The set $\mathcal{O}(D_{\rho_1,\ldots,\rho_n})[[t]]_s$ endowed with the usual algebraic operations and the usual derivations ∂_t and ∂_{x_ℓ} with $\ell = 1, \ldots, n$ is a $\mathbb{C}$-differential algebra. Moreover, it is stable under the anti-derivations ∂_t^{-1} and $\partial_{x_\ell}^{-1}$ with $\ell = 1, \ldots, n$.*

Proof Since $(\mathcal{O}(D_{\rho_1,\ldots,\rho_n})[[t]], +, \times, \partial_t, \partial_{x_1}, \ldots, \partial_{x_n})$ is a $\mathbb{C}$-differential algebra, it is sufficient to prove that $\mathcal{O}(D_{\rho_1,\ldots,\rho_n})[[t]]_s$ is a $\mathbb{C}$-differential subalgebra of $\mathcal{O}(D_{\rho_1,\ldots,\rho_n})[[t]]$ stable under the anti-derivations ∂_t^{-1} and $\partial_{x_\ell}^{-1}$ with $\ell = 1, \ldots, n$. To this end, let us consider two s-Gevrey formal series

$$\widetilde{u}(t,x) = \sum_{j \geqslant 0} u_j(x) t^j, \quad \widetilde{v}(t,x) = \sum_{j \geqslant 0} v_j(x) t^j \in \mathcal{O}(D_{\rho_1,\ldots,\rho_n})[[t]]_s.$$

We can always assume that the estimates of Definition 3.1 on the coefficients $u_j(x)$ and $v_j(x)$ are satisfied with the same positive constants $0 < r \leqslant \min(\rho_1, \ldots, \rho_n)$ and $C, K > 0$: for all $x \in D_{r,\ldots,r}$ and all $j \geqslant 0$,

$$\left|u_j(x)\right| \leqslant C K^j \Gamma(1+sj) \quad \text{and} \quad \left|v_j(x)\right| \leqslant C K^j \Gamma(1+sj). \tag{3.3}$$

• *Differential subalgebra.*

◁ *Stability under linear combination.* Let $\lambda, \mu \in \mathbb{C}$. Applying inequalities (3.3), it is clear that

$$\left|\lambda u_j(x) + \mu v_j(x)\right| \leqslant (|\lambda| + |\mu|) C K^j \Gamma(1+sj)$$

hold for all $x \in D_{r,\ldots,r}$ and all $j \geqslant 0$. Hence, $\lambda\widetilde{u} + \mu\widetilde{v} \in \mathcal{O}(D_{\rho_1,\ldots,\rho_n})[[t]]_s$.

◁ *Stability under multiplication.* The product $\widetilde{w} = \widetilde{u}\widetilde{v}$ is the formal power series

$$\widetilde{w}(t,x) = \sum_{j \geqslant 0} w_j(x) t^j \in \mathcal{O}(D_{\rho_1,\ldots,\rho_n})[[t]]$$

where

$$w_j(x) = \sum_{k=0}^{j} u_k(x) v_{j-k}(x)$$

for all $x \in D_{\rho_1,\dots,\rho_n}$ and all $j \geqslant 0$. From inequalities (3.3), we derive the inequalities

$$\left|w_j(x)\right| \leqslant C^2 K^j \sum_{k=0}^{j} \Gamma(1+sk)\Gamma(1+s(j-k)) = C^2 K^j \Gamma(1+sj) \sum_{k=0}^{j} \frac{1}{\binom{sj}{sk}}$$

for all $x \in D_{r,\dots,r}$ and all $j \geqslant 0$, where $\binom{sj}{sk}$ stands for a generalized binomial coefficient (see Chap. 13). Applying then Proposition 13.6 when $s > 0$, we get

$$\left|w_j(x)\right| \leqslant \begin{cases} C^2 K^j (j+1) \leqslant eC^2 (eK)^j & \text{if } s = 0 \\ C_s C^2 K^j \Gamma(1+sj) & \text{if } s > 0 \end{cases}$$

for all $x \in D_{r,\dots,r}$ and all $j \geqslant 0$, and, consequently, $\widetilde{w} \in \mathcal{O}(D_{\rho_1,\dots,\rho_n})[[t]]_s$.

◃ *Stability under the derivation* ∂_t. The derivative $\widetilde{w} = \partial_t \widetilde{u}$ is the formal power series

$$\widetilde{w}(t,x) = \sum_{j \geqslant 0} w_j(x) t^j \in \mathcal{O}(D_{\rho_1,\dots,\rho_n})[[t]]$$

with $w_j(x) = (j+1)u_{j+1}(x)$ for all $x \in D_{\rho_1,\dots,\rho_n}$ and all $j \geqslant 0$. From inequalities (3.3) and property (3.2), we get the relations

$$\left|w_j(x)\right| \leqslant (j+1) C K^{j+1} \Gamma(1+sj+s) \leqslant (j+1) C K^{j+1} \Gamma(1+sj+\lceil s \rceil)$$

for all $x \in D_{r,\dots,r}$ and all $j \geqslant 0$, where $\lceil s \rceil$ stands for the ceil of s. Applying then $\lceil s \rceil$ times the recurrence relation $\Gamma(1+z) = z\Gamma(z)$, we finally get

$$\begin{aligned} \left|w_j(x)\right| &\leqslant (j+1) \prod_{k=1}^{\lceil s \rceil} (sj+k) \times C K^{j+1} \Gamma(1+sj) \\ &\leqslant e^{1+\lceil s \rceil(\lceil s \rceil+1)/2} C K \left(e^{1+s\lceil s \rceil} K\right)^j \Gamma(1+sj) \end{aligned}$$

for all $x \in D_{r,\dots,r}$ and all $j \geqslant 0$, which proves that $\partial_t \widetilde{u}$ belongs to $\mathcal{O}(D_{\rho_1,\dots,\rho_n})[[t]]_s$.

◁ *Stability under the derivation* ∂_{x_ℓ}. Let us fix $\ell \in \{1, \dots, n\}$. The derivative $\widetilde{w} = \partial_{x_\ell}\widetilde{u}$ is the formal power series

$$\widetilde{w}(t, x) = \sum_{j \geqslant 0} w_j(x)t^j \in \mathcal{O}(D_{\rho_1,\dots,\rho_n})[[t]]$$

with $w_j(x) = \partial_{x_\ell} u_j(x)$ for all $x \in D_{\rho_1,\dots,\rho_n}$ and all $j \geqslant 0$. For a given radius $0 < r' < r$, the Cauchy Integral Formula implies the relation

$$w_j(x) = \frac{1}{(2i\pi)^n} \int_{\gamma(x)} \frac{u_j(x')}{(x'_\ell - x_\ell)^2 \prod_{\substack{k=1\\k\neq\ell}}^{n} (x'_k - x_k)} dx'$$

for all $x \in D_{r',\dots,r'}$ and all $j \geqslant 0$, where $\gamma(x)$ is the path defined by

$$\gamma(x) = \{x' = (x'_1, \dots, x'_n) \in \mathbb{C}^n;\ |x'_k - x_k| = r - r' \text{ for all } k \in \{1, \dots, n\}\}.$$

Observe that $x' \in D_{r,\dots,r}$ for all $x' \in \gamma(x)$. Therefore, applying (3.3), we get the inequalities

$$|w_j(x)| \leqslant C'K^j\Gamma(1 + sj) \quad \text{with } C' = \frac{C}{r - r'}$$

for all $x \in D_{r',\dots,r'}$ and all $j \geqslant 0$, which proves that $\partial_{x_\ell}\widetilde{u}$ belongs to $\mathcal{O}(D_{\rho_1,\dots,\rho_n})[[t]]_s$.

• *Anti-derivations.*

◁ *Stability under the anti-derivation* ∂_t^{-1}. The anti-derivative $\widetilde{w} = \partial_t^{-1}\widetilde{u}$ is the formal power series

$$\widetilde{w}(t, x) = \sum_{j \geqslant 0} w_j(x)t^j \in \mathcal{O}(D_{\rho_1,\dots,\rho_n})[[t]]$$

with $w_0(x) \equiv 0$ and $w_j(x) = u_{j-1}(x)/j$ for all $x \in D_{\rho_1,\dots,\rho_n}$ and all $j \geqslant 1$. Applying then inequalities (3.3) and property (3.2), we get the relations

$$\begin{aligned} |w_j(x)| \leqslant CK^{j-1}\Gamma(1 + s(j-1)) \leqslant CK^{j-1}\Gamma(2 + sj) &= C(1 + sj)K^j\Gamma(1 + sj) \\ &\leqslant \frac{eC}{K}(e^sK)^j\Gamma(1 + sj) \end{aligned}$$

for all $x \in D_{r,\dots,r}$ and all $j \geqslant 1$; hence, $\partial_t^{-1}\widetilde{u} \in \mathcal{O}(D_{\rho_1,\dots,\rho_n})[[t]]_s$.

◁ *Stability under the anti-derivation* $\partial_{x_\ell}^{-1}$. Let us fix $\ell \in \{1, \ldots, n\}$. The anti-derivative $\widetilde{w} = \partial_{x_\ell}^{-1}\widetilde{u}$ is the formal power series

$$\widetilde{w}(t, x) = \sum_{j \geqslant 0} w_j(x)t^j \in \mathcal{O}(D_{\rho_1,\ldots,\rho_n})[[t]]$$

with

$$w_j(x) = \partial_{x_\ell}^{-1} u_j(x) = \int_0^{x_\ell} u_j(x_1, \ldots, x_{\ell-1}, x'_\ell, x_{\ell+1}, \ldots, x_n)dx'_\ell$$

for all $x \in D_{\rho_1,\ldots,\rho_n}$ and all $j \geqslant 0$. Since $x \in D_{r,\ldots,r}$ implies

$$(x_1, \ldots, x_{\ell-1}, x'_\ell, x_{\ell+1}, \ldots, x_n) \in D_{r,\ldots,r},$$

we clearly derive from inequalities (3.3) the relations

$$\left|w_j(x)\right| \leqslant rCK^j\Gamma(1 + sj)$$

for all $x \in D_{r,\ldots,r}$ and all $j \geqslant 0$; hence, $\partial_{x_\ell}\widetilde{u} \in \mathcal{O}(D_{\rho_1,\ldots,\rho_n})[[t]]_s$. ■

Observe that the stability under the derivations ∂_{x_ℓ} is guaranteed by the condition "*there exists* $0 < r \leqslant \min(\rho_1, \ldots, \rho_n)$..." in Definition 3.1. Observe also that the stability under the anti-derivations $\partial_{x_\ell}^{-1}$ is due to the fact that inequalities (3.1) are uniform with respect to x.

Remark 3.5 Using the Stirling Formula (see Remark 3.2) and adapting the proof of [35] (see also [65, pp. 15–16]), one can also prove that $\mathcal{O}(D_{\rho_1,\ldots,\rho_n})[[t]]_s$ is stable under the composition. However, as this result is not useful to us in this work, we leave this proof to the reader.

3.2 Gevrey Formal Power Series and Partial Differential Equations

In this section, we are interested in the Gevrey regularity of the formal power series solutions of some partial differential equations. Before stating various general results, let us first consider in detail a simple example—the so-called *heat equation*—in order to introduce some theoretical tools that can be used in this kind of study.

Remark 3.6 In the literature, the formal power series solutions of partial differential equations are often written in the form

$$\widetilde{u}(t,x)=\sum_{j\geqslant 0}u_{j,*}(x)\frac{t^j}{j!}\quad\text{instead of}\quad \widetilde{u}(t,x)=\sum_{j\geqslant 0}u_{j,*}(x)t^j,$$

which we will also do in the following. Consequently, in Definition 3.1, the factor $\Gamma(1+sj)$ should be replaced by $\Gamma(1+(s+1)j)$ to study their Gevrey regularity. Notice, however, that the choice made in Definition 3.1 is in no way inconsistent with the choice made here. Indeed, we wanted to write a definition which can be adapted to any kind of equations, whether they are partial differential equations or moment partial differential equations, the solutions of the latter being still written in another form, namely

$$\widetilde{u}(t,x)=\sum_{j\geqslant 0}u_{j,*}(x)\frac{t^j}{m(j)}$$

with convenient sequences $(m(j))_{j\geqslant 0}$.

3.2.1 Example: The Heat Equation

Let us consider the inhomogeneous linear heat equation

$$\begin{cases}\partial_t u-a(t,x)\partial_x^2 u=\widetilde{f}(t,x)\\ u(0,x)=\varphi(x)\end{cases},\ (t,x)\in\mathbb{C}^2, \tag{3.4}$$

where $a(t,x)\in\mathcal{O}(D_{\rho_0}\times D_{\rho_1})$ satisfies $a(0,x)\not\equiv 0$, $\varphi(x)\in\mathcal{O}(D_{\rho_1})$ and $\widetilde{f}(t,x)\in\mathcal{O}(D_{\rho_1})[[t]]$. This equation describes heat propagation under thermodynamics and Fourier laws. The coefficient $a(t,x)$, named thermal diffusivity, is related to the thermal conductivity κ by the formula $a=\dfrac{\kappa}{c\rho}$, where c is the capacity and ρ the density of the medium. The inhomogeneity $\widetilde{f}(t,x)$, which stands for the internal heat input, may be smooth or not (hence, the notation with a tilde).

An important particular case of Eq. (3.4) is the case with uniform thermal diffusivity, that is with isotropic and homogeneous medium, and no internal heat generation, which corresponds to the homogeneous linear heat equation with constant coefficients

$$\begin{cases}\partial_t u-a\partial_x^2 u=0\\ u(0,x)=\varphi(x)\end{cases},\ (t,x)\in\mathbb{C}^2,\ a\in\mathbb{C}^*. \tag{3.5}$$

Lemma 3.7 *Equation (3.4) admits a unique formal solution* $\widetilde{u}(t,x) \in \mathcal{O}(D_{\rho_1})[[t]]$.

Proof Let us write the coefficient $a(t,x)$ and the inhomogeneity $\widetilde{f}(t,x)$ in the form

$$a(t,x) = \sum_{j \geqslant 0} a_{j,*}(x) \frac{t^j}{j!} \quad \text{and} \quad \widetilde{f}(t,x) = \sum_{j \geqslant 0} f_{j,*}(x) \frac{t^j}{j!}$$

with $a_{j,*}(x)$, $f_{j,*}(x) \in \mathcal{O}(D_{\rho_1})$ for all $j \geqslant 0$. Looking for $\widetilde{u}(t,x)$ on the same type:

$$\widetilde{u}(t,x) = \sum_{j \geqslant 0} u_{j,*}(x) \frac{t^j}{j!} \quad \text{with } u_{j,*}(x) \in \mathcal{O}(D_{\rho_1}) \text{ for all } j \geqslant 0,$$

one easily checks that the coefficients $u_{j,*}(x)$ are uniquely determined for all $j \geqslant 0$ by the recurrence relations

$$u_{j+1,*}(x) = f_{j,*}(x) + \sum_{k=0}^{j} \binom{j}{k} a_{j-k,*}(x) \partial_x^2 u_{k,*}(x) \tag{3.6}$$

together with the initial condition $u_{0,*}(x) = \varphi(x)$, which ends the proof. ∎

Remark 3.8 In the case of Eq. (3.5), relations (3.6) are reduced to the relations $u_{j+1,*}(x) = a\partial_x^2 u_{j,*}(x)$ and, therefore, the formal solution $\widetilde{u}(t,x)$ reads in the form

$$\widetilde{u}(t,x) = \sum_{j \geqslant 0} \partial_x^{2j} \varphi(x) \frac{(at)^j}{j!}. \tag{3.7}$$

Consequently, since the Cauchy inequalities imply, for any $0 < r < \rho_1$, the existence of two positive constants $C, K > 0$ such that the estimates

$$\left| \partial_x^{2j} \varphi(x) \right| \leqslant C K^j \Gamma(1+2j)$$

hold for all $x \in D_r$ and all $j \geqslant 0$, we can conclude in particular that this formal solution is 1-Gevrey. Moreover, as shown by the choice $\varphi(x) = 1/(1-x)$, hence $\partial_x^{2j} \varphi(x) = \Gamma(1+2j)$, this Gevrey estimate is optimal.

In the case of the general Eq. (3.4), the determination of the Gevrey regularity of the formal solution $\widetilde{u}(t,x)$ is more complicated. Indeed, it depends both on the structure of the equation, this is on its associated linear operator $\partial_t - a(t,x)\partial_x^2$, and on the Gevrey regularity of the inhomogeneity $\widetilde{f}(t,x)$. More precisely, we have the following.

Proposition 3.9 *Let $\widetilde{u}(t,x)$ be the formal solution in $\mathcal{O}(D_{\rho_1})[[t]]$ of the heat equation (3.4). Then,*

1. *$\widetilde{u}(t,x)$ and $\widetilde{f}(t,x)$ are simultaneously s-Gevrey for any $s \geqslant 1$.*
2. *$\widetilde{u}(t,x)$ is generically 1-Gevrey while $\widetilde{f}(t,x)$ is s-Gevrey with $s < 1$.*

Observe that, choosing $\widetilde{f}(t,x) \equiv 0$, we have in particular the following result:

Corollary 3.10 *Let $\widetilde{u}(t,x) \in \mathcal{O}(D_{\rho_1})[[t]]$ be the formal solution of the homogeneous linear heat equation*

$$\begin{cases} \partial_t u - a(t,x)\partial_x^2 u = 0 \\ u(0,x) = \varphi(x) \end{cases}, (t,x) \in \mathbb{C}^2,$$

where $a(t,x) \in \mathcal{O}(D_{\rho_0} \times D_{\rho_1})$ satisfies $a(0,x) \not\equiv 0$ and $\varphi(x) \in \mathcal{O}(D_{\rho_1})$. Then, $\widetilde{u}(t,x)$ is generically 1-Gevrey.

Remark 3.11 The expression "*generically 1-Gevrey*" means, on one hand, that $\widetilde{u}(t,x)$ is exactly s-Gevrey for some $s \in [0,1]$, but that this order can not be determined explicitly for general initial data, and, on the other hand, that the 1-Gevrey order is optimal, that is there are cases where $\widetilde{u}(t,x)$ is exactly 1-Gevrey (see, for instance, just above for the case of Eq. (3.5) and the proof of the second point of Proposition 3.9 for a more general case). Notice, however, that for certain initial data $\varphi(x)$, the Gevrey order of $\widetilde{u}(t,x)$ can also be calculated explicitly when it is strictly less than 1. For instance, if $\varphi(x) = e^{x^2}$, and, more generally, if $\varphi(x)$ is an entire function with an exponential growth of order at most 2 at infinity, then there exist, for any $r > 0$, two positive constants $C, K > 0$ such that the inequalities

$$\left|\partial_x^j \varphi(x)\right| \leqslant C K^j \Gamma\left(1 + \frac{j}{2}\right)$$

hold for all $x \in D_r$ and all $j \geqslant 0$ (write the Taylor expansion of φ at x and apply for instance [62, Lemma 1 page 5]), and, consequently, the formal solution (3.7) of Eq. (3.5) is 0-Gevrey; hence, convergent, and more precisely analytic on $D_R \times \mathbb{C}$ for a convenient radius $R > 0$ thanks to the Analytic Continuation Principle. We will show later that this condition is also necessary for the formal solution (3.7) to be convergent (see Corollary 7.22).

Remark 3.12 Proposition 3.9 shows in particular that the Gevrey regularity of $\widetilde{u}(t,x)$ follows a noteworthy dichotomy with respect to the s-Gevrey regularity of the inhomogeneity $\widetilde{f}(t,x)$: if $s > 1$, then $\widetilde{u}(t,x)$ inherits the Gevrey regularity of $\widetilde{f}(t,x)$; if $s \leqslant 1$, then $\widetilde{u}(t,x)$ keeps the 1-Gevrey regularity defined by the structure of Eq. (3.4), that is by its associated linear operator $\partial_t - a(t,x)\partial_x^2$ as shown by Corollary 3.10. The *"frontier"* value $s = 1$ is called *the critical value of Eq. (3.4)*.

The proof of Proposition 3.9 is detailed just below. The first point is the most technical and the most complicated. Its proof is based on the Nagumo norms [19, 94,

130] (see their definition and properties in Chap. 12) and on a technique of majorant series. As for the second point, it stems from the first one and from an explicit example in which $\widetilde{u}(t,x)$ is s'-Gevrey for no $s' < 1$ while $\widetilde{f}(t,x)$ is s-Gevrey with $s < 1$.

Proof of the First Point of Proposition 3.9 We adapt here the proof of W. Balser and M. Loday-Richaud [10] done in the case $s = 1$ and under the condition $a(t,x) = \alpha(x)$ independent of the variable t.

According to Proposition 3.4, it is clear that

$$\widetilde{u}(t,x) \in \mathcal{O}(D_{\rho_1})[[t]]_s \Rightarrow \widetilde{f}(t,x) \in \mathcal{O}(D_{\rho_1})[[t]]_s.$$

Reciprocally, let us fix $s \geqslant 1$ and let us assume that the inhomogeneity $\widetilde{f}(t,x)$ of Eq. (3.4) is s-Gevrey. By assumption, its coefficients $f_{j,*}(x) \in \mathcal{O}(D_{\rho_1})$ satisfy the following condition (see Definition 3.1 and Remark 3.6): there exist three positive constants $0 < r < \rho_1$, $C > 0$ and $K > 0$ such that the inequalities

$$\left|f_{j,*}(x)\right| \leqslant CK^j \Gamma(1 + (s+1)j) \tag{3.8}$$

hold for all $|x| \leqslant r$ and all $j \geqslant 0$, and we must prove that the coefficients $u_{j,*}(x) \in \mathcal{O}(D_{\rho_1})$ of the formal solution $\widetilde{u}(t,x)$ satisfy similar inequalities.

From the recurrence relations (3.6), we first derive, for all $|x| \leqslant r$ and all $j \geqslant 0$, the identities

$$\frac{u_{j+1,*}(x)}{\Gamma(1+(s+1)(j+1))} = \frac{f_{j,*}(x)}{\Gamma(1+(s+1)(j+1))} + \sum_{k=0}^{j} \binom{j}{k} \frac{a_{j-k,*}(x)\partial_x^2 u_{k,*}(x)}{\Gamma(1+(s+1)(j+1))}$$

together with the initial condition $u_{0,*}(x) = \varphi(x)$. Applying then the Nagumo norms of indices $((s+1)(j+1), r)$, we deduce successively from Property 1 and Properties 4–5 of Proposition 12.2 the inequalities

$$\begin{aligned}
\frac{\left\|u_{j+1,*}\right\|_{(s+1)(j+1),r}}{\Gamma(1+(s+1)(j+1))} &\leqslant \frac{\left\|f_{j,*}\right\|_{(s+1)(j+1),r}}{\Gamma(1+(s+1)(j+1))} \\
&\quad + \sum_{k=0}^{j} \binom{j}{k} \frac{\left\|a_{j-k,*}\partial_x^2 u_{k,*}\right\|_{(s+1)(j+1),r}}{\Gamma(1+(s+1)(j+1))} \\
&\leqslant \frac{\left\|f_{j,*}\right\|_{(s+1)(j+1),r}}{\Gamma(1+(s+1)(j+1))} +
\end{aligned}$$

$$\sum_{k=0}^{j} \binom{j}{k} \frac{\left\|a_{j-k,*}\right\|_{(s+1)(j-k)+s-1,r} \left\|\partial_x^2 u_{k,*}\right\|_{(s+1)k+2,r}}{\Gamma(1+(s+1)(j+1))}$$

$$\leqslant \frac{\left\|f_{j,*}\right\|_{(s+1)(j+1),r}}{\Gamma(1+(s+1)(j+1))} + \sum_{k=0}^{j} A_{s,j,k} \frac{\left\|u_{k,*}\right\|_{(s+1)k,r}}{\Gamma(1+(s+1)k)}$$

for all $j \geqslant 0$, where the terms $A_{s,j,k}$ are nonnegative and defined by

$$A_{s,j,k} = \frac{e^2 \left\|a_{j-k,*}\right\|_{(s+1)(j-k)+s-1,r}}{\Gamma(1+(s+1)(j-k))} \frac{\Gamma(1+(s+1)k+2)}{\Gamma(1+(s+1)(k+1))} \frac{\binom{j}{k}}{\binom{(s+1)(j+1)}{(s+1)(k+1)}}.$$

Observe that all the norms written in these inequalities, and especially the norms $\left\|a_{j-k,*}\right\|_{(s+1)(j-k)+s-1,r}$, are well-defined. Indeed, the assumption $s \geqslant 1$ implies $(s+1)(j-k)+s-1 \geqslant 0$ for all $k \in \{0, \ldots, j\}$. Observe also that

$$A_{s,j,k} \leqslant \frac{e^2 \left\|a_{j-k,*}\right\|_{(s+1)(j-k)+s-1,r}}{\Gamma(1+(s+1)(j-k))}.$$

Indeed,

- on one hand, the assumption $s \geqslant 1$ implies

$$1+(s+1)(k+1) \geqslant 1+(s+1)k+2 \geqslant 2;$$

  hence, the inequality

$$\frac{\Gamma(1+(s+1)k+2)}{\Gamma(1+(s+1)(k+1))} \leqslant 1$$

  by increasing of the Gamma function on $[2, +\infty[$;
- on the other hand, applying successively the Vandermonde Inequality (see Proposition 13.2) and Proposition 13.5, we get

$$\binom{(s+1)(j+1)}{(s+1)(k+1)} \geqslant \binom{j}{k} \binom{s(j+1)+1}{s(k+1)+1} \geqslant \binom{j}{k};$$

  hence, the inequality

$$\frac{\binom{j}{k}}{\binom{(s+1)(j+1)}{(s+1)(k+1)}} \leqslant 1.$$

This brings then us to the following inequalities

$$\frac{\left\|u_{j+1,*}\right\|_{(s+1)(j+1),r}}{\Gamma(1+(s+1)(j+1))} \leqslant g_{s,j} + \sum_{k=0}^{j} \alpha_{s,j-k} \frac{\left\|u_{k,*}\right\|_{(s+1)k,r}}{\Gamma(1+(s+1)k)},$$

for all $j \geqslant 0$, where we set

$$g_{s,j} = \frac{\left\|f_{j,*}\right\|_{(s+1)(j+1),r}}{\Gamma(1+(s+1)(j+1))} \quad \text{and} \quad \alpha_{s,j} = \frac{e^2 \left\|a_{j,*}\right\|_{(s+1)j+s-1,r}}{\Gamma(1+(s+1)j)}.$$

We shall now bound the Nagumo norms $\left\|u_{j,*}\right\|_{(s+1)j,r}$ for any $j \geqslant 0$ by using a technique of majorant series. To do that, let us consider the formal power series

$$v(X) = \sum_{j\geqslant 0} v_j X^j \in \mathbb{R}^+[[X]],$$

where the coefficients v_j are recursively determined from the initial condition $v_0 = \left\|u_{0,*}\right\|_{0,r} = \|\varphi\|_{0,r}$ by the relations

$$v_{j+1} = g_{s,j} + \sum_{k=0}^{j} \alpha_{s,j-k} v_k$$

for all $j \geqslant 0$. By construction, we clearly have

$$0 \leqslant \frac{\left\|u_{j,*}\right\|_{(s+1)j,r}}{\Gamma(1+(s+1)j)} \leqslant v_j \quad \text{for all } j \geqslant 0.$$

On the other hand, according to the assumption on the coefficients $f_{j,*}(x)$ (see inequality (3.8)) and the analyticity of the function $a(t,x)$ at the origin $(0,0) \in \mathbb{C}^2$, we derive from the definition of the Nagumo norms and from the classical property (3.2) the relations

$$0 \leqslant g_{s,j} \leqslant \frac{CK^j \Gamma(1+(s+1)j) r^{(s+1)(j+1)}}{\Gamma(1+(s+1)(j+1))} \leqslant Cr^{s+1}\left(Kr^{s+1}\right)^j$$

$$0 \leqslant \alpha_{s,j} \leqslant \frac{e^2 C' K'^j j! r^{(s+1)j+s-1}}{\Gamma(1+(s+1)j)} \leqslant e^2 C' r^{s-1}\left(K' r^{s+1}\right)^j$$

for all $j \geqslant 0$, with four convenient constants $C, K, C', K' > 0$ independent of j. Consequently, the series

$$g_s(X) = \sum_{j\geqslant 0} g_{s,j} X^j \quad \text{and} \quad \alpha_s(X) = \sum_{j\geqslant 0} \alpha_{s,j} X^j$$

are convergent. Thereby, the formal series $v(X)$ satisfying

$$(1 - X\alpha_s(X))v(X) = \|\varphi\|_{0,r} + Xg_s(X),$$

it is also convergent, and there exist two positive constants $C'', K'' > 0$ such that $v_j \leqslant C''K''^j$ for all $j \geqslant 0$. Hence,

$$\left\|u_{j,*}\right\|_{(s+1)j,r} \leqslant C''K''^j\Gamma(1 + (s + 1)j) \quad \text{for all } j \geqslant 0.$$

We deduce from this similar estimates on the sup-norm of the $u_{j,*}(x)$ by shrinking the closed disc $|x| \leqslant r$. Let $0 < r' < r$. Then, for all $x \in D_{r'}$ and all $j \geqslant 0$, we have

$$\begin{aligned}\left|u_{j,*}(x)\right| &= \left|u_{j,*}(x)d_r(x)^{(s+1)j}\frac{1}{d_r(x)^{(s+1)j}}\right| \\ &\leqslant \frac{\left|u_{j,*}(x)d_r(x)^{(s+1)j}\right|}{(r-r')^{(s+1)j}} \leqslant \frac{\left\|u_{j,*}\right\|_{(s+1)j,r}}{(r-r')^{(s+1)j}}\end{aligned}$$

and, consequently,

$$\sup_{x\in D_{r'}}\left|u_{j,*}(x)\right| \leqslant C''\left(\frac{K''}{(r-r')^{s+1}}\right)^j\Gamma(1+(s+1)j);$$

which ends the proof of the first point of Proposition 3.9. ■

Proof of the Second Point of Proposition 3.9 Let us fix $s < 1$. According to the filtration of the s-Gevrey spaces $\mathcal{O}(D_{\rho_1})[[t]]_s$ (see page 17) and the first point of Proposition 3.9, it is clear that the following implications hold:

$$\widetilde{f}(t,x) \in \mathcal{O}(D_{\rho_1})[[t]]_s \Rightarrow \widetilde{f}(t,x) \in \mathcal{O}(D_{\rho_1})[[t]]_1 \Rightarrow \widetilde{u}(t,x) \in \mathcal{O}(D_{\rho_1})[[t]]_1. \tag{3.9}$$

To conclude that we can not say better about the Gevrey order of $\widetilde{u}(t,x)$, that is $\widetilde{u}(t,x)$ is *generically* 1*-Gevrey*, we need to find an example for which the formal solution $\widetilde{u}(t,x)$ of Eq. (3.4) is s'-Gevrey for no $s' < 1$. Remark 3.8 already provides such an example in the case where the inhomogeneity $\widetilde{f}(t,x)$ is zero. More generally, let us consider the equation

$$\begin{cases}\partial_t u - a\partial_x^2 u = \widetilde{f}(t,x) & , a \in \mathbb{C}^* \\ u(0,x) = \varphi(x) = \dfrac{1}{1-x} \in \mathcal{O}(D_1)\end{cases} \tag{3.10}$$

where $\widetilde{f}(t,x) \in \mathcal{O}(D_1)[[t]]_s$ satisfies the condition $\partial_x^\ell f_{j,*}(0) \geqslant 0$ for all $\ell, j \geqslant 0$. According to the general relations (3.6), the coefficients $u_{j,*}(x) \in \mathcal{O}(D_1)$ of the formal solution $\widetilde{u}(t,x)$ of Eq. (3.10) are recursively determined by the initial condition $u_{0,*}(x) = \varphi(x)$ and, for all $j \geqslant 0$, by the relations

$$u_{j+1,*}(x) = f_{j,*}(x) + a\partial_x^2 u_{j,*}(x).$$

We derive straightaway from this that

$$u_{j,*}(x) = a^j \partial_x^{2j} \varphi(x) + \sum_{k=0}^{j-1} a^k \partial_x^{2k} f_{j-1-k,*}(x)$$

for all $x \in D_1$ and all $j \geqslant 0$, with the classical convention that the sum is 0 when $j = 0$. Hence, due to our assumption on the inhomogeneity $\widetilde{f}(t,x)$, the inequalities

$$u_{j,*}(0) \geqslant a^j \Gamma(1+2j) \quad \text{for all } j \geqslant 0. \tag{3.11}$$

Let us now suppose that $\widetilde{u}(t,x)$ is s'-Gevrey for some $s' < 1$. Then, Definition 3.1 and inequalities (3.11) above imply the relations

$$1 \leqslant C\left(\frac{K}{a}\right)^j \frac{\Gamma(1+(s'+1)j)}{\Gamma(1+2j)} \tag{3.12}$$

for all $j \geqslant 0$ and some convenient positive constants $C, K > 0$ independent of j. Applying then the Stirling's Formula, we get

$$C\left(\frac{K}{a}\right)^j \frac{\Gamma(1+(s'+1)j)}{\Gamma(1+2j)} \underset{j\to+\infty}{\sim} C'\left(\frac{K'}{j^{1-s'}}\right)^j$$

with

$$C' = C\sqrt{\frac{s'+1}{2}} \quad \text{and} \quad K' = \frac{Ke^{1-s'}(s'+1)^{s'+1}}{4a}.$$

We conclude that inequalities (3.12) are impossible since the right hand-side goes to 0 when j tends to infinity. Hence, $\widetilde{u}(t,x)$ is s'-Gevrey for no $s' < 1$ and, consequently, $\widetilde{u}(t,x)$ is exactly 1-Gevrey thanks to (3.9). This achieves the proof of the second point of Proposition 3.9. ■

As we shall see below, results similar to that of Proposition 3.9 can be given for much more general partial differential equations, including nonlinear ones, the critical value being determined by the structure of a suitable linear partial differential operator. Before stating these general results, let us start by studying the case of the homogeneous linear partial differential equations for which the Gevrey regularity of the formal solutions depends only on their structure, that is on their associated linear operator. This will allow us in particular to introduce another very important tool—the so-called *Newton polygon*—which will also be very useful in the study of the summability (see Part II).

3.2.2 Homogeneous Linear Partial Differential Equations and Newton Polygon

In this section, we are interested in the Gevrey regularity of the formal solutions of the homogeneous linear partial differential equations of the form

$$\begin{cases} L(u)=0, \quad L=\partial_t^\kappa-\sum_{i\in\mathcal{K}}\sum_{q\in Q_i} a_{i,q}(t,x)\partial_t^i\partial_x^q \\ \partial_t^j u(t,x)_{|t=0}=\varphi_j(x),\ j=0,\ldots,\kappa-1 \end{cases} \tag{3.13}$$

where

- $\kappa \geqslant 1$ is a positive integer;
- $\mathcal{K}$ is a nonempty subset of $\{0,\ldots,\kappa-1\}$;
- Q_i is a nonempty finite subset of $\mathbb{N}^n$ for all $i \in \mathcal{K}$;
- the coefficients $a_{i,q}(t,x)$ are analytic and not identically zero on a polydisc $D_{\rho_0,\rho_1,\ldots,\rho_n}$ centered at the origin of $\mathbb{C}^{n+1}$ for all $i \in \mathcal{K}$ and all $q \in Q_i$;
- the initial conditions $\varphi_j(x)$ are analytic on $D_{\rho_1,\ldots,\rho_n}$ for all $j=0,\ldots,\kappa-1$.

Following lemma tells us that Eq. (3.13) is formally well-posed.

Lemma 3.13 *Equation (3.13) admits a unique formal solution* $\widetilde{u}(t,x) \in \mathcal{O}(D_{\rho_1,\ldots,\rho_n})[[t]]$.

Proof As in the proof of Lemma 3.7, we write the coefficients $a_{i,q}(t,x)$ in the form

$$a_{i,q}(t,x)=\sum_{j\geqslant 0} a_{i,q;j,*}(x)\frac{t^j}{j!}$$

with $a_{i,q;j,*}(x) \in \mathcal{O}(D_{\rho_1,\ldots,\rho_n})$ for all $j \geqslant 0$. Looking for $\widetilde{u}(t,x)$ in the same type:

$$\widetilde{u}(t,x)=\sum_{j\geqslant 0} u_{j,*}(x)\frac{t^j}{j!} \quad \text{with } u_{j,*}(x)\in\mathcal{O}(D_{\rho_1,\ldots,\rho_n}) \text{ for all } j\geqslant 0,$$

one easily checks that the coefficients $u_{j,*}(x)$ are uniquely determined for all $j \geqslant 0$ by the recurrence relations

$$u_{j+\kappa,*}(x)=\sum_{i\in\mathcal{K}}\sum_{q\in Q_i}\sum_{\ell=0}^{j}\binom{j}{\ell}a_{i,q;\ell,*}(x)\partial_x^q u_{j-\ell+i,*}(x)$$

together with the initial conditions $u_{j,*}(x) = \varphi_j(x)$ for all $j = 0, \ldots, \kappa - 1$, which ends the proof. ■

As in the case of analytic linear ordinary differential equations [99, 101], the Gevrey regularity of the formal solution $\widetilde{u}(t, x)$ of Eq. (3.13) is closely related to the Newton polygon at $t = 0$ of the linear operator L [87, 96, 135]. The latter is defined as follows.

Definition 3.14 (Newton Polygon) For any $(a, b) \in \mathbb{R}^2$, we denote by $C(a, b)$ the domain

$$C(a, b) = \{(x, y) \in \mathbb{R}^2; x \leqslant a \text{ and } y \geqslant b\}.$$

1. Let $v = (v_{i,q})_{\substack{i \in \mathcal{K} \\ q \in Q_i}}$ be a tuple of nonnegative integers. The *Newton polygon* $\mathcal{N}_t(\mathcal{L}_{\kappa,v})$ *at* $t = 0$ *of the linear operator*

$$\mathcal{L}_{\kappa,v} = \partial_t^\kappa - \sum_{i \in \mathcal{K}} \sum_{q \in Q_i} t^{v_{i,q}} \partial_t^i \partial_x^q$$

is the convex hull of

$$C(\kappa, -\kappa) \cup \bigcup_{i \in \mathcal{K}} \bigcup_{q \in Q_i} C(\lambda(q) + i, v_{i,q} - i),$$

where $\lambda(q) = q_1 + \ldots + q_n$ denotes the length of $q = (q_1, \ldots, q_n) \in \mathbb{N}^n$.
2. The *Newton polygon* $\mathcal{N}_t(L)$ *at* $t = 0$ *of the linear operator* L associated with Eq. (3.13) is the Newton polygon $\mathcal{N}_t(\mathcal{L}_{\kappa,v})$, where, for all $i \in \mathcal{K}$ and all $q \in Q_i$, the nonnegative integer $v_{i,q}$ stands for the valuation at $t = 0$ of $a_{i,q}(t, x)$: $a_{i,q}(t, x) = t^{v_{i,q}} a'_{i,q}(t, x)$ with $a'_{i,q}(0, x) \not\equiv 0$.

The general geometric structure of the domain $\mathcal{N}_t(\mathcal{L}_{\kappa,v})$ is given in the following.

Proposition 3.15 *Let* $\mathcal{S} = \{(i, q)$ *such that* $i \in \mathcal{K}$, $q \in Q_i$ *and* $\lambda(q) > \kappa - i\}$ *be.*

1. *Suppose* $\mathcal{S} = \emptyset$. *Then,* $\mathcal{N}_t(\mathcal{L}_{\kappa,v}) = C(\kappa, -\kappa)$. *In particular,* $\mathcal{N}_t(\mathcal{L}_{\kappa,v})$ *has no side with a positive slope (see Fig. 3.1a).*
2. *Suppose* $\mathcal{S} \neq \emptyset$. *Then,* $\mathcal{N}_t(\mathcal{L}_{\kappa,v})$ *has at least one side with a positive slope. Moreover, its smallest positive slope* k *is given by*

$$k = \min_{(i,q) \in \mathcal{S}} \left(\frac{\kappa - i + v_{i,q}}{\lambda(q) - \kappa + i} \right) = \frac{\kappa - i^* + v_{i^*,q^*}}{\lambda(q^*) - \kappa + i^*},$$

where the pair $(i^*, q^*) \in \mathcal{S}$ *stands for any convenient pair, chosen and fixed once and for all, so that the edge of* $\mathcal{N}_t(\mathcal{L}_{\kappa,v})$ *with slope* k *be the segment with end points* $(\kappa, -\kappa)$ *and* $(\lambda(q^*) + i^*, v_{i^*,q^*} - i^*)$ *(see Fig. 3.1b).*

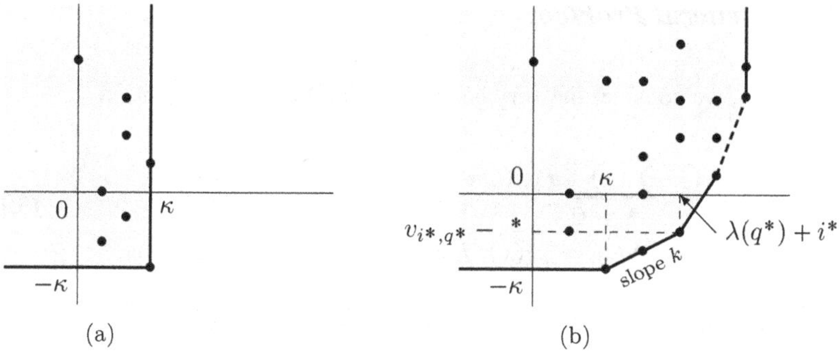

Fig. 3.1 The Newton polygon $\mathcal{N}_t(\mathcal{L}_{\kappa,v})$. **(a)** Case $\mathcal{S} = \emptyset$. **(b)** Case $\mathcal{S} \neq \emptyset$

Proof The first point stems obvious from the fact that the condition $\mathcal{S} = \emptyset$ implies $C(\lambda(q)+i, v_{i,q}-i) \subset C(\kappa, -\kappa)$ for all $i \in \mathcal{K}$ and $q \in Q_i$. As for the second point, it suffices to remark, on one hand, that $C(\lambda(q)+i, v_{i,q}-i) \subset C(\kappa, -\kappa)$ for all pairs $(i, q) \notin \mathcal{S}$, and, on the other hand, that the segment with the two end points $(\kappa, -\kappa)$ and $(\lambda(q)+i, v_{i,q}-i)$ has a positive slope equal to $(\kappa - i + v_{i,q})/(\lambda(q) - \kappa + i)$ for all pairs $(i, q) \in \mathcal{S}$. ∎

Using a similar approach to that developed in the previous section for the heat equation (see Sect. 3.2.1), we can prove the following.

Theorem 3.16 ([87, 96, 135]) *Let* $\mathcal{S} = \{(i, q)$ *such that* $i \in \mathcal{K},\ q \in Q_i$ *and* $\lambda(q) > \kappa - i\}$ *and* s_c *be the nonnegative rational number defined by*

$$s_c = \begin{cases} 0 & \text{if } \mathcal{S} = \emptyset \\ \dfrac{1}{k} = \dfrac{\lambda(q^*) - \kappa + i^*}{\kappa - i^* + v_{i^*,q^*}} & \text{if } \mathcal{S} \neq \emptyset \end{cases}.$$

Then, the formal solution $\widetilde{u}(t, x)$ *of Eq. (3.13) is generically* s_c*-Gevrey.*

Observe that Theorem 3.16 coincides with Corollary 3.10 in the case of the homogeneous heat equation since the Newton polygon of the operator $\partial_t - a(t, x)\partial_x^2$ with $a(0, x) \not\equiv 0$ admits only one positive slope, the latter being equal to 1.

Observe also that Theorem 3.16 yields a result similar to the Maillet-Ramis theorem for the ordinary linear differential equations [99, 101] (see also [65, Thm. 4.2.7]).

Corollary 3.17 *The formal solution* $\widetilde{u}(t, x)$ *of Eq. (3.13) is either convergent or* $1/k$*-Gevrey, where* k *stands for the smallest positive slope of the Newton polygon at* $t = 0$ *of the linear operator* L.

As we shall see in the next section, the value s_c defined in Theorem 3.16 will play a fundamental role in the study of the Gevrey regularity of formal solutions of much more general partial differential equations.

3.2.3 A General Problem

In this section, we consider the very general partial differential equation

$$\begin{cases} \partial_t^\kappa u - \sum\limits_{i\in\mathcal{K}}\sum\limits_{q\in Q_i} P_{i,q}(t,x,u)\partial_t^i\partial_x^q u = \widetilde{f}(t,x) \\ \partial_t^j u(t,x)_{|t=0} = \varphi_j(x),\ j=0,\ldots,\kappa-1 \end{cases}, \tag{3.14}$$

where κ, $\mathcal{K}$, Q_i and $\varphi_j(x)$ are as above (see Eq. (3.13)), and where

- $P_{i,q}(t,x,X)$ is a polynomial in X with analytic coefficients on the polydisc $D_{\rho_0,\rho_1,\ldots,\rho_n}$ centered at the origin of $\mathbb{C}^{n+1}$ for all $i\in\mathcal{K}$ and all $q\in Q_i$:

$$P_{i,q}(t,x,X) = \sum_{m\in E_{i,q}} a_{i,q,m}(t,x)X^m,$$

  where $E_{i,q}$ stands for a nonempty subset of $\mathbb{N}$, and where all the coefficients $a_{i,q,m}(t,x)\in\mathcal{O}(D_{\rho_0,\rho_1,\ldots,\rho_n})$ are not identically zero;
- the inhomogeneity $\widetilde{f}(t,x)$ belongs to $\mathcal{O}(D_{\rho_1,\ldots,\rho_n})[[t]]$.

Observe that, choosing $E_{i,q} = \{0\}$ for all (i,q) (resp. $E_{0,0}\cap(\mathbb{N}\backslash\{0\})\neq\emptyset$ and $E_{i,q}=\{0\}$ for all $(i,q)\neq(0,0)$), Eq. (3.14) is reduced to an inhomogeneous linear equation (resp. an inhomogeneous semilinear equation with a polynomial semilinearity), that is to an equation of the form

$$\begin{cases} L(u) - P(t,x,u) = \widetilde{f}(t,x) \\ \partial_t^j u(t,x)_{|t=0} = \varphi_j(x),\ j=0,\ldots,\kappa-1 \end{cases}, \tag{3.15}$$

where L is a linear partial differential operator of the same type as in Eq. (3.13), and where $P(t,x,X)$ is either zero, or a polynomial with valuation at least 2 in X and with analytic coefficients on $D_{\rho_0,\rho_1,\ldots,\rho_n}$.

Observe also that Eq. (3.14) is formally well-posed as shown by the following.

Lemma 3.18 *Equation (3.14) admits a unique formal solution* $\widetilde{u}(t,x)\in\mathcal{O}(D_{\rho_1,\ldots,\rho_n})[[t]]$.

Proof Writing as in the proof of Lemma 3.7 the coefficients $a_{i,q,m}(t,x)$ and the inhomogeneity $\widetilde{f}(t,x)$ in the form

$$a_{i,q,m}(t,x) = \sum_{j\geqslant 0} a_{i,q,m;j,*}(x)\frac{t^j}{j!} \quad\text{and}\quad \widetilde{f}(t,x) = \sum_{j\geqslant 0} f_{j,*}(x)\frac{t^j}{j!}$$

with $a_{i,q,m;j,*}(x)$, $f_{j,*}(x) \in \mathcal{O}(D_{\rho_1,\dots,\rho_n})$ for all $j \geqslant 0$, and looking for $\widetilde{u}(t,x)$ in the same type:

$$\widetilde{u}(t,x) = \sum_{j \geqslant 0} u_{j,*}(x)\frac{t^j}{j!} \quad \text{with } u_{j,*}(x) \in \mathcal{O}(D_{\rho_1,\dots,\rho_n}) \text{ for all } j \geqslant 0,$$

one easily checks that the coefficients $u_{j,*}(x)$ are uniquely determined for all $j \geqslant 0$ by the recurrence relations

$$\begin{aligned} u_{j+\kappa,*}(x) &= f_{j,*}(x) \\ &+ \sum_{i\in\mathcal{K}} \sum_{q\in Q_i} \sum_{m\in E_{i,q}} \sum_{j_0+\dots+j_{m+1}=j} \binom{j}{j_0,\dots,j_{m+1}} T_{i,q,m,j_0,\dots,j_{m+1}}(x) \end{aligned} \tag{3.16}$$

together with the initial conditions $u_{j,*}(x) = \varphi_j(x)$ for all $j = 0,\dots,\kappa-1$, where

$$T_{i,q,m,j_0,\dots,j_{m+1}}(x) = \begin{cases} a_{i,q,0;j_0,*}(x)\partial_x^q u_{j_1+i,*}(x) & \text{if } m = 0 \\ a_{i,q,m;j_0,*}(x)u_{j_1,*}(x)\dots u_{j_m,*}(x)\partial_x^q u_{j_{m+1}+i,*}(x) & \text{if } m \geqslant 1 \end{cases},$$

and where the notation $\binom{j}{j_0,\dots,j_{m+1}}$ stands for the multinomial coefficient. The proof is complete. ■

As for the heat equation (3.4) (see Proposition 3.9), the Gevrey regularity of the formal solution $\widetilde{u}(t,x)$ follows a noteworthy dichotomy with respect to the s-Gevrey regularity of the inhomogeneity $\widetilde{f}(t,x)$, the critical value being defined by the structure of a convenient linear partial differential operator.

In the particular case of the semilinear equation (3.15), this operator is the linear operator L and, consequently, the critical value is the value s_c defined in Theorem 3.16. More precisely, we have the following.

Proposition 3.19 ([106]) *Let s_c be the nonnegative rational number equal to the inverse of the smallest positive slope of the Newton polygon $\mathcal{N}_t(L)$ at $t = 0$ of the linear operator L if any exists, and equal to 0 otherwise.*

Let $\widetilde{u}(t,x)$ be the formal solution in $\mathcal{O}(D_{\rho_1,\dots,\rho_n})[[t]]$ of Eq. (3.15). Then,

1. *$\widetilde{u}(t,x)$ and $\widetilde{f}(t,x)$ are simultaneously s-Gevrey for any $s \geqslant s_c$;*
2. *$\widetilde{u}(t,x)$ is generically s_c-Gevrey while $\widetilde{f}(t,x)$ is s-Gevrey with $s < s_c$.*

In the much more general case of Eq. (3.14), the associated linear operator is defined in the following way.

Definition 3.20 (Associated Linear Operator [111]) We call *linear operator associated with Eq. (3.14)* the operator

$$\mathcal{L}_{\kappa,v} = \partial_t^{\kappa} - \sum_{i \in \mathcal{K}} \sum_{q \in Q_i} t^{v_{i,q}} \partial_t^i \partial_x^q$$

defined in Definition 3.14, where, for all $i \in \mathcal{K}$ and all $q \in Q_i$, the nonnegative integer $v_{i,q}$ stands for the smallest valuation at $t = 0$ of the analytic coefficients $a_{i,q,m}(t, x)$ of the polynomial $P_{i,q}(t, x, X)$.

The Gevrey regularity of the formal solution $\widetilde{u}(t, x)$ of Eq. (3.14) is then provided by the following.

Theorem 3.21 ([111]) *Let $\mathcal{L}_{\kappa,v}$ be the associated linear operator associated with Eq. (3.14) (see Definition 3.20).*

Let s_c be the nonnegative rational number equal to the inverse of the smallest positive slope of the Newton polygon $\mathcal{N}_t(\mathcal{L}_{\kappa,v})$ at $t = 0$ of $\mathcal{L}_{\kappa,v}$ if any exists, and equal to 0 otherwise (see Proposition 3.15).

Let $\widetilde{u}(t, x)$ be the formal solution in $\mathcal{O}(D_{\rho_1,\ldots,\rho_n})[[t]]$ of Eq. (3.14). Then,

1. *$\widetilde{u}(t, x)$ and $\widetilde{f}(t, x)$ are simultaneously s-Gevrey for any $s \geqslant s_c$;*
2. *$\widetilde{u}(t, x)$ is generically s_c-Gevrey while $\widetilde{f}(t, x)$ is s-Gevrey with $s < s_c$.*

Remark 3.22

- The linear operator $\mathcal{L}_{\kappa,v}$ associated with Eq. (3.15) does not necessarily coincide with the linear operator L, since their respective power $v_{0,0}$ may be different. However, since their Newton polygon at $t = 0$ are equal, Theorem 3.21 coincides well with Proposition 3.19 in the case of Eq. (3.15).
- In [106, 111], the approaches used to prove Proposition 3.19 and Theorem 3.21 are significantly different from the one outlined in Sect. 3.2.1 for the heat equation (3.4). However, an approach analogous to the latter can also be proposed. For this one, we will notice in particular that the nonlinear terms u^m naturally induce the appearance of multinomial coefficients (see for instance the relations (3.16)) and, consequently, require to adapt the previous proof by replacing in it the technical results on the generalized binomial coefficients by their analogues on the generalized multinomial coefficients (see Chap. 13).
- In the case of *real variables*, a result similar to that of Theorem 3.21 has been given by H. Tahara in [124] for slightly more general partial differential equations.

As a direct consequence of Theorem 3.21, we derive the following two corollaries. The first one provides, in the case where the Newton polygon $\mathcal{N}_t(\mathcal{L}_{\kappa,v})$ has no positive slope, a necessary and sufficient condition for the formal solution $\widetilde{u}(t, x)$ to be convergent; the second one extends to the general equation (3.14) the Maillet-Ramis type theorem stated in Corollary 3.17 for the homogeneous equation (3.13).

Corollary 3.23 *Assume that the Newton polygon $\mathcal{N}_t(\mathcal{L}_{\kappa,\upsilon})$ has no positive slope. Then, the formal solution $\widetilde{u}(t,x)$ of Eq. (3.14) is convergent if and only if the inhomogeneity $\widetilde{f}(t,x)$ is convergent.*

Corollary 3.24 *Assume that the inhomogeneity $\widetilde{f}(t,x)$ of Eq. (3.14) is convergent. Then, the formal solution $\widetilde{u}(t,x)$ is either convergent or $1/k$-Gevrey, where k stands for the smallest positive slope of the Newton polygon $\mathcal{N}_t(\mathcal{L}_{\kappa,\upsilon})$.*

As an application of Theorem 3.21, we end this section devoted to the Gevrey regularity of formal solutions of some partial differential equations with some classical examples, which will be also reconsidered later to illustrate our results of summability.

3.2.4 Some Classical Physical Applications

Equation (3.14) is fundamental in many physical, chemical, biological, and ecological problems. Among the many equations of this form that arise from these different problems, we are particularly interested in the following four classical equations in this section:

- the *n-dimensional heat equation*

$$\begin{cases} \partial_t u - a(t,x)\Delta_x u - u^2 P(t,x,u) = \widetilde{f}(t,x) \\ u(0,x) = \varphi(x) \end{cases}, (t,x) \in \mathbb{C}^{n+1} \qquad (3.17)$$

  which arises in problems involving diffusion and (possibly) nonlinear growth such as heat and mass transfer, combustion theory, and spread theory of animal or plant populations (the polynomial P may be zero);
- the *Klein-Gordon equation*

$$\begin{cases} \partial_t^2 u - a(t,x)\partial_x^2 u - u^2 P(t,x,u) = \widetilde{f}(t,x) \\ u(0,x) = \varphi_0(x), \partial_t u(t,x)_{|t=0} = \varphi_1(x) \end{cases}, (t,x) \in \mathbb{C}^2 \qquad (3.18)$$

  which describes the propagation of (nonlinear) waves in an inhomogeneous medium (the polynomial P may be zero);
- the *Euler-Lagrange equation*

$$\begin{cases} \partial_t^2 u - a(t,x)\partial_x^4 u - u^2 P(t,x,u) = \widetilde{f}(t,x) \\ u(0,x) = \varphi_0(x), \partial_t u(t,x)_{|t=0} = \varphi_1(x) \end{cases}, (t,x) \in \mathbb{C}^2 \qquad (3.19)$$

  which describes the relationship between the beam's deflection and an applied lateral (nonlinear) force (the polynomial P may be zero);

- the *generalized Burgers-Korteweg-de Vries equation* (in short, the gBKdV equation)

$$\begin{cases} \partial_t u - P_{q_1}(t,x,u)\partial_x^{q_1} u - P_{q_2}(t,x,u)\partial_x^{q_2} u = \widetilde{f}(t,x) \\ u(0,x) = \varphi(x), \quad q_1 \geqslant q_2 \end{cases}, (t,x) \in \mathbb{C}^2 \tag{3.20}$$

  which allows to model nonlinear waves in dispersive-dissipative media with instabilities, waves arising in thin films flowing down an inclined surface, moderate-amplitude shallow-water surface waves, changes of the concentration of substances in chemical reactions, etc.

Before applying the results of the previous section to these different equations, let us start by determining the Newton polygon at $t = 0$ of their associated linear operator.

◂ Associated Linear Operators

The Table 3.1 below lists the linear operators associated with Eqs. (3.17)–(3.20).

In the first three rows, v denotes the valuation at $t = 0$ of the coefficient $a(t,x)$ and w the smallest valuation at $t = 0$ of the coefficients of the polynomial $P(t,x,X)$ if $P \not\equiv 0$; in the last row, v_1 (resp. v_2) denotes the smallest valuation at $t = 0$ of the coefficients of the polynomial $P_{q_1}(t,x,X)$ (resp. $P_{q_2}(t,x,X)$).

◂ Newton Polygons

Figure 3.2 describes the Newton polygons of the linear operators associated with Eqs. (3.17)–(3.19) and Fig. 3.3 describes the one of the linear operator associated with Eq. (3.20) (see Table 3.1).

Table 3.1 Linear operators associated with Eqs. (3.17)–(3.20)

Equations	Associated linear operators
Heat equation (3.17)	$\begin{cases} \partial_t - t^v\partial_x^2 & \text{if } P \equiv 0 \\ \partial_t - t^v\partial_x^2 - t^w & \text{if } P \not\equiv 0 \end{cases}$
Klein-Gordon equation (3.18)	$\begin{cases} \partial_t^2 - t^v\partial_x^2 & \text{if } P \equiv 0 \\ \partial_t^2 - t^v\partial_x^2 - t^w & \text{if } P \not\equiv 0 \end{cases}$
Euler-Lagrange equation (3.19)	$\begin{cases} \partial_t^2 - t^v\partial_x^4 & \text{if } P \equiv 0 \\ \partial_t^2 - t^v\partial_x^4 - t^w & \text{if } P \not\equiv 0 \end{cases}$
gBKdV equation (3.20)	$\begin{cases} \partial_t - t^{v_1}\partial_x^{q_1} - t^{v_2}\partial_x^{q_2} & \text{if } q_1 > q_2 \\ \partial_t - t^{\min(v_1,v_2)}\partial_x^{q_1} & \text{if } q_1 = q_2 \end{cases}$

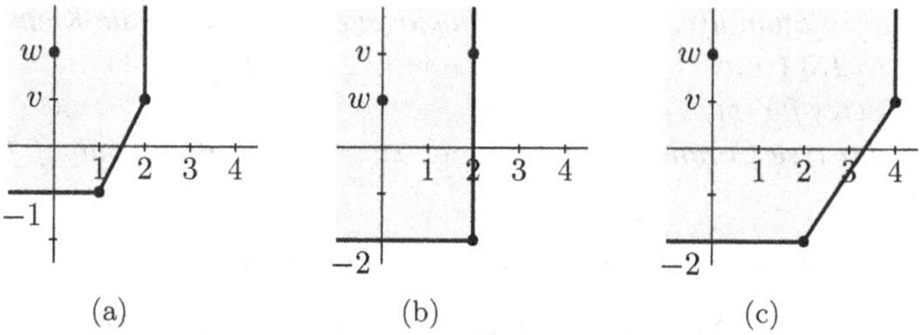

Fig. 3.2 The Newton polygons of the linear operators associated with (**a**) Eq. (3.17), (**b**) Eq. (3.18), (**c**) Eq. (3.19)

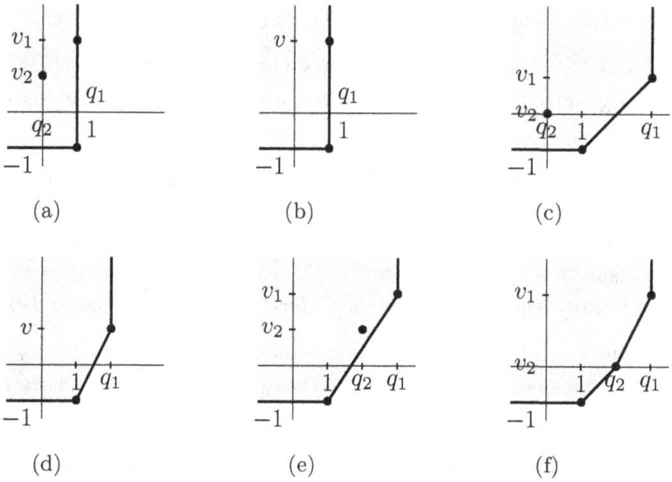

Fig. 3.3 The Newton polygon of the linear operator associated with Eq. (3.20). (**a**) Case $q_2 = 0$ and $q_1 = 1$, (**b**) Case $q_2 = q_1 \leqslant 1$; $v = \min(v_1, v_2)$, (**c**) Case $q_2 \leqslant 1 < q_1$, (**d**) Case $1 < q_2 = q_1$; $v = \min(v_1, v_2)$, (**e**) Case $1 < q_2 < q_1$ and $\dfrac{1+v_2}{q_2-1} \geqslant \dfrac{1+v_1}{q_1-1}$, (**f**) Case $1 < q_2 < q_1$ and $\dfrac{1+v_2}{q_2-1} < \dfrac{1+v_1}{q_1-1}$

◂ Applications of Theorem 3.21

We are now able to apply Theorem 3.21 in view to study the Gevrey regularity of the formal solutions of the various equations (3.17)–(3.20). Observe that several cases appear depending on whether the Newton polygon of the associated linear operator has no, exactly one or at least one positive slope (see Figs. 3.2 and 3.3).

Application 3.25 (Klein-Gordon and gBKdV Equations with $q_1 \leqslant 1$) Since the Newton polygons of the linear operators associated with these equations have no positive slope (see Figs. 3.2b, 3.3a and b), the critical value is zero. Consequently, applying Theorem 3.21, we get the following result :

- *The formal solution $\widetilde{u}(t,x)$ and the inhomogeneity $\widetilde{f}(t,x)$ of the Klein-Gordon equation (3.18) (resp. the gBKdV equation (3.20) with $q_1 \leqslant 1$) are simultaneously s-Gevrey for any $s \geqslant 0$.*
- *In particular (see Corollary 3.23), $\widetilde{u}(t,x)$ is convergent if and only if $\widetilde{f}(t,x)$ is convergent.*

Application 3.26 (Heat and Euler-Lagrange Equations)

1. *Heat equation.* The Newton polygon of the associated linear operator admits a unique positive slope, which is equal to $1+v$ (see Fig. 3.2a). Thereby, the critical value of the equation is $1/(1+v)$; hence, applying Theorem 3.21, the following result:
Let $\widetilde{u}(t,x)$ be the formal solution of the heat equation (3.17). Then,
    - *$\widetilde{u}(t,x)$ and $\widetilde{f}(t,x)$ are simultaneously s-Gevrey for any $s \geqslant 1/(1+v)$.*
    - *$\widetilde{u}(t,x)$ is generically $1/(1+v)$-Gevrey while $\widetilde{f}(t,x)$ is s-Gevrey with $s < 1/(1+v)$.*

2. *Euler-Lagrange equation.* The Newton polygon of the associated linear operator admits a unique positive slope, which is equal to $(2+v)/2$ (see Fig. 3.2). Thereby, the critical value of the equation is $2/(2+v)$; hence, applying Theorem 3.21, the following result:
Let $\widetilde{u}(t,x)$ be the formal solution of the Euler-Lagrange equation (3.19). Then,
    - *$\widetilde{u}(t,x)$ and $\widetilde{f}(t,x)$ are simultaneously s-Gevrey for any $s \geqslant 2/(2+v)$.*
    - *$\widetilde{u}(t,x)$ is generically $2/(2+v)$-Gevrey while $\widetilde{f}(t,x)$ is s-Gevrey with $s < 2/(2+v)$.*

Application 3.27 (gBKdV Equation with $q_1 > 1$) Depending on the values of the derivation order q_2 and of the valuations v_1 and v_2, the Newton polygon of the associated linear operator admits either a single or two positive slopes (see Fig. 3.3c–f). Considering then the smaller one, we deduce that the critical value s_c of the equation is defined by

$$s_c = \begin{cases} \dfrac{q_1-1}{1+v_1} & \text{if } q_2 \leqslant 1 < q_1 \text{ or } \left(1 < q_2 < q_1 \text{ and } \dfrac{1+v_1}{q_1-1} \leqslant \dfrac{1+v_2}{q_2-1}\right) \\ \dfrac{q_1-1}{1+\min(v_1,v_2)} & \text{if } 1 < q_2 = q_1 \\ \dfrac{q_2-1}{1+v_2} & \text{if } 1 < q_2 < q_1 \text{ and } \dfrac{1+v_2}{q_2-1} \leqslant \dfrac{1+v_1}{q_1-1} \end{cases}.$$

Hence, applying Theorem 3.21, the following result:
Let $\widetilde{u}(t,x)$ be the formal solution of the gBKdV equation (3.20) with $q_1 > 1$. Then,

- *$\widetilde{u}(t,x)$ and $\widetilde{f}(t,x)$ are simultaneously s-Gevrey for any $s \geqslant s_c$.*
- *$\widetilde{u}(t,x)$ is generically s_c-Gevrey while $\widetilde{f}(t,x)$ is s-Gevrey with $s < s_c$.*

In particular, this result provides the Gevrey regularity of the formal solution of the classical inhomogeneous Korteweg-de Vries equation

$$\partial_t u + \partial_x^3 u - 6u\partial_x u = \widetilde{f}(t,x),$$

and the Gevrey regularity of the formal solution of the classical inhomogeneous Burgers equation

$$\partial_t u - \partial_x^2 u - 2u\partial_x u = \widetilde{f}(t,x).$$

Indeed, these two equations both correspond to the case presented in Fig. 3.3c and admit respectively $s_c = 2$ and $s_c = 1$ as critical value.

Remark 3.28 The approaches developed in Sect. 3.2.1 and in [111] to describe the Gevrey regularity of the formal power series solutions of the heat equation (3.4) and of the general equation (3.14) can also be applied to other types of partial differential equations. For instance, an approach similar to that of [111] was used by the author in [105] to determine the Gevrey regularity of the formal power series solutions of some inhomogeneous linear Cauchy-Goursat problems. More recently, adapting the approach presented in Sect. 3.2.1, the author proved that the formal power series solution of the inhomogeneous generalized Boussinesq equation

$$\begin{cases} \partial_t^2 u - a(t,x)\partial_x^4 u - P(t,x,u)\partial_x^2 u - Q(t,x,u)(\partial_x u)^2 = \widetilde{f}(t,x) \\ \partial_t^j u(t,x)_{|t=0} = \varphi_j(x), \ j = 0,1 \end{cases}, (t,x) \in \mathbb{C}^2$$

with $a(0,0) \neq 0$ admits the same Gevrey regularity as the formal power series solution of the Euler-Lagrange equation [110].

Chapter 4
Gevrey Asymptotics

As said at the beginning of Chap. 3, Poincaré asymptotics is not sufficient to study the summability of the formal power series solutions of partial differential equations, since the conditions required on the constants $C_{x_0,\Sigma',J}$ given in Definition 2.1 are too weak. In particular, they are not uniform with respect to x and their dependence in J is not known. To get around these problems, one strengthens Poincaré asymptotics into Gevrey asymptotics in which the asymptotic functions admit a uniform Gevrey condition in x.

In the sequel of this chapter, s denotes a positive real number and Σ an open sector with vertex 0 either in $\mathbb{C}^*$ or in $\widetilde{\mathbb{C}}$.

4.1 Algebras of s-Gevrey Asymptotic Functions

Definition 4.1 (*s*-**Gevrey Asymptotic Expansion**) Let $r > 0$, $u(t,x) \in \mathcal{O}(\Sigma \times D_{r,\ldots,r})$ be a holomorphic function on $\Sigma \times D_{r,\ldots,r}$ and

$$\widetilde{u}(t,x) = \sum_{j \geqslant 0} u_j(x)t^j \in \mathcal{O}(D_{r,\ldots,r})[[t]]$$

a formal power series with analytic coefficients in $D_{r,\ldots,r}$.

The function u is said to *admit* $\widetilde{u}$ *as* s*-Gevrey asymptotic expansion at* 0 *on* Σ (or to *be* s*-Gevrey asymptotic to* $\widetilde{u}$ *at* 0 *on* Σ) if for all proper subsector $\Sigma' \Subset \Sigma$, there exist two positive constants $C_{\Sigma'}, K_{\Sigma'} > 0$ such that the following estimate holds for all $J \in \mathbb{N}^*$ and all $(t,x) \in \Sigma' \times D_{r,\ldots,r}$:

$$\left| u(t,x) - \sum_{j=0}^{J-1} u_j(x)t^j \right| \leqslant C_{\Sigma'} K_{\Sigma'}^J \Gamma(1+sJ)\,|t|^J\,. \tag{4.1}$$

P. Remy, *Asymptotic Expansions and Summability*, Lecture Notes in Mathematics 2351, https://doi.org/10.1007/978-3-031-59094-8_4

Observe that the s-Gevrey asymptotic expansion $\widetilde{u}(t,x)$ at 0 on Σ of a given function $u(t,x) \in \mathcal{O}(\Sigma \times D_{r,\ldots,r})$ is unique, if any exists. Indeed, inequality (4.1) implies that u belongs to $\mathcal{T}_r(\Sigma)$ with $T_{r,\Sigma}(u) = \widetilde{u}$.

Observe also that the coefficients $u_j(x)$ of such a s-Gevrey asymptotic expansion may be holomorphic on a common polydisc $D_{\rho_1,\ldots,\rho_n}$ "greater" than $D_{r,\ldots,r}$, that is satisfying $\min(\rho_1,\ldots,\rho_n) \geqslant r$.

Definition 4.2 (s-Gevrey Asymptotic Series) A formal power series which is the s-Gevrey asymptotic expansion at 0 on Σ of a function $u(t,x) \in \mathcal{O}(\Sigma \times D_{r,\ldots,r})$ for some $r > 0$ is said to be *a s-Gevrey asymptotic series at* 0 *on* Σ.

Proposition 4.3 *A s-Gevrey asymptotic series at* 0 *on* Σ *is a s-Gevrey formal power series.*

Proof Let $\widetilde{u}(t,x) = \sum_{j\geqslant 0} u_j(x)t^j \in \mathcal{O}(D_{r,\ldots,r})[[t]]$ the s-Gevrey asymptotic expansion at 0 on Σ of a function $u(t,x) \in \mathcal{O}(\Sigma \times D_{r,\ldots,r})$. Let us fix a radius $0 < r' < r$ and a proper subsector $\Sigma' \Subset \Sigma$. By assumption (see Definition 4.1), there exist two positive constants $C', K > 0$ such that the following estimate holds for all $J \in \mathbb{N}^*$ and all $(t,x) \in \Sigma' \times D_{r,\ldots,r}$:

$$\left| u(t,x) - \sum_{j=0}^{J-1} u_j(x)t^j \right| \leqslant C'K^J\Gamma(1+sJ)\,|t|^J .$$

Applying this twice to the relation

$$u_J(x)t^J = \left(u(t,x) - \sum_{j=0}^{J-1} u_j(x)t^j \right) - \left(u(t,x) - \sum_{j=0}^{J} u_j(x)t^j \right)$$

and making t tend to 0, we get

$$|u_J(x)| \leqslant C'K^J\Gamma(1+sJ)$$

for all $J \geqslant 1$ and all $x \in D_{r,\ldots,r}$; hence, the inequalities

$$|u_J(x)| \leqslant CK^J\Gamma(1+sJ)$$

for all $J \geqslant 0$ and all $x \in D_{r',\ldots,r'}$ by setting $C = \max\left(C', \sup_{\|x\|\leqslant r'} |u_0(x)| \right)$, which achieves the proof. ■

Notation 4.4 For any $r > 0$, we denote by

- $\overline{\mathcal{A}}_{s,r}(\Sigma)$ the set of all the functions $u(t,x) \in \mathcal{O}(\Sigma \times D_{r,\dots,r})$ admitting a s-Gevrey asymptotic expansion at 0 on Σ;
- $T_{s,r,\Sigma} : \overline{\mathcal{A}}_{s,r}(\Sigma) \longrightarrow \mathcal{O}(D_{r,\dots,r})[[t]]_s$ the map assigning to each $u \in \overline{\mathcal{A}}_{s,r}(\Sigma)$ its s-Gevrey asymptotic expansion at 0 on Σ.

Observe that the map $T_{s,r,\Sigma}$ is the restriction of the map $T_{r,\Sigma}$ to $\overline{\mathcal{A}}_{s,r}(\Sigma)$, and that the inclusion $T_{s,r,\Sigma}(\overline{\mathcal{A}}_{s,r}(\Sigma)) \subset \mathcal{O}(D_{r,\dots,r})[[t]]_s$ is due to Proposition 4.3 above.

Observe also that the inclusion $\overline{\mathcal{A}}_{s,r}(\Sigma) \subset \overline{\mathcal{A}}_{s,r'}(\Sigma)$ and the identity $T_{s,r,\Sigma}(u) = T_{s,r',\Sigma}(u)$ hold for all $u \in \overline{\mathcal{A}}_{s,r}(\Sigma)$ and all $0 < r' \leqslant r$.

4.1.1 Characterization of s-Gevrey Asymptotic Functions

As said just above, a function $u \in \overline{\mathcal{A}}_{s,r}(\Sigma)$ belongs to $\mathcal{T}_r(\Sigma)$ and $T_{s,r,\Sigma}(u) = T_{r,\Sigma}(u)$. Reciprocally, the following proposition provides a characterization for a function $u \in \mathcal{T}_r(\Sigma)$ to belong to $\overline{\mathcal{A}}_{s,r}(\Sigma)$.

Proposition 4.5 *A function $u \in \mathcal{T}_r(\Sigma)$ belongs to $\overline{\mathcal{A}}_{s,r}(\Sigma)$ if and only if for all proper subsector $\Sigma' \Subset \Sigma$, there exist two positive constants $C'_{\Sigma'} > 0$ and $K'_{\Sigma'} > 0$ such that the following estimate holds for all $\ell \in \mathbb{N}$ and all $(t,x) \in \Sigma' \times D_{r,\dots,r}$:*

$$\left|\partial_t^\ell u(t,x)\right| \leqslant C'_{\Sigma'} K'^{\ell}_{\Sigma'} \Gamma(1+(s+1)\ell). \tag{4.2}$$

Proof Let $u \in \mathcal{T}_r(\Sigma)$ and

$$\widetilde{u}(t,x) = \sum_{j\geqslant 0} u_j(x)t^j \in \mathcal{O}(D_{r,\dots,r})[[t]]$$

its Taylor expansion at 0 on Σ.

◁ *The* only if *part.* Let us assume $u \in \overline{\mathcal{A}}_{s,r}(\Sigma)$ and let us fix a proper subsector $\Sigma' \Subset \Sigma$. Like in the proof of Proposition 2.4, p. 9, we consider a sector Σ'' such that $\Sigma' \Subset \Sigma'' \Subset \Sigma'$ and we attach to any $t \in \Sigma'$ the closed disc $\overline{D}(t, |t|\,\delta)$ with center t and radius $|t|\,\delta$, the constant δ being chosen so small that $\overline{D}(t, |t|\,\delta)$ be contained in Σ'' (see Fig. 2.1). To prove the existence of the constants $C'_{\Sigma'}, K'_{\Sigma'} > 0$ so that inequality (4.2) holds, we proceed as follows. Let us fix $x \in D_{r,\dots,r}$ and let us start by observing that

$$\partial_t^\ell u(t,x) = \partial_t^\ell \left(u(t,x) - \sum_{j=0}^{\ell-1} u_j(x)t^j \right)$$

for all $\ell \geqslant 0$ and all $t \in \Sigma'$. Indeed, the sum is 0 when $\ell = 0$ and the ℓ-th derivative of a polynomial of degree $\ell - 1$ is also 0 when $\ell \geqslant 1$. Then, applying successively the Cauchy Integral Formula and inequality (4.1), we derive for all $\ell \geqslant 0$ and all $t \in \Sigma'$ the relations

$$\begin{aligned}
\left|\partial_t^\ell u(t,x)\right| &= \left|\frac{\ell!}{2i\pi}\int_{|t'-t|=|t|\delta}\left(u(t',x) - \sum_{j=0}^{\ell-1} u_j(x)t'^j\right)\frac{dt'}{(t'-t)^{\ell+1}}\right| \\
&\leqslant C_{\Sigma''}K_{\Sigma''}^{\ell}\Gamma(1+s\ell)\ell!\frac{|t|^\ell(1+\delta)^\ell}{|t|^\ell\delta^\ell} \\
&= C_{\Sigma''}\left(\frac{K_{\Sigma''}(1+\delta)}{\delta}\right)^\ell \Gamma(1+(s+1)\ell) \times \frac{1}{\binom{(s+1)\ell}{\ell}}
\end{aligned}$$

Hence, using $\binom{(s+1)\ell}{\ell} \geqslant 1$ (see Proposition 13.5) and setting $C'_{\Sigma'} = C_{\Sigma''}$ and $K'_{\Sigma'} = K_{\Sigma''}\left(1 + \frac{1}{\delta}\right)$, the following inequality

$$\left|\partial_t^\ell u(t,x)\right| \leqslant C'_{\Sigma'}K'^{\ell}_{\Sigma'}\Gamma(1+(s+1)\ell)$$

for all $\ell \geqslant 0$ and all $t \in \Sigma'$, which proves inequality (4.2) since the positive constants $C'_{\Sigma'}$ and $K'_{\Sigma'}$ are both independent of x by construction.

◁ *The* if *part.* Let us now assume that inequalities (4.2) on the successive derivatives $\partial_t^\ell u$ hold. To prove that $u \in \overline{\mathcal{A}}_{s,r}(\Sigma)$, that is for any proper subsector $\Sigma' \Subset \Sigma$, there exist two positive constants $C_{\Sigma'}, K_{\Sigma'} > 0$ such that inequality (4.1) holds, we proceed like in the proof of Proposition 2.7. For any fixed $x \in D_{r,\ldots,r}$, we first derive from the Taylor Formula with integral remainder the identity

$$\begin{aligned}
u(t,x) - \sum_{j=0}^{J-1} u_j(x)t^j &= \int_0^t \frac{(t-t')^{J-1}}{(J-1)!}\frac{\partial^J u}{\partial t^J}(t',x)dt' \\
&= -\frac{1}{J!}\int_0^t \frac{\partial^J u}{\partial t^J}(t',x)d(t-t')^J
\end{aligned}$$

for all $J \geqslant 1$ and all $t \in \Sigma'$. Hence, applying inequality (4.2) and using Proposition 13.5, the relations

$$\begin{aligned}
\left|u(t,x) - \sum_{j=0}^{J-1} u_j(x)t^j\right| &\leqslant C'_{\Sigma'}K'^{J}_{\Sigma'}\Gamma(1+sJ)\,|t|^J \times \binom{(s+1)J}{J} \\
&\leqslant C'_{\Sigma'}K'^{J}_{\Sigma'}\Gamma(1+sJ)\,|t|^J \times 4^{(\lceil s\rceil+1)J}
\end{aligned}$$

for all $J \geqslant 1$ and all $t \in \Sigma'$, where $\lceil s \rceil$ stands for the ceil of s. Setting then $C_{\Sigma'} = C'_{\Sigma'}$ and $K_{\Sigma'} = 4^{\lceil s \rceil + 1} K'_{\Sigma'}$, we finally get the inequality

$$\left| u(t,x) - \sum_{j=0}^{J-1} u_j(x) t^j \right| \leqslant C_{\Sigma'} K_{\Sigma'}^J \Gamma(1+sJ)\, |t|^J$$

for all $J \geqslant 1$ and all $t \in \Sigma'$, which proves inequality (4.1) since the positive constants $C_{\Sigma'}$ and $K_{\Sigma'}$ are both independent of x by construction. ■

In Chap. 2, Corollary 2.9, we saw that a function $u(t,x) \in \mathcal{O}(\Sigma \times D_{r,\dots,r})$ which is bounded, as well as all its derivatives $\partial_t^\ell u$, on $\Sigma \times D_{r,\dots,r}$ belongs to $\mathcal{T}_r(\Sigma')$ for any proper subsector $\Sigma' \Subset \Sigma$. Combining this with Proposition 4.5 above, we straightaway derive the following more precise result which will be very useful later for the study of the summability of the formal power series solutions of the partial differential equations.

Corollary 4.6 *Let $u(t,x) \in \mathcal{O}(\Sigma \times D_{r,\dots,r})$ for some $r > 0$.*

Let us assume there exist two positive constants $C, K > 0$ such that the following estimate holds for all $\ell \in \mathbb{N}$ and all $(t,x) \in \Sigma \times D_{r,\dots,r}$:

$$\left| \partial_t^\ell u(t,x) \right| \leqslant C K^\ell \Gamma(1+(s+1)\ell).$$

Then, $u \in \overline{\mathcal{A}}_{s,r}(\Sigma')$ for any proper subsector $\Sigma' \Subset \Sigma$.

4.1.2 Algebraic Structure of s-Gevrey Asymptotic Functions

Let us now turn to the study of the algebraic structure of the s-Gevrey asymptotic functions. We have the following first result.

Proposition 4.7 *Let $r > 0$.*

- *The set $\overline{\mathcal{A}}_{s,r}(\Sigma)$ endowed with the usual algebraic operations and the usual derivation ∂_t is a $\mathbb{C}$-differential algebra. Moreover, it is stable under the anti-derivations ∂_t^{-1} and $\partial_{x_\ell}^{-1}$ for all $\ell = 1, \dots, n$.*
- *The map $T_{s,r,\Sigma} : \overline{\mathcal{A}}_{s,r}(\Sigma) \longrightarrow \mathcal{O}(D_{r,\dots,r})[[t]]_s$ is a morphism of $\mathbb{C}$-differential algebras which commutes with ∂_t^{-1} and $\partial_{x_\ell}^{-1}$ for all $\ell = 1, \dots, n$.*

Proof Since $\overline{\mathcal{A}}_{s,r}(\Sigma) \subset \mathcal{T}_r(\Sigma)$ and since $(\mathcal{T}_r(\Sigma), +, \times, \partial_t)$ is a $\mathbb{C}$-differential algebra (see Proposition 2.4), it is sufficient to prove that $\overline{\mathcal{A}}_{s,r}(\Sigma)$ is a $\mathbb{C}$-differential subalgebra of $\mathcal{T}_r(\Sigma)$ stable under the anti-derivations ∂_t^{-1} and $\partial_{x_\ell}^{-1}$ for all $\ell = 1, \dots, n$.

• *Differential subalgebra and stability under ∂_t^{-1}.* We use here the characterization of $\overline{\mathcal{A}}_{s,r}(\Sigma)$ stated in Proposition 4.5. Let $u, v \in \overline{\mathcal{A}}_{s,r}(\Sigma)$ and $\Sigma' \Subset \Sigma$ be a proper subsector of Σ. According to Proposition 4.5, there exist four positive constants $C_{\Sigma'}, C'_{\Sigma'}, K_{\Sigma'}, K'_{\Sigma'} > 0$ such that

$$\begin{aligned}\left|\partial_t^\ell u(t,x)\right| &\leqslant C_{\Sigma'} K_{\Sigma'}^\ell \Gamma(1+(s+1)\ell)\\ \text{and} \quad \left|\partial_t^\ell v(t,x)\right| &\leqslant C'_{\Sigma'} K'^{\ell}_{\Sigma'} \Gamma(1+(s+1)\ell)\end{aligned} \tag{4.3}$$

for all $\ell \geqslant 0$ and all $(t,x) \in \Sigma' \times D_{r,\dots,r}$.

$\triangleleft$ *Stability under linear combination.* Let $\lambda, \mu \in \mathbb{C}$. According to Proposition 2.4, $\lambda u + \mu v \in \mathcal{T}_r(\Sigma)$ and

$$T_{r,\Sigma}(\lambda u + \mu v) = \lambda T_{r,\Sigma}(u) + \mu T_{r,\Sigma}(v).$$

On the other hand, we straightaway derive from inequalities (4.3) that

$$\begin{aligned}\left|\partial_t^\ell(\lambda u+\mu v)(t,x)\right| &\leqslant |\lambda|\, C_{\Sigma'} K_{\Sigma'}^\ell \Gamma(1+(s+1)\ell) + |\mu|\, C'_{\Sigma'} K'^{\ell}_{\Sigma'} \Gamma(1+(s+1)\ell)\\ &\leqslant (|\lambda|\, C_{\Sigma'} + |\mu|\, C'_{\Sigma'})(K_{\Sigma'} + K'_{\Sigma'})^\ell \Gamma(1+(s+1)\ell)\end{aligned}$$

for all $\ell \geqslant 0$ and all $(t,x) \in \Sigma' \times D_{r,\dots,r}$. Hence, applying Proposition 4.5:

$$\lambda u + \mu v \in \overline{\mathcal{A}}_{s,r}(\Sigma) \quad \text{and} \quad T_{s,r,\Sigma}(\lambda u + \mu v) = \lambda T_{s,r,\Sigma}(u) + \mu T_{s,r,\Sigma}(v).$$

$\triangleleft$ *Stability under multiplication.* According to Proposition 2.4, the product uv belongs to $\mathcal{T}_r(\Sigma)$ and $T_{r,\Sigma}(uv) = T_{r,\Sigma}(u) T_{r,\Sigma}(v)$. Moreover, applying successively the Leibniz Formula, inequalities (4.3) and Proposition 13.5, its derivatives satisfy

$$\begin{aligned}\left|\partial_t^\ell(uv)(t,x)\right| &\leqslant \sum_{k=0}^{\ell} \binom{\ell}{k} \left|\partial_t^k u(t,x)\right| \left|\partial_t^{\ell-k} v(t,x)\right|\\ &\leqslant C_{\Sigma'} C'_{\Sigma'} \left(\sum_{k=0}^{\ell} \frac{\binom{\ell}{k}}{\binom{(s+1)\ell}{(s+1)k}} K_{\Sigma'}^k K'^{\ell-k}_{\Sigma'} \right) \Gamma(1+(s+1)\ell)\\ &\leqslant C_{\Sigma'} C'_{\Sigma'} (K_{\Sigma'} + K'_{\Sigma'})^\ell \Gamma(1+(s+1)\ell)\end{aligned}$$

for all $\ell \geqslant 0$ and all $(t,x) \in \Sigma' \times D_{r,\dots,r}$. Hence, applying Proposition 4.5:

$$uv \in \overline{\mathcal{A}}_{s,r}(\Sigma) \quad \text{and} \quad T_{s,r,\Sigma}(uv) = T_{s,r,\Sigma}(u) T_{s,r,\Sigma}(v).$$

$\triangleleft$ *Stability under the derivation* ∂_t. As previously, Proposition 2.4 tells us that the derivative $\partial_t u$ belongs to $\mathcal{T}_r(\Sigma)$ and $T_{r,\Sigma}(\partial_t u) = \partial_t T_{r,\Sigma}(u)$. Besides, from inequalities (4.3) and Proposition 13.5, we derive the relations

$$\begin{aligned}\left|\partial_t^\ell(\partial_t u)(t,x)\right| &\leqslant C_{\Sigma'}K_{\Sigma'}^{\ell+1}\Gamma(1+(s+1)(\ell+1))\\ &= \Gamma(2+s)C_{\Sigma'}K_{\Sigma'}^{\ell+1}\binom{(s+1)(\ell+1)}{s+1}\Gamma(1+(s+1)\ell)\\ &\leqslant \Gamma(2+s)4^{\lceil s\rceil+1}C_{\Sigma'}K_{\Sigma'}\left(4^{\lceil s\rceil+1}K_{\Sigma'}\right)^\ell\Gamma(1+(s+1)\ell)\end{aligned}$$

for all $\ell \geqslant 0$ and all $(t,x) \in \Sigma' \times D_{r,\dots,r}$, where $\lceil s\rceil$ stands for the ceil of s. Hence, applying Proposition 4.5:

$$\partial_t u \in \overline{\mathcal{A}}_{s,r}(\Sigma) \quad\text{and}\quad T_{s,r,\Sigma}(\partial_t u) = \partial_t T_{s,r,\Sigma}(u).$$

◃ *Stability under the anti-derivation* ∂_t^{-1}. Once again, Proposition 2.4 tells us that the anti-derivative $\partial_t^{-1}u$ belongs to $\mathcal{T}_r(\Sigma)$ and $T_{r,\Sigma}(\partial_t^{-1}u) = \partial_t^{-1}T_{r,\Sigma}(u)$. Let us now denote by R the radius of Σ. Applying inequalities (4.3) and Proposition 13.5, the derivatives of $\partial_t^{-1}u$ satisfy, for all $(t,x) \in \Sigma' \times D_{r,\dots,r}$,

$$\left|\partial_t^\ell(\partial_t^{-1}u)(t,x)\right| \leqslant C_{\Sigma'}|t| \leqslant C_{\Sigma'}R$$

when $\ell = 0$ and

$$\begin{aligned}\left|\partial_t^\ell(\partial_t^{-1}u)(t,x)\right| &\leqslant C_{\Sigma'}K_{\Sigma'}^{\ell-1}\Gamma(1+(s+1)(\ell-1))\\ &= \frac{C_{\Sigma'}}{\Gamma(2+s)K_{\Sigma'}}K_{\Sigma'}^\ell\Gamma(1+(s+1)\ell)\times\frac{1}{\binom{(s+1)\ell}{s+1}}\\ &\leqslant \frac{C_{\Sigma'}}{\Gamma(2+s)K_{\Sigma'}}K_{\Sigma'}^\ell\Gamma(1+(s+1)\ell)\end{aligned}$$

when $\ell \geqslant 1$. Hence, applying Proposition 4.5:

$$\partial_t^{-1}u \in \overline{\mathcal{A}}_{s,r}(\Sigma) \quad\text{and}\quad T_{s,r,\Sigma}(\partial_t^{-1}u) = \partial_t^{-1}T_{s,r,\Sigma}(u).$$

• *Stability under the anti-derivation* $\partial_{x_\ell}^{-1}$. Let us fix $\ell \in \{1,\dots,n\}$. Since $\mathcal{T}_r(\Sigma)$ is not stable under the anti-derivation $\partial_{x_\ell}^{-1}$ (see Remark 2.6), we can no longer reason as before. To get around this problem, we directly use here the definition of the Gevrey asymptotics. Let $u \in \overline{\mathcal{A}}_{s,r}(\Sigma)$ and

$$\widetilde{u}(t,x) = \sum_{j\geqslant 0} u_j(x)t^j$$

its s-Gevrey asymptotic expansion at 0 on Σ. Let us also consider a proper subsector $\Sigma' \Subset \Sigma$ and, for any $J \geqslant 1$, let us denote by U_J the function defined on $\Sigma \times D_{r,\ldots,r}$ by

$$U_J(t,x) = u(t,x) - \sum_{j=0}^{J-1} u_j(x)t^j.$$

By assumption (see Definition 4.1), there exist two positive constants $C_{\Sigma'}, K_{\Sigma'} > 0$ such that, for all $J \geqslant 1$ and all $(t,x) \in \Sigma' \times D_{r,\ldots,r}$:

$$|U_J(t,x)| \leqslant C_{\Sigma'} K_{\Sigma'}^J \Gamma(1+sJ)\, |t|^J \tag{4.4}$$

We must prove similar inequalities for $\partial_{x_\ell}^{-1} U_J$. To this end, we first observe that

$$\partial_{x_\ell}^{-1} U_J(t,x) = \int_0^{x_\ell} U_J(t,x_1,\ldots,x_{\ell-1},x'_\ell,x_{\ell+1},\ldots,x_n)dx'_\ell$$

and that $x \in D_{r,\ldots,r}$ implies $(x_1,\ldots,x_{\ell-1},x'_\ell,x_{\ell+1},\ldots,x_n) \in D_{r,\ldots,r}$. Thereby, we clearly derive from inequalities (4.4) the relations

$$\left|\partial_{x_\ell}^{-1} U_J(t,x)\right| \leqslant C_{\Sigma'} K_{\Sigma'}^J \Gamma(1+sJ)\, |t|^J\, |x_\ell| \leqslant r C_{\Sigma'} K_{\Sigma'}^J \Gamma(1+sJ)\, |t|^J$$

for all $J \geqslant 1$ and all $(t,x) \in \Sigma' \times D_{r,\ldots,r}$ and, consequently, $\partial_{x_\ell}^{-1} u \in \overline{\mathcal{A}}_{s,r}(\Sigma)$ with $T_{s,r,\Sigma}(\partial_{x_\ell}^{-1} u) = \partial_{x_\ell}^{-1} T_{s,r,\Sigma}(u)$. ■

Observe that the stability of $\overline{\mathcal{A}}_{s,r}(\Sigma)$ under the anti-derivations $\partial_{x_\ell}^{-1}$ is due to the fact that conditions (4.1) which are required in the Gevrey asymptotics are uniform in both t and x (which was not the case for conditions (2.1) required in the Taylor expansions).

Observe also that, despite these strong conditions, the sets $\overline{\mathcal{A}}_{s,r}(\Sigma)$ are stable under the derivations ∂_{x_ℓ} for no $r > 0$ since the Cauchy Integral Formula requires to shrink the polydisc $D_{r,\ldots,r}$. More precisely, we have the following.

Proposition 4.8 *Let $\ell \in \{1,\ldots,n\}$, $r > 0$, $u \in \overline{\mathcal{A}}_{s,r}(\Sigma)$ and $\widetilde{u} = T_{s,r,\Sigma}(u)$.*
Then, $\partial_{x_\ell} u \in \overline{\mathcal{A}}_{s,r'}(\Sigma)$ with $T_{s,r',\Sigma}(\partial_{x_\ell} u) = \partial_{x_\ell} \widetilde{u}$ for all $0 < r' < r$.

Proof We proceed like in the proof of the stability of $\overline{\mathcal{A}}_{s,r}(\Sigma)$ under the anti-derivation $\partial_{x_\ell}^{-1}$.

Let us consider a radius $r' \in]0,r[$ and a proper subsector $\Sigma' \Subset \Sigma$, and, for any $J \geqslant 1$, let us denote by U_J the function defined on $\Sigma \times D_{r,\ldots,r}$ by

$$U_J(t,x) = u(t,x) - \sum_{j=0}^{J-1} u_j(x)t^j,$$

where the $u_j(x)$'s stand as before for the coefficients of $\widetilde{u}(t,x)$.

We must prove there exist two positive constants $C_{\Sigma'}, K_{\Sigma'} > 0$ such that

$$\left|\partial_{x_\ell} U_J(t,x)\right| \leqslant C_{\Sigma'} K_{\Sigma'}^J \Gamma(1+sJ)\,|t|^J$$

for all $J \geqslant 1$ and all $(t,x) \in \Sigma' \times D_{r',\dots,r'}$. To this end, let us fix, like in the proof of Proposition 2.4, p. 9, a sector Σ'' such that $\Sigma' \Subset \Sigma'' \Subset \Sigma$, and let us choose $\delta > 0$ so small that, for all $t \in \Sigma'$, the closed disc with center t and radius $|t|\,\delta$ be contained in Σ'' (see Fig. 2.1). By assumption (see Definition 4.1), there exist two positive constants $C_{\Sigma''}, K_{\Sigma''} > 0$ such that, for all $J \geqslant 1$ and all $(t,x) \in \Sigma'' \times D_{r,\dots,r}$:

$$|U_J(t,x)| \leqslant C_{\Sigma''} K_{\Sigma''}^J \Gamma(1+sJ)\,|t|^J\,. \tag{4.5}$$

Let us now apply the Cauchy Integral Formula: for all $J \geqslant 1$ and all $(t,x) \in \Sigma' \times D_{r',\dots,r'}$,

$$\partial_{x_\ell} U_J(t,x) = \frac{1}{(2i\pi)^{n+1}} \int_{\gamma(t,x)} \frac{U_j(t',x')}{(t'-t)(x'_\ell - x_\ell)^2 \prod\limits_{\substack{k=1\\k\neq\ell}}^{n} (x'_k - x_k)} dt'dx',$$

where $\gamma(t,x)$ is the path defined as

$$\{(t',x') \in \mathbb{C}^{n+1};\ \left|t'-t\right| = |t|\,\delta \text{ and } \left|x'_k - x_k\right| = r - r' \text{ for all } k \in \{1,\dots,n\}\}.$$

Since $(t',x') \in \Sigma'' \times D_{r,\dots,r}$ for any $(t',x') \in \gamma(t,x)$, we deduce from (4.5) the relation

$$\left|\partial_{x_\ell} U_J(t,x)\right| \leqslant C_{\Sigma''} K_{\Sigma''}^J \Gamma(1+sJ) \frac{|t|^J (1+\delta)^J}{r-r'};$$

hence, by setting $C_{\Sigma'} = \dfrac{C_{\Sigma''}}{r-r'}$ and $K_{\Sigma'} = (1+\delta)K_{\Sigma''}$, the inequality

$$\left|\partial_{x_\ell} U_J(t,x)\right| \leqslant C_{\Sigma'} K_{\Sigma'}^J \Gamma(1+sJ)\,|t|^J$$

for all $J \geqslant 1$ and all $(t,x) \in \Sigma' \times D_{r',\dots,r'}$, which completes the proof. ■

Combining then Propositions 4.7 and 4.8 and using the inclusion $\overline{\mathcal{A}}_{s,r}(\Sigma) \subset \overline{\mathcal{A}}_{s,r'}(\Sigma)$ for any $0 < r' < r$, we finally get the following result which provides a *"good"* $\mathbb{C}$-differential algebra for studying the summability of the formal power series solutions of the partial differential equations.

Proposition 4.9 *Let* $\mathbb{C}\{x\}[[t]]_s = \bigcup_{r>0} \mathcal{O}(D_{r,\dots,r})[[t]]_s$ *denote the set of all the* s*-Gevrey formal series.*

- *The set $\overline{\mathcal{A}}_s(\Sigma) = \bigcup_{r>0} \overline{\mathcal{A}}_{s,r}(\Sigma)$ endowed with the usual algebraic operations and the usual derivations ∂_t and ∂_{x_ℓ} for all $\ell = 1, \dots, n$ is a $\mathbb{C}$-differential algebra. Moreover, it is stable under the anti-derivations ∂_t^{-1} and $\partial_{x_\ell}^{-1}$ for all $\ell = 1, \dots, n$.*
- *The map $T_{s,\Sigma} : \overline{\mathcal{A}}_s(\Sigma) \longrightarrow \mathbb{C}\{x\}[[t]]_s$ assigning to each $u \in \overline{\mathcal{A}}_s(\Sigma)$ its s-Gevrey asymptotic expansion at 0 on Σ is a morphism of $\mathbb{C}$-differential algebras which commutes with ∂_t^{-1} and $\partial_{x_\ell}^{-1}$ for all $\ell = 1, \dots, n$.*

Observe that the fact that the set $\mathbb{C}\{x\}[[t]]_s$ be a $\mathbb{C}$-differential algebra follows from Proposition 3.4 and from the inclusion $\mathcal{O}(D_{r,\dots,r})[[t]]_s \subset \mathcal{O}(D_{r',\dots,r'})[[t]]_s$ for any $0 < r' < r$.

Observe also that the map $T_{s,\Sigma}$ is well-defined since $T_{s,r,\Sigma}(u) = T_{s,r',\Sigma}(u)$ for any $u \in \overline{\mathcal{A}}_{s,r}(\Sigma)$ and any $0 < r' < r$.

4.1.3 Gevrey Asymptotics and Change of Variable

We investigate here the effect of a change of variable of the form $t = z^p$ with $p \in \mathbb{N}^*$.

Clearly, if a formal power series $\widetilde{u}(t,x) \in \mathbb{C}\{x\}[[t]]_s$ is s-Gevrey, then the formal power series $\widetilde{u}(z^p, x)$ is s/p-Gevrey (see Definition 3.1). But, what about the asymptotics?

Given a sector $\Sigma = \Sigma_{\alpha,\beta}(R)$ (see page 5 for the notation), we denote by $\Sigma_{/p}$ the sector $\Sigma_{\alpha/p,\beta/p}(R^{1/p})$ so that as the variable z runs over $\Sigma_{/p}$, the variable $t = z^p$ runs over Σ. Then, from Definition 4.1, we can state the following.

Proposition 4.10 (Gevrey Asymptotics in an Extension of the Variable) *The following two assertions are equivalent:*

1. *the function $u(t,x) \in \overline{\mathcal{A}}_s(\Sigma)$ is s-Gevrey asymptotic to $\widetilde{u}(t,x)$ at 0 on Σ;*
2. *the function $v(z,x) = u(z^p, x) \in \overline{\mathcal{A}}_{s/p}(\Sigma_{/p})$ is s/p-Gevrey asymptotic to $\widetilde{v}(z,x) = \widetilde{u}(z^p, x)$ at 0 on $\Sigma_{/p}$.*

Way back, that is, given a s'-Gevrey formal power series $\widetilde{v}(z,x) \in \mathbb{C}\{x\}[[z]]_{s'}$, the series $\widetilde{u}(t,x) = \widetilde{v}(t^{1/p}, x)$ exhibits in general fractional powers in t. To get around this problem and to keep working with series of integer powers of t, one may extend the classical method of the rank reduction for the elements in $\mathbb{C}[[z]]$ as follows [64].

The formal series $\widetilde{v}(z,x)$ can be uniquely decomposed as the sum

$$\widetilde{v}(z,x) = \sum_{j=0}^{p-1} z^j \widetilde{v}_j(z^p, x)$$

where all the terms $\widetilde{v}_j(z^p,x)$ are power series of integer powers in z^p. Setting then $\omega = e^{2i\pi/p}$ and $t = z^p$, one can easily check that the formal series $\widetilde{v}_j(t,x)$ are given, for all $j = 0,\dots,p-1$, by the relations

$$pz^j\widetilde{v}_j(z^p,x) = \sum_{\ell=0}^{p-1} \omega^{\ell(p-j)}\widetilde{v}(\omega^\ell z,x).$$

For all $\ell = 0,\dots,p-1$, let us denote by $\Sigma^\ell_{/p}$ the sector

$$\Sigma^\ell_{/p} = \Sigma_{\alpha/p+2\ell\pi/p,\beta/p+2\ell\pi/p}(R^{1/p})$$

so that as the variable z runs over $\Sigma_{/p} = \Sigma^0_{/p}$, the variable $\omega^\ell z$ runs over $\Sigma^\ell_{/p}$ and the variable $t = z^p$ runs over Σ.

From the two relations above and Proposition 4.10, we can state the following.

Corollary 4.11 (Gevrey Asymptotics and Rank Reduction) *The following two assertions are equivalent:*

1. *for $\ell = 0,\dots,p-1$, the formal power series $\widetilde{v}(z,x)$ is a s'-Gevrey asymptotic series at* 0 *on $\Sigma^\ell_{/p}$ (in the variable z);*
2. *for $j = 0,\dots,p-1$, the p-rank reduced series $\widetilde{v}_j(t,x)$ is a $s'p$-Gevrey asymptotic series at* 0 *on Σ (in the variable $t = z^p$).*

These results allow us to limit the study of Gevrey asymptotics to small value of s (say $s \leqslant s_0$) or to large ones (say $s \geqslant s_1$) at convenience.

4.2 s-Gevrey Asymptotic Expansion in $\mathcal{O}(D_{\rho_1,\dots,\rho_n})[[t]]$

In the definition of the set $\overline{\mathcal{A}}_{s,r}(\Sigma)$ (see Definition 4.1), there is no control of the domain of holomorphy of the coefficients of the s-Gevrey asymptotic expansions, except that it must contain the polydisc $D_{r,\dots,r}$. However, in order to prepare the following part on the summability, it is also interesting to consider s-Gevrey asymptotic expansions in a fixed space of formal power series, say $\mathcal{O}(D_{\rho_1,\dots,\rho_n})[[t]]$.

Notation 4.12 In addition to Notation 4.4, we denote, for all $\rho_1,\dots,\rho_n > 0$ and all $0 < r \leqslant \min(\rho_1,\dots,\rho_n)$, by

- $\overline{\mathcal{A}}^{(\rho_1,\dots,\rho_n)}_{s,r}(\Sigma)$ the set of all the functions $u \in \overline{\mathcal{A}}_{s,r}(\Sigma)$ such that $T_{s,r,\Sigma}(u) \in \mathcal{O}(D_{\rho_1,\dots,\rho_n})[[t]]$;
- $T^{(\rho_1,\dots,\rho_n)}_{s,r,\Sigma} : \overline{\mathcal{A}}^{(\rho_1,\dots,\rho_n)}_{s,r}(\Sigma) \longrightarrow \mathcal{O}(D_{\rho_1,\dots,\rho_n})[[t]]_s$ the restriction of the map $T_{s,r,\Sigma}$ to $\overline{\mathcal{A}}^{(\rho_1,\dots,\rho_n)}_{s,r}(\Sigma)$, that is the map assigning to each $u \in \overline{\mathcal{A}}^{(\rho_1,\dots,\rho_n)}_{s,r}(\Sigma)$ its s-Gevrey asymptotic expansion in $\mathcal{O}(D_{\rho_1,\dots,\rho_n})[[t]]$.

Observe that, we have as before the inclusion $\overline{\mathcal{A}}_{s,r}^{(\rho_1,\dots,\rho_n)}(\Sigma) \subset \overline{\mathcal{A}}_{s,r'}^{(\rho_1,\dots,\rho_n)}(\Sigma)$ and the identity $T_{s,r,\Sigma}^{(\rho_1,\dots,\rho_n)}(u) = T_{s,r',\Sigma}^{(\rho_1,\dots,\rho_n)}(u)$ for all $u \in \overline{\mathcal{A}}_{s,r}^{(\rho_1,\dots,\rho_n)}(\Sigma)$ and all $0 < r' \leqslant r$.

The following three results are the direct consequences of Proposition 4.7, 4.8 and 4.9. The first one yields the algebraic structure of the set $\overline{\mathcal{A}}_{s,r}^{(\rho_1,\dots,\rho_n)}(\Sigma)$, the second one describes the effect of the derivation ∂_{x_ℓ} on $\overline{\mathcal{A}}_{s,r}^{(\rho_1,\dots,\rho_n)}(\Sigma)$, and the last one provides the *"good"* $\mathbb{C}$-differential algebra that will allow us to define the summability of formal power series in $\mathcal{O}(D_{\rho_1,\dots,\rho_n})[[t]]$, and thereby to study the summability of the formal power series solutions of partial differential equations.

Proposition 4.13 *Let $\rho_1, \dots, \rho_n > 0$ and $0 < r \leqslant \min(\rho_1, \dots, \rho_n)$.*

- *The set $\overline{\mathcal{A}}_{s,r}^{(\rho_1,\dots,\rho_n)}(\Sigma)$ endowed with the usual algebraic operations and the usual derivation ∂_t is a $\mathbb{C}$-differential algebra. Moreover, it is stable under the anti-derivations ∂_t^{-1} and $\partial_{x_\ell}^{-1}$ for all $\ell = 1, \dots, n$.*
- *The map $T_{s,r,\Sigma}^{(\rho_1,\dots,\rho_n)} : \overline{\mathcal{A}}_{s,r}^{(\rho_1,\dots,\rho_n)}(\Sigma) \longrightarrow \mathcal{O}(D_{\rho_1,\dots,\rho_n})[[t]]_s$ is a morphism of $\mathbb{C}$-differential algebras which commutes with ∂_t^{-1} and $\partial_{x_\ell}^{-1}$ for all $\ell = 1, \dots, n$.*

Proposition 4.14 *Let $\rho_1, \dots, \rho_n > 0$ and $0 < r \leqslant \min(\rho_1, \dots, \rho_n)$.*
Let $\ell \in \{1, \dots, n\}$, $u \in \overline{\mathcal{A}}_{s,r}^{(\rho_1,\dots,\rho_n)}(\Sigma)$ and $\widetilde{u} = T_{s,r,\Sigma}^{(\rho_1,\dots,\rho_n)}(u)$.
Then, $\partial_{x_\ell} u \in \overline{\mathcal{A}}_{s,r'}^{(\rho_1,\dots,\rho_n)}(\Sigma)$ with $T_{s,r',\Sigma}^{(\rho_1,\dots,\rho_n)}(\partial_{x_\ell} u) = \partial_{x_\ell}\widetilde{u}$ for all $0 < r' < r$.

Proposition 4.15 *Let $\rho_1, \dots, \rho_n > 0$.*

- *The set $\overline{\mathcal{A}}_{s,(\rho_1,\dots,\rho_n)}(\Sigma) = \bigcup_{0<r\leqslant\min(\rho_1,\dots,\rho_n)} \overline{\mathcal{A}}_{s,r}^{(\rho_1,\dots,\rho_n)}(\Sigma)$ endowed with the usual algebraic operations and the usual derivations ∂_t and ∂_{x_ℓ} for all $\ell = 1, \dots, n$ is a $\mathbb{C}$-differential algebra. Moreover, it is stable under the anti-derivations ∂_t^{-1} and $\partial_{x_\ell}^{-1}$ for all $\ell = 1, \dots, n$.*
- *The map $T_{s,(\rho_1,\dots,\rho_n),\Sigma} : \overline{\mathcal{A}}_{s,(\rho_1,\dots,\rho_n)}(\Sigma) \longrightarrow \mathcal{O}(D_{\rho_1,\dots,\rho_n})[[t]]_s$ assigning to each $u \in \overline{\mathcal{A}}_{s,(\rho_1,\dots,\rho_n)}(\Sigma)$ its s-Gevrey asymptotic expansion at 0 on Σ is a morphism of $\mathbb{C}$-differential algebras which commutes with ∂_t^{-1} and $\partial_{x_\ell}^{-1}$ for all $\ell = 1, \dots, n$.*

4.3 Flat s-Gevrey Asymptotic Functions

In this section, we are interested in the *flat s-Gevrey asymptotic functions*, that is in the kernels of the maps $T_{s,r,\Sigma}$ and $T_{s,\Sigma}$, and their restrictions $T_{s,r,\Sigma}^{(\rho_1,\dots,\rho_n)}$ and $T_{s,(\rho_1,\dots,\rho_n),\Sigma}$.

To this end, we first introduce the notion of exponential flatness.

Definition 4.16 Let $r, k > 0$. A function $u(t,x) \in \mathcal{O}(\Sigma \times D_{r,\dots,r})$ is said to be *exponentially flat of order k* (or *k-exponentially flat*) *on* Σ if, for any proper subsector $\Sigma' \Subset \Sigma$, there exist two positive constants $A_{\Sigma'}, C_{\Sigma'} > 0$ such that the following estimate holds for all $(t, x) \in \Sigma' \times D_{r,\dots,r}$:

$$|u(t,x)| \leqslant C_{\Sigma'} \exp\left(-\frac{A_{\Sigma'}}{|t|^k}\right).$$

Notation 4.17 Given $r, k > 0$, we denote by $\overline{\mathcal{A}}_r^{\leqslant -k}(\Sigma)$ the set of all the functions $u \in \mathcal{O}(\Sigma \times D_{r,\dots,r})$ which are k-exponentially flat on Σ. We also denote by $\overline{\mathcal{A}}^{\leqslant -k}(\Sigma)$ and $\overline{\mathcal{A}}_{(\rho_1,\dots,\rho_n)}^{\leqslant -k}(\Sigma)$ the sets

$$\overline{\mathcal{A}}^{\leqslant -k}(\Sigma) = \bigcup_{r>0} \overline{\mathcal{A}}_r^{\leqslant -k}(\Sigma) \quad \text{and} \quad \overline{\mathcal{A}}_{(\rho_1,\dots,\rho_n)}^{\leqslant -k}(\Sigma) = \bigcup_{0<r\leqslant \min(\rho_1,\dots,\rho_n)} \overline{\mathcal{A}}_r^{\leqslant -k}(\Sigma).$$

Observe that the inclusion $\overline{\mathcal{A}}_r^{\leqslant -k}(\Sigma) \subset \overline{\mathcal{A}}_{r'}^{\leqslant -k}(\Sigma)$ holds for any $0 < r' \leqslant r$.

Example 4.18 Let $r > 0$ and Σ be the sector defined by

$$\Sigma = \left\{t; 0 < |t| < R \text{ and } |\arg(t)| < \frac{\pi s}{2}\right\}, \qquad R, s > 0.$$

Then, setting $k = 1/s$, the function

$$u(t,x) = \exp\left(-1/t^k + x_1 + \dots + x_n\right) \in \mathcal{O}(\Sigma \times D_{r,\dots,r})$$

belongs to $\overline{\mathcal{A}}_r^{\leqslant -k}(\Sigma)$. In particular, $u \in \overline{\mathcal{A}}_r^{\leqslant -k}(\Sigma')$ for any subsector $\Sigma' \subset \Sigma$. On the other hand, observe that $u \in \overline{\mathcal{A}}_r^{\leqslant -k}(\mathfrak{s})$ for no sector $\mathfrak{s}$ containing Σ.

We are now able to state the result in view in this section.

Proposition 4.19 *Let $r > 0$ and $k = 1/s$. Then,*

$$\text{Ker } T_{s,r,\Sigma} = \overline{\mathcal{A}}_r^{\leqslant -k}(\Sigma) \quad \textit{and} \quad \text{Ker } T_{s,\Sigma} = \overline{\mathcal{A}}^{\leqslant -k}(\Sigma).$$

Moreover, if $r \leqslant \min(\rho_1, \dots, \rho_n)$ for some $\rho_1, \dots, \rho_n > 0$, then

$$\text{Ker } T_{s,r,\Sigma}^{(\rho_1,\dots,\rho_n)} = \overline{\mathcal{A}}_r^{\leqslant -k}(\Sigma) \quad \textit{and} \quad \text{Ker } T_{s,(\rho_1,\dots,\rho_n),\Sigma} = \overline{\mathcal{A}}_{(\rho_1,\dots,\rho_n)}^{\leqslant -k}(\Sigma).$$

Proof It is sufficient to prove the first identity $\text{Ker } T_{s,r,\Sigma} = \overline{\mathcal{A}}_r^{\leqslant -k}(\Sigma)$.

Before starting the calculations, let us first observe that, taking into account the Stirling Formula, there exist four positive constants $a, b, a', b' > 0$ such that the following estimate holds for all $J \geqslant 1$:

$$ab^J J^{sJ} \leqslant \Gamma(1 + sJ) \leqslant a'b'^J J^{sJ}. \tag{4.6}$$

$\triangleleft$ *Let* $u \in \operatorname{Ker} T_{s,r,\Sigma}$ *and prove that* $u \in \overline{\mathcal{A}}_r^{\leqslant -k}(\Sigma)$. Let $\Sigma' \Subset \Sigma$ a proper subsector of Σ. From the hypothesis (see Definition 4.1) and the second inequality of (4.6), we know that there exist two positive constants $C_{\Sigma'}, K_{\Sigma'} > 0$ such that the estimate

$$|u(t,x)| \leqslant C_{\Sigma'} K_{\Sigma'}^J J^{sJ} |t|^J = C_{\Sigma'} \exp\left(\frac{J}{k} \ln\left(J\left(K_{\Sigma'}|t|\right)^k\right)\right)$$

holds for all $J \geqslant 1$ and all $(t,x) \in \Sigma' \times D_{r,\ldots,r}$ (recall that $s = 1/k$).

For (t,x) fixed, we look for a lower bound of the right-hand side of this estimate as J runs over $\mathbb{N}^*$. The derivative $\varphi'(J) = \ln\left(J\left(K_{\Sigma'}|t|\right)^k\right) + 1$ of the function $\varphi(J) = J\ln\left(J\left(K_{\Sigma'}|t|\right)^k\right)$ seen as a function of a real variable $J > 0$ vanishes at $J_0 = 1/\left(e\left(K_{\Sigma'}|t|\right)^k\right)$ and φ reaches its minimal value at that point. In general, J_0 is not an integer and the minimal value of $\varphi(J)$ on integer values of J is larger than $\varphi(J_0)$. However, taking into account the increasing of φ on $[J_0, J_0+1]$ and the fact that there is at least one integer number in this interval, we can assert that

$$\inf_{J\in\mathbb{N}^*} \varphi(J) \leqslant \varphi(J_0+1) = -\frac{1}{e\left(K_{\Sigma'}|t|\right)^k} + \psi(t)$$

with

$$\psi(t) = \frac{\left(1 + e\left(K_{\Sigma'}|t|\right)^k\right)\ln\left(1 + e\left(K_{\Sigma'}|t|\right)^k\right)}{e\left(K_{\Sigma'}|t|\right)^k}.$$

Since the function ψ is bounded on Σ', there exists a positive constant $C'_{\Sigma'} > 0$ such that $\psi(t) \leqslant C'_{\Sigma'}$ for all $t \in \Sigma'$, and consequently, it follows that the following estimate

$$|u(t,x)| \leqslant C''_{\Sigma'} \exp\left(-\frac{A_{\Sigma'}}{|t|^k}\right)$$

with $C''_{\Sigma'} = C_{\Sigma'} \exp\left(C'_{\Sigma'}/k\right) > 0$ and $A_{\Sigma'} = -1/\left(keK_{\Sigma'}^k\right) > 0$ independent of $(t,x) \in \Sigma' \times D_{r,\ldots,r}$. This proves that u belongs to $\overline{\mathcal{A}}_r^{\leqslant -k}(\Sigma)$.

$\triangleleft$ *Let* $u \in \overline{\mathcal{A}}_r^{\leqslant -k}(\Sigma)$ *and prove that* $u \in \operatorname{Ker} T_{s,r,\Sigma}$. Let $\Sigma' \Subset \Sigma$ a proper subsector of Σ. From the hypothesis, we know now that there exist two positive constants $A_{\Sigma'}, C_{\Sigma'} > 0$ such that the following estimate

$$|u(t,x)| \leqslant C_{\Sigma'} \exp\left(-\frac{A_{\Sigma'}}{|t|^k}\right)$$

holds for all $(t, x) \in \Sigma' \times D_{r,\dots,r}$. Hence,

$$|u(t,x)| \cdot |t|^{-J} \leqslant C_{\Sigma'} \exp\left(-\frac{A_{\Sigma'}}{|t|^k}\right) |t|^{-J}$$

for all $J \geqslant 1$ and all $(t, x) \in \Sigma' \times D_{r,\dots,r}$.

For J fixed, we look for an upper bound of the right-hand side of this estimate as $|t|$ runs over $\mathbb{R}_+^* =]0, +\infty[$. To do that, let us consider the function ψ defined on $\mathbb{R}_+^*$ by $\psi(t) = \exp\left(-A_{\Sigma'}/t^k\right) t^{-J}$. Its logarithmic derivative

$$\frac{\psi'(t)}{\psi(t)} = -\frac{J}{t} + \frac{kA_{\Sigma'}}{t^{k+1}}$$

vanishes for $kA_{\Sigma'}/t^k = J$ and ψ reaches its maximal value at that point.

It follows that

$$\max_{|t|>0} \psi(|t|) = \exp\left(-\frac{J}{k}\right)\left(\frac{J}{kA_{\Sigma'}}\right)^{J/k}$$

and, consequently, the following estimate

$$|u(t,x)| \leqslant C_{\Sigma'} K_{\Sigma'}^J J^{sj} |t|^J$$

with $K_{\Sigma'} = (keA_{\Sigma'})^{-1/k}$ holds for all $J \geqslant 1$ and all $(t, x) \in \Sigma' \times D_{r,\dots,r}$ (recall that $s = 1/k$). Hence, u belongs to Ker $T_{s,r,\Sigma}$ by applying the first inequality of (4.6). This achieves the proof of Proposition 4.19. ■

4.4 s-Gevrey Borel-Ritt Theorem

In Sect. 4.1, Proposition 4.9, we saw that with any asymptotic function $u \in \overline{\mathcal{A}}_s(\Sigma)$, the map $T_{s,\Sigma}$ associates a s-Gevrey formal series $\widetilde{u} = T_{s,\Sigma}(u)$. We address now the converse problem:

Is any s-Gevrey formal series the s-Gevrey asymptotic expansion at 0 of an asymptotic function on a given sector?

We address also the same problem in restriction to the map $T_{s,(\rho_1,\dots,\rho_n),\Sigma}$ defined in Sect. 4.2, Proposition 4.15.

Theorem 4.20 below states the answer is yes for any open sector Σ with small enough opening in $\mathbb{C}^*$ or $\widetilde{\mathbb{C}}$. Notice that the s-Gevrey asymptotic expansion at 0 of a function $u \in \overline{\mathcal{A}}_s(\mathbb{C}^*)$ is necessarily convergent by the Removable Singularity Theorem. Thus, when the s-Gevrey asymptotic expansion at 0 is divergent, the sector Σ can not be a full neighborhood of 0 in $\mathbb{C}^*$.

Theorem 4.20 (Borel-Ritt) *Let $\Sigma \neq \mathbb{C}^*$ be an open sector of $\mathbb{C}^*$ or of the Riemann surface of the logarithm $\widetilde{\mathbb{C}}$ with opening $|\Sigma| \leqslant \pi s$.*

1. *The map $T_{s,\Sigma} : \overline{\mathcal{A}}_s(\Sigma) \longrightarrow \mathbb{C}\{x\}[[t]]_s$ is onto.*
2. *The map $T_{s,(\rho_1,\dots,\rho_n),\Sigma} : \overline{\mathcal{A}}_{s,(\rho_1,\dots,\rho_n)}(\Sigma) \longrightarrow \mathcal{O}(D_{\rho_1,\dots,\rho_n})[[t]]_s$ is onto.*

Proof It is sufficient to consider a sector Σ with opening πs. Moreover, by means of a rotation, we can besides assume that Σ is bisected by the direction $\theta = 0$.

Let $\widetilde{u}(t,x) = \sum_{j\geqslant 0} u_j(x)t^j \in \mathbb{C}\{x\}[[t]]_s$ or $\mathcal{O}(D_{\rho_1,\dots,\rho_n})[[t]]_s$ be a s-Gevrey formal series. By assumption (see the definition of $\mathbb{C}\{x\}[[t]]_s$ just above and Definition 3.1), there exist a radius $r > 0$ and two positive constants $C, K > 0$ such that

- $u_j(x) \in \mathcal{O}(D_{r,\dots,r})$ for all $j \geqslant 0$;
- $\left|u_j(x)\right| \leqslant CK^j\Gamma(1+sj)$ for all $j \geqslant 0$ and all $x \in D_{r,\dots,r}$.

Therefore, the series

$$\widehat{u}(\tau,x) = \sum_{j\geqslant 0} \frac{u_j(x)}{\Gamma(1+sj)}\tau^j$$

converges for all $(\tau,x) \in D_{1/K} \times D_{r,\dots,r}$.

Let us now fix $b \in \left]0, \frac{1}{K}\right[$ and let us consider the holomorphic function $u(t,x) \in \mathcal{O}(\Sigma \times D_{r,\dots,r})$ defined by the truncated k-Laplace transform

$$u(t,x) = t^{-k}\int_0^{b^k} \widehat{u}(\xi^s,x)e^{-\xi/t^k}d\xi, \quad \text{where } s = \frac{1}{k} \text{ and } \xi = \tau^k.$$

We shall prove below that $u(t,x)$ is s-Gevrey asymptotic to $\widetilde{u}$ at 0 on Σ.

Let $\Sigma'' \Subset \Sigma$ a proper subsector of Σ and let $R > 0$ denote the radius of Σ. By assumption, there exist two positive constants $\delta \in \left]0, \frac{\pi}{2}\right[$ and $R' \in]0, R[$ so that Σ'' be included in the sector

$$\Sigma' = \left\{t \text{ such that } |\arg(t)| < \frac{\pi}{2k} - \frac{\delta}{k} \text{ and } 0 < |t| < R'\right\}.$$

Let us now fix $J \geqslant 1$ and $(t,x) \in \Sigma' \times D_{r,\dots,r}$ be. From the relation

$$t^j = t^{-k}\int_0^{+\infty} \frac{\xi^{sj}}{\Gamma(1+sj)}e^{-\xi/t^k}d\xi \quad \text{for all } j \geqslant 0$$

(see [6, pp. 78–79] for instance), we first have

$$u(t,x) - \sum_{j=0}^{J-1} u_j(x)t^j = t^{-k} \int_0^{b^k} \left(\sum_{j\geqslant 0} \frac{u_j(x)}{\Gamma(1+sj)} \xi^{sj} e^{-\xi/t^k} \right) d\xi$$
$$- \sum_{j=0}^{J-1} u_j(x) t^{-k} \int_0^{+\infty} \frac{\xi^{sj}}{\Gamma(1+sj)} e^{-\xi/t^k} d\xi.$$

Since

$$t \in \Sigma' \Rightarrow |\arg(t)| < \frac{\pi}{2} \Rightarrow \mathrm{Re}(t) > 0 \Rightarrow \left|\xi^{sj} e^{-\xi/t^k}\right| = |\xi|^{sj} e^{-\xi \frac{\mathrm{Re}(t^k)}{|t|^{2k}}} \leqslant b^j$$

for all $\xi \in [0, b^k]$, the series

$$\sum_{j\geqslant 0} \frac{u_j(x)}{\Gamma(1+sj)} \xi^{sj} e^{-\xi/t^k}$$

converges normally on $[0, b^k]$. Therefore, we can permute the sum and the integral. Hence,

$$u(t,x) - \sum_{j=0}^{J-1} u_j(x)t^j = \sum_{j\geqslant J} \frac{u_j(x)}{\Gamma(1+sj)} t^{-k} \int_0^{b^k} \xi^{sj} e^{-\xi/t^k} d\xi$$
$$- \sum_{j=0}^{J-1} \frac{u_j(x)}{\Gamma(1+sj)} t^{-k} \int_{b^k}^{+\infty} \xi^{sj} e^{-\xi/t^k} d\xi.$$

Let us now observe that the inequalities $(\xi/b^k)^{sj} \leqslant (\xi/b^k)^{Js}$ hold both when $\xi \leqslant b^k$ and $j \geqslant J$ and when $\xi \geqslant b^k$ and $j < J$. This brings then us to the following

$$\left| u(t,x) - \sum_{j=0}^{J-1} u_j(x)t^j \right| \leqslant \sum_{j\geqslant J} \frac{b^{j-J} \left|u_j(x)\right|}{\Gamma(1+sj)} |t|^{-k} \int_0^{b^k} \xi^{sJ} e^{-\xi \mathrm{Re}(1/t^k)} d\xi$$
$$+ \sum_{j=0}^{J-1} \frac{b^{j-J} \left|u_j(x)\right|}{\Gamma(1+sj)} |t|^{-k} \int_{b^k}^{+\infty} \xi^{sJ} e^{-\xi \mathrm{Re}(1/t^k)} d\xi$$

and thereby to the inequalities

$$\left| u(t,x) - \sum_{j=0}^{J-1} u_j(x) t^j \right| \leqslant \sum_{j\geqslant 0} \frac{b^{j-J}\,|u_j(x)|}{\Gamma(1+sj)} |t|^{-k} \int_0^{+\infty} \xi^{sJ} e^{-\xi \mathrm{Re}(1/t^k)} d\xi$$

$$\leqslant \sum_{j\geqslant 0} \frac{b^{j-J}\,|u_j(x)|}{\Gamma(1+sj)} |t|^{-k} \int_0^{+\infty} \xi^{sJ} e^{-\xi \sin(\delta)/|t|^k} d\xi.$$

Observe that the last inequality stems from the fact that $t \in \Sigma'$ implies

$$\mathrm{Re}\left(\frac{1}{t^k}\right) = \frac{\cos\left(\arg(t^k)\right)}{|t|^k} \geqslant \frac{\cos\left(\frac{\pi}{2} - \delta\right)}{|t|^k} = \frac{\sin(\delta)}{|t|^k}.$$

Setting then $\eta = \xi \sin(\delta)/\,|t|^k$, we obtain

$$\left| u(t,x) - \sum_{j=0}^{J-1} u_j(x) t^j \right| \leqslant \sum_{j\geqslant 0} \frac{b^{j-J}\,|u_j(x)|\,|t|^J}{\Gamma(1+sj)(\sin(\delta))^{sJ+1}} \int_0^{+\infty} \eta^{sJ} e^{-\eta} d\eta$$

$$= \sum_{j\geqslant 0} \frac{b^{j-J}\,|u_j(x)|}{\Gamma(1+sj)(\sin(\delta))^{sJ+1}} \Gamma(1+sJ)\,|t|^J,$$

where, according to the choice of b (see the beginning of the proof), we have

$$\sum_{j\geqslant 0} \frac{|u_j(x)|\,b^j}{\Gamma(1+sj)} \leqslant C \sum_{j\geqslant 0} (Kb)^j = \frac{C}{1-Kb}.$$

Consequently, we finally get

$$\left| u(t,x) - \sum_{j=0}^{J-1} u_j(x) t^j \right| \leqslant C' K'^J \Gamma(1+sJ) |t|^J,$$

with $C' = \dfrac{C}{(1-Kb)\sin(\delta)}$ and $K' = \dfrac{1}{b(\sin(\delta))^s}$. The constants C' and K' depending on Σ' and on the choice of b, but not on t and x, this achieves the proof. ■

Taking into account Propositions 4.9, 4.15 and 4.19, we can reformulate the Borel-Ritt Theorem 4.20 as follows:

Corollary 4.21 *Let $\Sigma \neq \mathbb{C}^*$ be an open sector of $\mathbb{C}^*$ or of the Riemann surface of the logarithm $\widetilde{\mathbb{C}}$ with opening $|\Sigma| \leqslant \pi s$. Then, the sequences*

$$0 \longrightarrow \overline{\mathcal{A}}^{\leqslant -k}(\Sigma) \longrightarrow \overline{\mathcal{A}}_s(\Sigma) \xrightarrow{T_{s,\Sigma}} \mathbb{C}\{x\}[[t]]_s \longrightarrow 0$$

and

$$0 \longrightarrow \overline{\mathcal{A}}^{\leqslant -k}_{(\rho_1,\dots,\rho_n)}(\Sigma) \longrightarrow \overline{\mathcal{A}}_{s,(\rho_1,\dots,\rho_n)}(\Sigma) \xrightarrow{T_{s,(\rho_1,\dots,\rho_n),\Sigma}} \mathcal{O}(D_{\rho_1,\dots,\rho_n})[[t]]_s \longrightarrow 0$$

are exact sequences of morphisms of differential algebras.

4.5 Watson's Lemma

When Σ is a "small" sector, that is with opening $|\Sigma| \leqslant \pi s$, we saw in the previous section that the maps $T_{s,\Sigma}$ and $T_{s,(\rho_1,\dots,\rho_n),\Sigma}$ are onto. On the other hand, when Σ is a "wide" sector, that is with opening $|\Sigma| > \pi s$, the situation is very different. In this case, the two maps $T_{s,\Sigma}$ and $T_{s,(\rho_1,\dots,\rho_n),\Sigma}$ are no longer onto, but become injective. This is due to the following result.

Theorem 4.22 (Watson's Lemma [131]) *Let Σ be an open sector with opening $|\Sigma| = \pi s$ and a radius $r > 0$. Suppose that $u(t,x) \in \mathcal{O}(\Sigma \times D_{r,\dots,r})$ satisfies a global estimate of exponentially decreasing of order $k = 1/s$ on Σ, that is there exist two positive constants $C, A > 0$ such that the following estimate holds for all $(t,x) \in \Sigma \times D_{r,\dots,r}$:*

$$|u(t,x)| \leqslant C \exp\left(-\frac{A}{|t|^k}\right).$$

Then, u is identically equal to 0 on $\Sigma \times D_{r,\dots,r}$.

Proof Without loss of generality, we can always assume that Σ is bisected by the direction $\theta = 0$, that is Σ reads in the form

$$\Sigma = \left\{t; 0 < |t| < R \text{ and } |\arg(t)| < \frac{\pi}{2k}\right\}$$

with a convenient radius $R > 0$.

Let us fix $x \in D_{r,\dots,r}$. For any positive real number $a > k$, we consider the function U defined on Σ by

$$U(t,x,a) = u(t,x)\exp\left(\frac{A}{\cos\left(\frac{k\pi}{2a}\right)t^k}\right). \tag{4.7}$$

By assumption, U is holomorphic on Σ and the following estimate holds for all $t \in \Sigma$:

$$|U(t,x,a)| \leqslant C\exp\left(\frac{A}{|t|^k}\left(\frac{\cos(k\arg(t))}{\cos\left(\frac{k\pi}{2a}\right)} - 1\right)\right). \tag{4.8}$$

Let us now denote by $\Sigma_{a,r}$ the sector

$$\Sigma_{a,r} = \left\{t;\, 0 < |t| < r \text{ and } |\arg(t)| < \frac{\pi}{2a}\right\}, \quad 0 < r < R.$$

The function U is holomorphic on $\Sigma_{a,r}$ and continuous up to the boundary

$$\mathcal{B}(\Sigma_{a,r}) = \left\{t;\, 0 < |t| \leqslant r \text{ and } |\arg(t)| = \frac{\pi}{2a}\right\}.$$

of $\Sigma_{a,r}$. Moreover, inequality (4.8) implies that

$$\begin{cases} |U(t,x,a)| \leqslant C & \text{for all } t \in \mathcal{B}(\Sigma_{a,r}) \\ |U(t,x,a)| \leqslant C\exp\left(\frac{A}{|t|^k}\left(\frac{1}{\cos\left(\frac{k\pi}{2a}\right)} - 1\right)\right) & \text{for all } t \in \Sigma_{a,r} \end{cases}$$

with $A\left(\frac{1}{\cos\left(\frac{k\pi}{2a}\right)} - 1\right) > 0$. Applying then the Phragmén-Lindelöf Principle, we derive that

$$|U(t,x,a)| \leqslant C$$

for all $t \in \Sigma_{a,r}$ and, consequently, for all $t \in \Sigma$ with $|\arg(t)| < \frac{\pi}{2a}$. Hence,

$$|U(t,x,a)| \leqslant C\exp\left(\frac{A}{R^k}\left(\frac{1}{\cos\left(\frac{k\pi}{2a}\right)} - 1\right)\right)$$

for all $t \in \Sigma$ with $|\arg(t)| < \frac{\pi}{2a}$.

Using identity (4.7), we finally get the inequality

$$|u(t,x)| \leqslant C \exp\left(-\frac{A}{R^k} + \frac{A}{\cos\left(\frac{k\pi}{2a}\right)}\left(\frac{1}{R^k} - \frac{1}{t^k}\right)\right)$$

for all $t \in \Sigma$ with $\arg(t) = 0$ and all $a > k$. Then, by making a tend to k and by using the fact that $0 < t < R$, we derive that $u(t,x) = 0$ for all $t \in \Sigma$ with $\arg(t) = 0$, and we conclude by the Analytic Continuation Principle. ■

In other words, as soon as a sector Σ admits a proper subsector Σ' with opening $|\Sigma'| = \pi s$, that is $|\Sigma| > \pi s$, then $\overline{\mathcal{A}}_r^{\leqslant -k}(\Sigma) = \{0\}$. Consequently, we can reformulate the Watson's Lemma as follows:

Corollary 4.23 *Let $\Sigma \neq \mathbb{C}^*$ be an open sector of $\mathbb{C}^*$ or of the Riemann surface of the logarithm $\widetilde{\mathbb{C}}$ with opening $|\Sigma| > \pi s$. Then, the sequences*

$$0 \longrightarrow \overline{\mathcal{A}}_s(\Sigma) \xrightarrow{T_{s,\Sigma}} \mathbb{C}\{x\}[[t]]_s \longrightarrow 0$$

and

$$0 \longrightarrow \overline{\mathcal{A}}_{s,(\rho_1,\ldots,\rho_n)}(\Sigma) \xrightarrow{T_{s,(\rho_1,\ldots,\rho_n),\Sigma}} \mathcal{O}(D_{\rho_1,\ldots,\rho_n})[[t]]_s \longrightarrow 0$$

are exact sequences of morphisms of differential algebras.

Part II
Summability

The aim of a summation theory on a given germ of sector (there might be some constraints on the size and the position of the sector) is to associate with any formal power series an asymptotic function uniquely determined in a way as much natural as possible. What natural means depends on the category we want to consider. There is no known operator of summation applying to the algebra of all power series at one time and very little hope towards such a universal tool. For instance, to apply such a theory to the formal power series solutions of ordinary or partial differential equations, an eligible request is that the summation operator be a morphism of differential algebras from an algebra of power series (containing the series under consideration) into an algebra of asymptotic functions (containing the corresponding asymptotic solutions). Both algebras must be chosen carefully and correspondingly.

In the present work, we are only interested in the simplest case of the summation theories, the so-called *k-summability* ($k > 0$ a positive real number). In the case of formal power series in $\mathbb{C}[[t]]$, this theory is perfectly established since the end of the twentieth century and several equivalent approaches have been given, mainly by J.-P. Ramis, J. Martinet, Y. Sibuya and B. Malgrange (see [65, Chapter 5] and the references therein).

Based on the results established in the first part of this work, we propose here to show how this classical theory of the k-summability in $\mathbb{C}[[t]]$ can be extended to the formal power series in $\mathcal{O}(D_{\rho_1,\ldots,\rho_n})[[t]]$. In particular, we present two characterizations of these k-summable formal power series and, for each of them, we attach some applications to partial differential equations fitting especially that point of view.

All along this part, k denotes a positive real number, $s = 1/k$ the inverse of k, and θ a direction in $\mathbb{R}/2\pi\mathbb{Z}$.

Chapter 5
k-Summability: Definition and First Algebraic Properties

As said at the beginning of Part I, we consider t as the variable and $x = (x_1, \dots, x_n)$ as a parameter. Thereby, to define the notion of k-summability of formal power series in $\mathcal{O}(D_{\rho_1,\dots,\rho_n})[[t]]$, one extends the classical notion of k-summability of formal power series in $\mathbb{C}[[t]]$ to families parametrized by x in requiring similar conditions, the estimates being however uniform with respect to x.

Among the many equivalent definitions of the k-summability in a given direction $\arg(t) = \theta$ at $t = 0$ [65, Chapter 5], we choose here a generalization of Ramis' definition which states that a formal power series $\widetilde{f}(t) \in \mathbb{C}[[t]]$ is k-summable in the direction θ if there exists a holomorphic function f which is s-Gevrey asymptotic to $\widetilde{f}$ in an open sector $\Sigma_{\theta,>\pi s}$ bisected by θ and with opening larger than πs [100, Def. 3.1] (recall that $s = 1/k$).

Notice that such a sector $\Sigma_{\theta,>\pi s}$ is drawn in $\mathbb{C}^*$ when $k > 1/2$, and on the Riemann surface $\widetilde{\mathbb{C}}$ of the logarithm otherwise.

Definition 5.1 (k-Summability) A formal power series $\widetilde{u}(t,x) \in \mathcal{O}(D_{\rho_1,\dots,\rho_n})[[t]]$ is said to be *k-summable in the direction* $\arg(t) = \theta$ if there exist a sector $\Sigma_{\theta,>\pi s}$ and a function $u \in \overline{\mathcal{A}}_{s,(\rho_1,\dots,\rho_n)}(\Sigma_{\theta,>\pi s})$ such that $T_{s,(\rho_1,\dots,\rho_n),\Sigma_{\theta,>\pi s}}(u) = \widetilde{u}$, that is if there exist a sector $\Sigma_{\theta,>\pi s}$, a radius $0 < r \leqslant \min(\rho_1, \dots, \rho_n)$ and a holomorphic function $u(t,x) \in \mathcal{O}(\Sigma_{\theta,>\pi s} \times D_{r,\dots,r})$ which is s-Gevrey asymptotic to $\widetilde{u}$ at 0 on $\Sigma_{\theta,>\pi s}$.

Definition 5.2 (k-Sum) The function u above, which is uniquely determined when it exists accordingly to Watson's Lemma 4.22, is called the *k-sum of $\widetilde{u}$ in the direction θ*.

Notation 5.3 We denote by $\mathcal{O}(D_{\rho_1,\dots,\rho_n})\{t\}_{k;\theta}$ the subset of $\mathcal{O}(D_{\rho_1,\dots,\rho_n})[[t]]$ made of all the k-summable formal power series in the direction $\arg(t) = \theta$.

Notice that the k-sum of a k-summable formal series $\widetilde{u}(t,x) \in \mathcal{O}(D_{\rho_1,\dots,\rho_n})\{t\}_{k;\theta}$ may be analytic with respect to x on a polydisc $D_{r,\dots,r}$ smaller than the common polydisc $\mathcal{O}(D_{\rho_1,\dots,\rho_n})$ of analyticity of the coefficients $u_j(x)$ of $\widetilde{u}(t,x)$.

P. Remy, *Asymptotic Expansions and Summability*, Lecture Notes in Mathematics 2351, https://doi.org/10.1007/978-3-031-59094-8_5

Notice also that the inclusion $\mathcal{O}(D_{\rho_1,\dots,\rho_n})\{t\}_{k;\theta} \subset \mathcal{O}(D_{\rho'_1,\dots,\rho'_n})\{t\}_{k;\theta}$ holds for any radii $0 < \rho'_j \leqslant \rho_j$ with $j = 1, \dots, n$.

The following proposition, which specifies the algebraic structure of $\mathcal{O}(D_{\rho_1,\dots,\rho_n})\{t\}_{k;\theta}$, is straightforward from Proposition 4.15.

Proposition 5.4 *The set $\mathcal{O}(D_{\rho_1,\dots,\rho_n})\{t\}_{k;\theta}$ endowed with the usual algebraic operations and the usual derivations ∂_t and ∂_{x_ℓ} for all $\ell = 1, \dots, n$ is a $\mathbb{C}$-differential subalgebra of the Gevrey space $\mathcal{O}(D_{\rho_1,\dots,\rho_n})[[t]]_s$. Moreover, it is stable under the anti-derivations ∂_t^{-1} and $\partial_{x_\ell}^{-1}$ for all $\ell = 1, \dots, n$.*

With respect to t, the k-sum $u(t,x)$ of a k-summable series $\widetilde{u}(t,x) \in \mathcal{O}(D_{\rho_1,\dots,\rho_n})\{t\}_{k;\theta}$ is analytic on an open sector for which there is no control on the angular opening except that it must be larger than πs (hence, it contains a closed sector $\overline{\Sigma}_{\theta,\pi s}$ bisected by θ and with opening πs) and no control on the radius except that it must be positive. Thereby, the k-sum $u(t,x)$ is well-defined as a section of the sheaf of analytic functions in (t,x) on a germ of closed sector of opening πs (that is, a closed interval $\overline{I}_{\theta,\pi s}$ of length πs on the circle S^1 of directions issuing from 0; see [72, 1.1] or [63, I.2]) times $\{(0,\dots,0)\}$ (in the space $\mathbb{C}^n$ of the variable x). We denote by $\mathcal{O}_{\overline{I}_{\theta,\pi s}\times\{(0,\dots,0)\}}$ the space of such sections.

Corollary 5.5 *The operator of k-summation*

$$\begin{aligned} \mathcal{S}_{k;\theta} : \mathcal{O}(D_{\rho_1,\dots,\rho_n})\{t\}_{k;\theta} &\longrightarrow \mathcal{O}_{\overline{I}_{\theta,\pi/k}\times\{(0,\dots,0)\}} \\ \widetilde{u}(t,x) &\longmapsto u(t,x) \end{aligned}$$

is a homomorphism of $\mathbb{C}$-differential algebras for the derivations ∂_t and ∂_{x_ℓ} with $\ell = 1, \dots, n$. Moreover, it commutes with the anti-derivations ∂_t^{-1} and $\partial_{x_\ell}^{-1}$ for all $\ell = 1, \dots, n$.

As in Sect. 4.1.3 (cf. Proposition 4.10 and Corollary 4.11), let us now observe the effect of a change of variable of the form $t = z^p$ with $p \in \mathbb{N}^*$.

Given a sector $\Sigma = \Sigma_{\theta,>\pi s}$, we keep the notation $\Sigma^\ell_{/p}$ introduced page 51. Recall that when the variable z runs over $\Sigma_{/p} = \Sigma^0_{/p}$, the variable $\omega^\ell z$ with $\omega = e^{2i\pi/p}$ runs over $\Sigma^\ell_{/p}$ and the variable $t = z^p$ runs over Σ.

Observing then that $\Sigma^\ell_{/p}$ is a sector bisected by the direction $\theta/p + 2\ell\pi/p$ and with opening larger than $\pi s/p$ for all $\ell = 0, \dots, p-1$, we immediately derive from Definition 5.1 and Proposition 4.10 the following.

Proposition 5.6 (k-Summability in an Extension of the Variable) *The following two assertions are equivalent:*

1. *the formal power series $\widetilde{u}(t,x) \in \mathcal{O}(D_{\rho_1,\dots,\rho_n})[[t]]$ is k-summable in the direction θ with k-sum $u(t,x) \in \mathcal{O}(\Sigma \times D_{r,\dots,r})$;*
2. *the formal power series $\widetilde{v}(z,x) = \widetilde{u}(z^p,x) \in \mathcal{O}(D_{\rho_1,\dots,\rho_n})[[z]]$ is kp-summable in the direction θ/p with kp-sum $v(z,x) = u(z^p,x) \in \mathcal{O}(\Sigma_{/p} \times D_{r,\dots,r})$.*

Way back, given a formal power series $\widetilde{v}(z,x) \in \mathcal{O}(D_{\rho_1,\dots,\rho_n})[[z]]$, recall (cf. Sect. 4.1.3) that the p-rank reduction consists in replacing $\widetilde{v}(z,x)$ by the p formal power series $\widetilde{v}_j(t,x)$, $j = 0,\dots,p-1$ defined by the unique decomposition

$$\widetilde{v}(z,x) = \sum_{j=0}^{p-1} z^j \widetilde{v}_j(z^p,x)$$

and that the formal series $\widetilde{v}_j(t,x)$ are given, for all $j = 0,\dots,p-1$, by the relations

$$pz^j \widetilde{v}_j(z^p,x) = \sum_{\ell=0}^{p-1} \omega^{\ell(p-j)} \widetilde{v}(\omega^\ell z,x).$$

From Definition 5.1 and Corollary 4.11, we can state the following.

Corollary 5.7 (k-Summability and Rank Reduction) *The following two assertions are equivalent:*

1. *for $\ell = 0,\dots,p-1$, the formal power series $\widetilde{v}(z,x) \in \mathcal{O}(D_{\rho_1,\dots,\rho_n})[[z]]$ is kp-summable in the direction $\theta/p+2\ell\pi/p$ with kp-sum $v(z,x) \in \mathcal{O}(\Sigma^\ell_{/p} \times D_{r,\dots,r})$;*
2. *for $j = 0,\dots,p-1$, the p-rank reduced series $\widetilde{v}_j(t,x) \in \mathcal{O}(D_{\rho_1,\dots,\rho_n})[[t]]$ is k-summable in the direction θ with k-sum $v_j(t,x) \in \mathcal{O}(\Sigma \times D_{r,\dots,r})$ defined by the relation*

$$pt^{j/p} v_j(t,x) = \sum_{\ell=0}^{p-1} \omega^{\ell(p-j)} v(\omega^\ell t^{1/p},x), \quad t^{1/p} \in \Sigma^0_{/p}.$$

If the definition of the k-summability given above is natural given the results of Part I (Gevrey regularity of the formal power series solutions of partial differential equations, algebraic structures of the s-Gevrey formal power series and of the s-Gevrey asymptotic functions, Watson's Lemma, etc.), it is in general difficult to apply in practice because of inequalities (4.1) which can be very complicated to verify. To get around this problem, it is therefore interesting to replace Definition 5.1 by various equivalent definitions in order to have more flexibility in the tools allowing the study of the k-summability.

In the next two chapters, we focus on two characterizations of the k-summability: the first one based on a property related to the successive derivatives of the k-sum; the second one based on the Borel-Laplace method and which allows to construct the k-sum in an efficient way. For each of these characterizations, we attach some applications to partial differential equations fitting especially that point of view.

Chapter 6
First Characterization of the k-Summability: The Successive Derivatives

In Proposition 4.5, we saw a characterization of the s-Gevrey asymptotic functions in terms of their successive derivatives. Combining this with Definition 5.1, we immediately obtain the following characterization of the k-summability.

Proposition 6.1 *A formal power series* $\widetilde{u}(t,x) \in \mathcal{O}(D_{\rho_1,\dots,\rho_n})[[t]]$ *is* k*-summable in the direction* $\arg(t) = \theta$ *if and only if there exist a sector* $\Sigma_{\theta,>\pi s}$*, a radius* $0 < r \leqslant \min(\rho_1,\dots,\rho_n)$ *and a function* $u(t,x)$ *such that*

1. u *is defined and holomorphic on* $\Sigma_{\theta,>\pi s} \times D_{r,\dots,r}$ *;*
2. *for any* $x \in D_{r,\dots,r}$*, the map* $t \longmapsto u(t,x)$ *admits* $\widetilde{u}(t,x)$ *as Taylor expansion at* 0 *on* $\Sigma_{\theta,>\pi s}$*;*
3. *for any proper subsector* $\Sigma \Subset \Sigma_{\theta,>\pi s}$*, there exist two positive constants* $C, K > 0$ *such that the following estimate holds for all* $\ell \in \mathbb{N}$ *and all* $(t,x) \in \Sigma \times D_{r,\dots,r}$*:*

$$\left|\partial_t^\ell u(t,x)\right| \leqslant CK^\ell \Gamma(1+(s+1)\ell).$$

In Sect. 6.2 below, we shall show how this characterization allows to study the k-summability of the formal power series solutions of some partial differential equations. Before that, let us start by proving some additional algebraic properties on the k-summable formal power series.

6.1 Some Additional Algebraic Properties

The two following propositions investigate the k-summability of the analytic functions at the origin of $\mathbb{C}^{n+1}$, and the k-summability of the inverse of the invertible k-summable formal series $\widetilde{g}(t,x) \in \mathcal{O}(D_{\rho_1,\dots,\rho_n})\{t\}_{k;\theta}$.

P. Remy, *Asymptotic Expansions and Summability*, Lecture Notes in Mathematics 2351, https://doi.org/10.1007/978-3-031-59094-8_6

Proposition 6.2 *Let $a(t,x)$ be an analytic function on a polydic $D_{\rho_0,\rho_1,\dots,\rho_n}$. Then, $a(t,x) \in \mathcal{O}(D_{\rho_1,\dots,\rho_n})\{t\}_{k;\theta}$ for any $k > 0$ and any direction $\theta \in \mathbb{R}/2\pi\mathbb{Z}$.*

Moreover, $S_{k;\theta}(a) = a$.

Proof Let us fix $k > 0$, a direction $\theta \in \mathbb{R}/2\pi\mathbb{Z}$, and $2n+2$ radii $0 < \rho''_j < \rho'_j < \rho_j$ with $j = 0, \dots, n$. Let us first start by observing that $a(t,x)$ is its own Taylor expansion at 0 on $D_{\rho_0,\rho_1,\dots,\rho_n}$. On the other hand, we derive from the Cauchy Integral Formula

$$\partial_t^\ell a(t,x) = \frac{\ell!}{(2i\pi)^{n+1}} \int_{\substack{|t'-t|=\rho'_0-\rho''_0 \\ |x'_j-x_j|=\rho'_j-\rho''_j \\ j\in\{1,\dots,n\}}} \frac{a(t',x')}{(t'-t)^{\ell+1}(x'_1-x_1)\dots(x'_n-x_n)} dt'dx'$$

the inequalities

$$|\partial_t^\ell a(t,x)| \leqslant \alpha \left(\frac{1}{\rho'_0-\rho''_0}\right)^\ell \ell!$$

for all $\ell \geqslant 0$ and all $(t,x) \in D_{\rho''_0,\rho''_1,\dots,\rho''_n}$, where α stands for the maximum of $|a(t,x)|$ on the closed polydisc $\overline{D}_{\rho'_0,\rho'_1,\dots,\rho'_n}$ ($\overline{D}_\rho$ denotes the closed disc with center $0 \in \mathbb{C}$ and radius $\rho > 0$). Observing then that $\ell! = \Gamma(1+\ell) \leqslant \Gamma(1+(s+1)\ell)$ for all $\ell \geqslant 0$ (it is clear for $\ell = 0$ and stems from the increase of the Gamma function on $[2, +\infty[$ for $\ell \geqslant 1$), we finally get the inequalities

$$|\partial_t^\ell a(t,x)| \leqslant \alpha \left(\frac{1}{\rho'_0-\rho''_0}\right)^\ell \Gamma(1+(s+1)\ell)$$

for all $\ell \geqslant 0$ and all $(t,x) \in D_{\rho''_0,\rho''_1,\dots,\rho''_n}$. According to Proposition 6.1, the choice of a sector $\Sigma_{\theta,>\pi s} \subset D_{\rho''_0}$ and of a radius $0 < r \leqslant \min(\rho''_1, \dots, \rho''_n)$ ends the proof of Proposition 6.2. ■

Proposition 6.3 *Let $k > 0$, $\theta \in \mathbb{R}/2\pi\mathbb{Z}$ and $\widetilde{g}(t,x) \in \mathcal{O}(D_{\rho_1,\dots,\rho_n})\{t\}_{k;\theta}$.*

Assume $\widetilde{g}(0,0) \neq 0$. Then, $\widetilde{g}(t,x)$ is invertible and its inverse $\widetilde{g}^{-1}(t,x)$ belongs to $\mathcal{O}(D_{\rho'_1,\dots,\rho'_n})\{t\}_{k;\theta}$ for some convenient radii $0 < \rho'_j \leqslant \rho_j$ for all $j = 1, \dots, n$.

Moreover, $S_{k;\theta}(\widetilde{g}^{-1}) = (S_{k;\theta}(\widetilde{g}))^{-1}$, wherever the right-hand side is defined.

Proof Let us write $\widetilde{g}(t,x)$ in the form

$$\widetilde{g}(t,x) = \sum_{j\geqslant 0} g_j(x)t^j \quad \text{with } g_j(x) \in \mathcal{O}(D_{\rho_1,\dots,\rho_n}) \text{ for all } j \geqslant 0.$$

The condition $\widetilde{g}(0,0) \neq 0$ implying $g_0(0) \neq 0$, the function $g_0^{-1}(x)$ is well-defined and analytic on a convenient polydisc $D_{\rho'_1,\dots,\rho'_n}$ with $0 < \rho'_j \leqslant \rho_j$ for all $j = 1, \dots, n$. Consequently, the formal series $\widetilde{g}(t,x)$ is invertible and its inverse

$\widetilde{g}^{-1}(t,x)$ is defined by

$$\widetilde{g}^{-1}(t,x) = \sum_{j\geqslant 0} h_j(x)t^j$$

with

$$\begin{cases} h_0 = g_0^{-1} \quad \text{and} \\ h_j = -g_0^{-1} \sum_{j_0=0}^{j-1} g_{j-j_0} h_{j_0} \text{ for all } j \geqslant 1. \end{cases} \tag{6.1}$$

In particular, $\widetilde{g}^{-1}(t,x) \in \mathcal{O}(D_{\rho'_1,\dots,\rho'_n})[[t]]$.

We shall now prove that $\widetilde{g}^{-1}(t,x)$ is k-summable in the direction θ. A natural candidate for its k-sum is the inverse, if any exists, of the k-sum $g(t,x)$ of $\widetilde{g}(t,x)$.

Let us denote by $\Sigma_{\theta,>\pi s} \times D_{\rho,\dots,\rho}$ the domain of analyticity of $g(t,x)$. By continuity, we have $g(0,0) = \widetilde{g}(0,0) \neq 0$. Consequently, there exist a proper subsector $\Sigma'_{\theta,>\pi s} \Subset \Sigma_{\theta,>\pi s}$ and a radius $0 < r < \min(\rho, \rho'_1, \dots, \rho'_n)$ such that $g(t,x) \neq 0$ for all $(t,x) \in \overline{\Sigma}'_{\theta,>\pi s} \times \overline{D}_{r,\dots,r}$ (otherwise, we can construct a convergent sequence of zeros of $g(t,x)$, which implies $g(t,x) \equiv 0$ by the Isolated Zeros Theorem; hence, in particular, $g(0,0) = 0$). Thereby, $g^{-1}(t,x)$ exists for all $(t,x) \in \overline{\Sigma}'_{\theta,>\pi s} \times \overline{D}_{r,\dots,r}$ and is holomorphic on $\Sigma'_{\theta,>\pi s} \times D_{r,\dots,r}$, which proves the first condition of Proposition 6.1.

Let us now fix $x \in D_{r,\dots,r}$ and let us prove that the map $t \longmapsto g^{-1}(t,x)$ admits $\widetilde{g}^{-1}(t,x)$ as Taylor expansion at 0 on $\Sigma'_{\theta,>\pi s}$ (see Definition 2.1). To do that, we shall prove that the remainders

$$r_{g^{-1}}(t,x,J) = t^{-J}\left(g^{-1}(t,x) - \sum_{j=0}^{J-1} h_j(x)t^j\right), \quad J \geqslant 1$$

are bounded on $\overline{\Sigma}'_{\theta,>\pi s}$. First of all, the map $t \longmapsto g(t,x)$ admitting $\widetilde{g}(t,x)$ as Taylor expansion at 0 on $\Sigma_{\theta,>\pi s}$, there exists a constant $C'_{x,J,\Sigma_{\theta,>\pi s}} > 0$ such that $|r_g(t,x,J)| \leqslant C'_{x,J,\Sigma_{\theta,>\pi s}}$ for all $J \geqslant 1$ and all $t \in \overline{\Sigma}'_{\theta,>\pi s}$. On the other hand, the definition of the functions $h_j(x)$ and $g(t,x)$ implies there exist two positive constants C''_j and δ independent of $t \in \overline{\Sigma}'_{\theta,>\pi s}$ and $x \in \overline{D}_{r,\dots,r}$ such that $|h_j(x)| \leqslant C''_j$ and $|g(t,x)| \geqslant \delta$. Consequently, using the identity

$$\sum_{j=0}^{J-1}\sum_{j'=0}^{J-j-1} h_j(x)g_{j'}(x)t^{j+j'} = \sum_{\ell=0}^{J-1}\left(\sum_{j_0+j_1=\ell} h_{j_0}(x)g_{j_1}(x)\right)t^\ell = 1$$

derived from (6.1), we get the relation

$$r_{g^{-1}}(t,x,J)g(t,x) = -\sum_{j=0}^{J-1} h_j(x) r_g(t,x,J-j)$$

and finally the bound

$$|r_{g^{-1}}(t,x,J)| \leqslant \delta^{-1} \sum_{j=0}^{J-1} C''_j C'_{x,J-j,\Sigma_{\theta,>\pi s}} = C_{x,J,\Sigma'_{\theta,>\pi s}}$$

for all $J \geqslant 1$ and all $t \in \overline{\Sigma}'_{\theta,>\pi s}$, which completes the proof of the second condition of Proposition 6.1.

We are left to prove the third condition of Proposition 6.1. By assumption, there exist two positive constants $C, K > 0$ such that

$$|\partial_t^\ell g(t,x)| \leqslant C K^\ell \Gamma(1+(s+1)\ell)$$

for all $\ell \geqslant 0$ and all $(t,x) \in \Sigma'_{\theta,>\pi s} \times D_{r,\dots,r}$. Let K' denote the maximum of K and $CK\delta^{-1}C_s$, where $\delta > 0$ is as above and where C_s is the positive real number defined by

$$C_s = s'(2+\Gamma(ss')) \quad \text{with} \quad s' = \begin{cases} 1 & \text{if } s \geqslant 1 \\ \left\lfloor \frac{1}{s} \right\rfloor + 1 & \text{if } s < 1 \end{cases},$$

the notation $\lfloor a \rfloor$ refering to the floor of $a \in \mathbb{R}$. We shall prove by induction on $\ell \geqslant 0$ the properties

$$(\mathcal{P}_\ell) : |\partial_t^\ell g^{-1}(t,x)| \leqslant \delta^{-1} K'^\ell \Gamma(1+(s+1)\ell) \quad \text{for all } (t,x) \in \Sigma'_{\theta,>\pi s} \times D_{r,\dots,r}.$$

The result is obvious for $\ell = 0$ by definition of δ. Let us now suppose that the properties $(\mathcal{P}_k)$ hold for all $k \in \{0,\dots,\ell\}$ for a certain $\ell \geqslant 0$. Applying the Leibniz Formula to the identity $g^{-1}(t,x)g(t,x) = 1$, we get

$$\partial_t^{\ell+1} g^{-1}(t,x) = -g^{-1}(t,x) \sum_{j=0}^{\ell} \binom{\ell+1}{j} \partial_t^{\ell+1-j} g(t,x) \partial_t^j g^{-1}(t,x);$$

hence, the inequality

$$|\partial_t^{\ell+1} g^{-1}(t,x)| \leqslant \delta^{-1} C K \delta^{-1} K'^{\ell} \Gamma(1+(s+1)(\ell+1)) \sum_{j=0}^{\ell} \frac{\binom{\ell+1}{j}}{\binom{(s+1)(\ell+1)}{(s+1)j}}$$

for all $(t,x) \in \Sigma'_{\theta,>\pi s} \times D_{r,\dots,r}$ (use in particular the relation $K \leqslant K'$), the notation $\binom{(s+1)(\ell+1)}{(s+1)j}$ standind for the generalized binomial coefficient (see Chapter 13). Applying then the Vandermonde Inequality (see Proposition 13.2, (1))

$$\binom{(s+1)(\ell+1)}{(s+1)j} \geqslant \binom{s(\ell+1)}{sj}\binom{\ell+1}{j},$$

we finally get the relations

$$\begin{aligned}
|\partial_t^{\ell+1} g^{-1}(t,x)| &\leqslant \delta^{-1} C K \delta^{-1} K'^{\ell} \Gamma(1+(s+1)(\ell+1)) \sum_{j=0}^{\ell} \frac{1}{\binom{s(\ell+1)}{sj}} \\
&\leqslant \delta^{-1} C K \delta^{-1} C_s K'^{\ell} \Gamma(1+(s+1)(\ell+1)) \quad \text{(Proposition 13.6)} \\
&\leqslant \delta^{-1} K'^{\ell+1} \Gamma(1+(s+1)(\ell+1))
\end{aligned}$$

for all $(t,x) \in \Sigma'_{\theta,>\pi s} \times D_{r,\dots,r}$, which proves the third condition of Proposition 6.1 and completes thereby the proof of Proposition 6.3. ■

6.2 Summability and Partial Differential Equations

In this section, we investigate the summability of the formal power series solutions $\widetilde{u}(t,x) \in \mathcal{O}(D_{\rho_1})[[t]]$ of some inhomogeneous partial differential equations in two complex variables $(t,x) \in \mathbb{C}^2$. As we shall see, the approach used here, which is based on the characterization of the k-summability stated in Proposition 6.1, allows to obtain a characterization of the summability of these solutions both in terms of the inhomogeneity and of a finite number of its formal coefficients $\partial_x^n \widetilde{u}(t,x)_{|x=0}$.

Before stating various general results, let us start by detailing, as in Sect. 3.2 devoted to the study of the Gevrey regularity, the case of the heat equation in order to introduce some theoretical and technical tools that can be used in this kind of study.

Notation 6.4 In the sequel of this section, we write any element $\widetilde{g}(t,x) \in \mathcal{O}(D_{\rho_1})[[t]]$ in the form

$$\widetilde{g}(t,x) = \sum_{n\geqslant 0} \widetilde{g}_{*,n}(t)\frac{x^n}{n!}$$

with $\widetilde{g}_{*,n}(t) \in \mathbb{C}[[t]]$ for all $n \geqslant 0$. As before, the coefficients $\widetilde{g}_{*,n}(t)$ are denoted with a tilde to emphasize their possible divergence. However, when $\widetilde{g}(t,x)$ is analytic at the origin of $\mathbb{C}^2$, we omit it and we simply write $g_{*,n}(t)$ instead of $\widetilde{g}_{*,n}(t)$.

6.2.1 *Example: The Heat Equation*

Let us consider the inhomogeneous linear heat equation

$$\begin{cases} \partial_t u - a(t,x)\partial_x^2 u = \widetilde{f}(t,x) \\ u(0,x) = \varphi(x) \end{cases}, (t,x) \in \mathbb{C}^2, \tag{6.2}$$

where $a(t,x) \in \mathcal{O}(D_{\rho_0} \times D_{\rho_1})$ satisfies $a(0,x) \not\equiv 0$, $\varphi(x) \in \mathcal{O}(D_{\rho_1})$ and $\widetilde{f}(t,x) \in \mathcal{O}(D_{\rho_1})[[t]]$. In Sect. 3.2.1, Proposition 3.9, we saw that the Gevrey regularity of the unique formal power series solution $\widetilde{u}(t,x) \in \mathcal{O}(D_{\rho_1})[[t]]$ of Eq. (6.2) depends both on the structure of the equation, that is on its associated linear operator $\partial_t - a(t,x)\partial_x^2$, which induces a 1-Gevrey regularity, and on the s-Gevrey regularity of the inhomogeneity $\widetilde{f}(t,x)$:

- if $s \leqslant 1$, then $\widetilde{u}(t,x)$ keeps the 1-Gevrey regularity defined by the structure of Eq. (6.2);
- if $s > 1$, then $\widetilde{u}(t,x)$ inherits the s-Gevrey regularity of the inhomogeneity $\widetilde{f}(t,x)$.

In particular, when $\widetilde{f}(t,x) \equiv 0$, that is when Eq. (6.2) is homogeneous, the formal solution $\widetilde{u}(t,x)$ is generically 1-Gevrey.

Here below, we focus on the critical value $s = 1$ and we investigate the following natural question: *how to characterize the* 1*-summability of* $\widetilde{u}(t,x)$?

A first answer to this question was provided in [67] by D. A. Lutz, M. Miyake and R. Schäfke in the case $a(t,x) \equiv 1$ and $\widetilde{f}(t,x) \equiv 0$ using an approach based on the Borel-Laplace method to which we will return in more detail in Sect. 7.5.1.

In the 2009 article [10], W. Balser and M. Loday-Richaud studied the more general case of Eq. (6.2) with the condition $a(t,x) = \alpha(x)$ independent of t. Using the approach we consider in the present section, they gave a necessary and sufficient condition both on the inhomogeneity $\widetilde{f}(t,x)$ and on the two formal power series $\widetilde{u}_{*,0}(t)$ and $\widetilde{u}_{*,1}(t)$ for $\widetilde{u}(t,x)$ to be 1-summable in a fixed direction $\arg(t) = \theta$.

By adapting this approach, we propose here to prove the following general result.

Proposition 6.5 *Let $\widetilde{u}(t,x)$ be the formal solution in $\mathcal{O}(D_{\rho_1})[[t]]$ of the heat equation (6.2). Let* $\arg(t) = \theta \in \mathbb{R}/2\pi\mathbb{Z}$ *be a direction issuing from* 0.

Let us assume that either $a(0,0) \neq 0$ or $a(t,0) \equiv 0$ and $\partial_x a(0,0) \neq 0$. Then,

1. *$\widetilde{u}(t,x)$ is 1-summable in the direction θ if and only if the inhomogeneity $\widetilde{f}(t,x)$ and the two formal series $\widetilde{u}_{*,0}(t), \widetilde{u}_{*,1}(t) \in \mathbb{C}[[t]]$ are 1-summable in the direction θ.*
2. *Moreover, the 1-sum $u(t,x)$, if any exists, satisfies Eq. (6.2) in which $\widetilde{f}(t,x)$ is replaced by its 1-sum $f(t,x)$ in the direction θ.*

Remark 6.6 If Point 1 allows us to determine whether the formal solution $\widetilde{u}(t,x)$ is 1-summable, Point 2 shows the full value of the summability theory, at least when applied to the heat equation. Indeed, assuming that $\widetilde{u}(t,x)$ is 1-summable in a given direction θ, this one proves that its 1-sum $u(t,x)$ defines an analytic solution of the equation on a convenient domain of the form $\Sigma_{\theta,>\pi} \times D_r$. Furthermore, this solution being by definition 1-Gevrey asymptotic to $\widetilde{u}$ at 0 on $\Sigma_{\theta,>\pi}$, we can deduce from this some useful information about its behaviour in the neighborhood of the origin $t = 0$.

Proof of Proposition 6.5 Let us first observe that the necessary condition of the first point is straightforward from Proposition 5.4 and that the second point stems obvious from Corollary 5.5 and Proposition 6.2. Consequently, we are left to prove the sufficient condition of the first point.

From now on, we fix a direction θ and we suppose that the inhomogeneity $\widetilde{f}(t,x)$ and the two formal power series $\widetilde{u}_{*,0}(t), \widetilde{u}_{*,1}(t) \in \mathbb{C}[[t]]$ are all 1-summable in the direction θ. To prove that the formal solution $\widetilde{u}(t,x)$ is also 1-summable in this direction, we shall proceed through a fixed point method similar to the one already used by W. Balser and M. Loday-Richaud in [10].

$\triangleleft$ *The case* $a(0,0) \neq 0$.

• *A preliminary remark.* Before starting the calculations, let us begin with a preliminary remark. Writing the formal solution $\widetilde{u}(t,x)$, the coefficient $a(t,x)$ and the inhomogeneity $\widetilde{f}(t,x)$ as in Notation 6.4, we first notice that, by identification of the terms in x^n in Eq. (6.2), we get the identities

$$a_{*,0}(t)\widetilde{u}_{*,n+2}(t) + \sum_{n_0=1}^{n} \binom{n}{n_0} a_{*,n_0}(t)\widetilde{u}_{*,n+2-n_0}(t) = \partial_t \widetilde{u}_{*,n}(t) - \widetilde{f}_{*,n}(t) \tag{6.3}$$

for all $n \geqslant 0$. The coefficient $a_{*,0}(t)$ being invertible in $\mathbb{C}[[t]]$ according to the assumption $a(0,0) \neq 0$ which implies $a_{*,0}(0) \neq 0$, we derive then from relations (6.3) that each coefficient $\widetilde{u}_{*,n}(t)$ is uniquely determined from the inhomogeneity $\widetilde{f}(t,x)$ and from the two formal series $\widetilde{u}_{*,0}(t)$ and $\widetilde{u}_{*,1}(t)$. In particular, the same applies to $\widetilde{u}(t,x)$.

• *An auxiliary equation.* Let us now introduce the formal power series $\widetilde{w}(t,x) \in \mathcal{O}(D_{\rho_1})[[t]]$ defined by the relation

$$\widetilde{u}(t,x) = \widetilde{u}_{*,0}(t) + x\widetilde{u}_{*,1}(t) + \partial_x^{-2}\widetilde{w}(t,x).$$

With this notation, Eq. (6.2) becomes

$$\widetilde{w}(t,x) - A(t,x)\partial_t\partial_x^{-2}\widetilde{w}(t,x) = \widetilde{g}(t,x) \tag{6.4}$$

with

$$\widetilde{g}(t,x) = A(t,x)\left(\partial_t\widetilde{u}_{*,0}(t) + x\partial_t\widetilde{u}_{*,1}(t) - \widetilde{f}(t,x)\right),$$

where, thanks to the assumption $a(0,0) \neq 0$, the coefficient $A(t,x) = 1/a(t,x)$ is well-defined and holomorphic on a convenient polydisc $D_{\rho_0'} \times D_{\rho_1'}$ with $0 < \rho_0' \leqslant \rho_0$ and $0 < \rho_1' \leqslant \rho_1$.

Observe that $\widetilde{w}(t,x)$ is the unique formal power series solution of Eq. (6.4) (reason as in Lemma 3.7 by exchanging the role of t and x). Observe also that, according to our assumption on the 1-summability of the inhomogeneity $\widetilde{f}(t,x)$ and of the two formal power series $\widetilde{u}_{*,0}(t)$ and $\widetilde{u}_{*,1}(t)$, the formal series $\widetilde{g}(t,x) \in \mathcal{O}(D_{\rho_1'})[[t]]$ is also 1-summable in the direction θ (see Propositions 5.4 and 6.2). Consequently, the identity (6.4) above tells us that it is sufficient to prove that it is the same for the formal power series $\widetilde{w}(t,x)$. To do that, we shall proceed through a fixed point method as follows.

• *The fixed point procedure.* Let us set

$$\widetilde{W}(t,x) = \sum_{\mu \geqslant 0} \widetilde{w}_\mu(t,x)$$

and let us choose the solution of Eq. (6.4) recursively determined by the relations

$$\begin{cases} \widetilde{w}_0(t,x) = \widetilde{g}(t,x) \\ \widetilde{w}_{\mu+1} = A(t,x)\partial_t\partial_x^{-2}\widetilde{w}_\mu(t,x) \quad \text{for all } \mu \geqslant 0 \end{cases}. \tag{6.5}$$

Observe that $\widetilde{w}_\mu(t,x) \in \mathcal{O}(D_{\rho_1'})[[t]]$ for all $\mu \geqslant 0$. Observe also that the $\widetilde{w}_\mu(t,x)$'s are of order $O(x^{2\mu})$ in x for all $\mu \geqslant 0$. Thereby, the series $\widetilde{W}(t,x)$ itself makes sense as a formal power series in t and x and, consequently, $\widetilde{W}(t,x) = \widetilde{w}(t,x)$ by unicity.

Let us now denote by w_0 the 1-sum of $\widetilde{w}_0$ in the direction θ and, for all $\mu > 0$, let w_μ be determined by the relations (6.5) in which all the $\widetilde{w}_\mu$ are replaced by w_μ. By construction, all the functions $w_\mu(t,x)$ are defined and holomorphic on a common domain $\Sigma_{\theta,>\pi} \times D_{\rho_1''}$ with a convenient radius $0 < \rho_1'' < \rho_1'$.

To end the proof, it remains to prove that the series $\sum_{\mu \geqslant 0} w_\mu(t,x)$ is convergent and that its sum $w(t,x)$ is the 1-sum of $\widetilde{w}(t,x)$ in the direction θ. To do that, let us start by making some estimates on the functions $w_\mu(t,x)$.

• *Some estimates on the* $w_\mu(t,x)$*'s.* According to Propositions 6.1 and 6.2, there exists a radius $0 < r_1 < \rho_1''$ such that, for any proper subsector $\Sigma \Subset \Sigma_{\theta,>\pi}$, there exist two positive constants $C, K > 0$ such that, for all $\ell \geqslant 0$ and all $(t,x) \in \Sigma \times D_{r_1}$, the functions w_0 and A satisfy the inequalities

$$\left|\partial_t^\ell w_0(t,x)\right| \leqslant CK^\ell \Gamma(1+2\ell) \quad \text{and} \quad \left|\partial_t^\ell A(t,x)\right| \leqslant CK^\ell \Gamma(1+2\ell). \tag{6.6}$$

Let us fix a proper subsector $\Sigma \Subset \Sigma_{\theta,>\pi}$ and let us start by proving by recursion on $\mu \geqslant 0$ that the following estimate holds for all $\ell, \mu \geqslant 0$ and all $(t,x) \in \Sigma \times D_{r_1}$:

$$\left|\partial_t^\ell w_\mu(t,x)\right| \leqslant C(3C)^\mu K^{\mu+\ell} \Gamma(1+2\mu+2\ell) \frac{|x|^{2\mu}}{(2\mu)!}. \tag{6.7}$$

The case $\mu = 0$ is straightforward from the first inequality of (6.6) and we now suppose that the inequalities (6.7) hold for all the functions w_j with $j = 0, \ldots, \mu$ for a certain $\mu \geqslant 0$. According to the relations (6.5), we first derive from the Leibniz Formula the identities

$$\partial_t^\ell w_{\mu+1}(t,x) = \sum_{\ell_0=0}^{\ell} \binom{\ell}{\ell_0} \partial_t^{\ell-\ell_0} A(t,x) \partial_t^{\ell_0+1} \partial_x^{-2} w_\mu(t,x)$$

for all $\ell \geqslant 0$ and all $(t,x) \in \Sigma \times D_{r_1}$. Applying then the second inequality of (6.6) and the inequalities (6.7) to the function w_μ, we finally get the inequalities

$$\begin{aligned} \left|\partial_t^\ell w_{\mu+1}(t,x)\right| \leqslant C3^\mu C^{\mu+1} K^{\mu+1+\ell} \Gamma(1+2(\mu+1)+2\ell) \frac{|x|^{2(\mu+1)}}{(2(\mu+1))!} \\ \times \sum_{\ell_0=0}^{\ell} \frac{\binom{\ell}{\ell_0}}{\binom{2(\mu+1)+2\ell}{2(\mu+1)+2\ell_0}} \end{aligned} \tag{6.8}$$

for all $\ell \geqslant 0$ and all $(t,x) \in \Sigma \times D_{r_1}$. The Vandermonde Inequality (see Proposition 13.2, (1))

$$\binom{2(\mu+1)+2\ell}{2(\mu+1)+2\ell_0} \geqslant \binom{\ell}{\ell_0}\binom{\ell}{\ell_0}\binom{2(\mu+1)}{2(\mu+1)} = \binom{\ell}{\ell_0}^2$$

and Proposition 13.6, (1) applied with $s = 1$ complete the proof of the inequalities (6.7) for $w_{\mu+1}$.

Let us now improve the estimates (6.7) so that we can more easily use the characterization of the 1-summability given in Proposition 6.1. To do that, let us observe that

$$\Gamma(1+2\mu+2\ell) = \Gamma(1+2\mu)\Gamma(1+2\ell)\binom{2\mu+2\ell}{2\mu} \leqslant 4^{\mu+\ell}(2\mu)!\Gamma(1+2\ell) \tag{6.9}$$

and let us set $K_1 = 4K$ and $c = 12CK$. Then, the following estimate holds for all $\ell, \mu \geqslant 0$ and all $(t,x) \in \Sigma \times D_{r_1}$:

$$\left|\partial_t^\ell w_\mu(t,x)\right| \leqslant CK_1^\ell \Gamma(1+2\ell)(c\,|x|^2)^\mu. \tag{6.10}$$

• *Conclusion.* Let us now choose for Σ a sector containing a proper subsector Σ' bisected by the direction θ and opening larger than π (such a choice is already possible by definition of a proper subsector, see page 5). Let us also choose a radius $0 < r < \min(r_1, 1/\sqrt{c})$ and let us set $C_1 := C\sum_{\mu\geqslant 0}(cr^2)^\mu \in \mathbb{R}_+^*$.

Thanks to the estimates (6.10), the series $\sum_{\mu\geqslant 0} \partial_t^\ell w_\mu(t,x)$ are normally convergent on $\Sigma \times D_r$ for all $\ell \geqslant 0$ and satisfy the inequalities

$$\sum_{\mu\geqslant 0}\left|\partial_t^\ell w_\mu(t,x)\right| \leqslant C_1 K_1^\ell \Gamma(1+2\ell)$$

for all $(t,x) \in \Sigma \times D_r$. In particular, the sum $w(t,x)$ of the series $\sum_{\mu\geqslant 0} w_\mu(t,x)$ is well-defined, holomorphic on $\Sigma \times D_r$ and satisfies the inequalities

$$\left|\partial_t^\ell w(t,x)\right| \leqslant C_1 K_1^\ell \Gamma(1+2\ell)$$

for all $\ell \geqslant 0$ and all $(t,x) \in \Sigma \times D_r$. Hence, Conditions 1 and 3 of Proposition 6.1 hold, since $\Sigma' \Subset \Sigma$.

To prove the second condition of Proposition 6.1, we proceed as follows. Thanks to the definition of the function w, we first derive from Corollary 4.6 that w admits a 1-Gevrey asymptotic expansion $\widetilde{\omega}$ at 0 on Σ'. On the other hand, considering recurrence relations (6.5) with $w_\mu(t,x)$ and the 1-sum $g(t,x)$ instead of $\widetilde{w}_\mu(t,x)$ and $\widetilde{g}(t,x)$, it is clear that $w(t,x)$ satisfies Eq. (6.4) with right-hand side $g(t,x)$ in place of $\widetilde{g}(t,x)$ and, consequently, so does its Gevrey asymptotic expansion $\widetilde{\omega}$ according to Proposition 4.9. Then, since Eq. (6.4) admits $\widetilde{w}(t,x)$ as the unique formal power series, we then conclude that $\widetilde{\omega} \equiv \widetilde{w}$. Hence, Condition 2 of Proposition 6.1 holds.

This achieves the proof of the sufficient condition when $a(0,0) \neq 0$.

◁ *The case* $a(t,0) \equiv 0$ *and* $\partial_x a(0,0) \neq 0$. Let us first observe that, under this assumption, the identities (6.3) become

$$\partial_t \widetilde{u}_{*,0}(t) - \widetilde{f}_{*,0}(t) = 0 \tag{6.11}$$

for $n = 0$, and

$$na_{*,1}(t)\widetilde{u}_{*,n+1}(t) + \sum_{n_0=2}^{n} \binom{n}{n_0} a_{*,n_0}(t)\widetilde{u}_{*,n+2-n_0}(t) = \partial_t^{\kappa}\widetilde{u}_{*,n}(t) - \widetilde{f}_{*,n}(t) \tag{6.12}$$

with $a_{*,1}(0) \neq 0$ for all $n \geqslant 1$. In particular, the identities (6.12) tell us, as in the case $a(0,0) \neq 0$, that each coefficient $\widetilde{u}_{*,n}(t)$ (hence, the formal solution $\widetilde{u}(t,x)$ too) is uniquely determined from the inhomogeneity $\widetilde{f}(t,x)$ and from the two formal series $\widetilde{u}_{*,0}(t)$ and $\widetilde{u}_{*,1}(t)$.

Observe also that our assumption allows us to write the function $a(t,x)$ in the form $a(t,x) = xa_1(t,x)$ with $a_1(0,0) \neq 0$. Thereby, the function $A_1 = 1/a_1$ is well-defined and holomorphic on a convenient polydisc centered at the origin $(0,0) \in \mathbb{C}^2$, say $D_{\rho_0'} \times D_{\rho_1'}$ with $0 < \rho_0' \leqslant \rho_0$ and $0 < \rho_1' \leqslant \rho_1$ to use the same notations as the case $a(0,0) \neq 0$.

Setting as before $\widetilde{u}(t,x) = \widetilde{u}_{*,0}(t) + x\widetilde{u}_{*,1}(t) + \partial_x^{-2}\widetilde{w}(t,x)$ with $\widetilde{w}(t,x) \in \mathcal{O}(D_{\rho_1})[[t]]$, Eq. (6.2) becomes now

$$\widetilde{w} - \frac{A_1(t,x)}{x}\partial_t\partial_x^{-2}\widetilde{w} = \widetilde{g}(t,x) \tag{6.13}$$

where

$$\widetilde{g}(t,x) = A_1(t,x)\left(\frac{\partial_t\widetilde{u}_{*,0}(t) - \widetilde{f}(t,x)}{x} + \partial_t\widetilde{u}_{*,1}(t)\right)$$

is again a formal power series in t and x (indeed, due to the identity (6.11), the term $\partial_t\widetilde{u}_{*,0}(t) - \widetilde{f}(t,x)$ is of order $O(x)$ in x). Moreover, $\widetilde{w}(t,x)$ is again the unique formal power series of Eq. (6.13).

Assuming then $\widetilde{g}(t,x)$ to be 1-summable in the direction θ, we can prove as in the case $a(0,0) \neq 0$ that $\widetilde{w}(t,x)$ is also 1-summable in the direction θ. The proof being similar to the one developed just above, we give below only the key points to modify. In particular, we keep all the notations on the choices of the sectors Σ and Σ', and on the choices of the various radii. We keep also the estimates (6.6), but replacing A by A_1.

First of all, let us start by observing that the $\widetilde{w}_\mu(t,x)$'s are now recursively determined for all $\mu \geqslant 0$ by the relations

$$\widetilde{w}_{\mu+1}(t,x) = \frac{A_1(t,x)}{x}\partial_t\partial_x^{-2}\widetilde{w}_\mu(t,x)$$

together with the initial condition $\widetilde{w}_0 = \widetilde{g}$. In particular, the operator $\frac{1}{x}\partial_x^{-2}$ in place of ∂_x^{-2} implies that the $\widetilde{w}_\mu(t,x)$'s are of order $O(x^\mu)$ in x for all $\mu \geqslant 0$, instead of $O(x^{2\mu})$ as in the case $a(0,0) \neq 0$. Still, $\widetilde{W}(t,x)$ is again a formal power series in t and in x, and we have again $\widetilde{W}(t,x) = \widetilde{w}(t,x)$.

In doing so, the estimates on the derivatives $\partial_t^\ell w_\mu$ given in (6.7) are modified as follows: for all $\ell, \mu \geqslant 0$ and all $(t,x) \in \Sigma \times D_{r_1}$:

$$\left|\partial_t^\ell w_\mu(t,x)\right| \leqslant C(3C)^\mu K^{\mu+\ell}\Gamma(1+2\mu+2\ell)\frac{|x|^\mu}{(\mu!)^2}.$$

These inequalities are obtained as before by replacing the inequality (6.8) by the inequality

$$\left|\partial_t^\ell w_{\mu+1}(t,x)\right| \leqslant C3^\mu C^{\mu+1}K^{\mu+1+\ell}\Gamma(1+2(\mu+1)+2\ell)\frac{|x|^{\mu+1}}{((\mu+1)!)^2}$$
$$\times\sum_{\ell_0=0}^{\ell}\frac{\binom{\ell}{\ell_0}}{\binom{2(\mu+1)+2\ell}{2(\mu+1)+2\ell_0}}\frac{((\mu+1)!)^2}{(\mu!)^2(\mu+1)(\mu+2)}.$$

and by observing that

$$\frac{((\mu+1)!)^2}{(\mu!)^2(\mu+1)(\mu+2)} \leqslant \frac{((\mu+1)!)^2}{(\mu!)^2(\mu+1)^2} = 1.$$

Using then the inequality (6.9), we get

$$\left|\partial_t^\ell w_\mu(t,x)\right| \leqslant C(4K)^\ell\Gamma(1+2\ell)\binom{2\mu}{\mu}(12CK\,|x|)^\mu$$

for all $\ell, \mu \geqslant 0$ and all $(t,x) \in \Sigma \times D_{r_1}$. Hence, setting $K_1 = 4K$ and $c = 48CK$, the following estimate holds for all $\ell, \mu \geqslant 0$ and all $(t,x) \in \Sigma \times D_{r_1}$:

$$\left|\partial_t^\ell w_\mu(t,x)\right| \leqslant CK_1^\ell\Gamma(1+2\ell)(c\,|x|)^\mu.$$

The end of the proof is similar to the one of the case $a(0,0) \neq 0$ and is left to the reader. This completes the proof of Proposition 6.5. ∎

It is interesting to note here that Proposition 6.5 may fail in the case where the coefficient $a(t,x)$ is of order $O(x^2)$ in x. Indeed, the appearance of the operator $\frac{1}{x^2}\partial_x^{-2}$ instead of ∂_x^{-2} or $\frac{1}{x}\partial_x^{-2}$ implies that no power of x remains in the estimates (6.7) and we cannot guaranty the convergence of the estimate for $\partial_t^\ell w$.

Proposition 6.7 below provides in this case a counterexample showing that the 1-summability of $\widetilde{u}(t,x)$ may fail.

Proposition 6.7 ([10]) *Let us consider the linear heat equation (6.2) with*

$$a(t,x) = x^2, \quad \widetilde{f}(t,x) \equiv 0 \quad \textit{and} \quad \varphi(x) = \frac{1}{1-x}.$$

Then, the formal solution $\widetilde{u}(t,x)$ is 1-summable in no direction.

Proof Let us first observe that the relations (3.6) allow to write the formal solution $\widetilde{u}(t,x)$ in the form

$$\widetilde{u}(t,x) = \varphi(x) + O(tx^2) = \sum_{n\geqslant 0} n!\frac{x^n}{n!} + O(tx^2).$$

Thereby, $\widetilde{u}_{*,0}(t) \equiv 0! = 1$ and $\widetilde{u}_{*,1}(t) \equiv 1! = 1$ are both 1-summable in any direction. On the other hand, applying also the relations (6.3), it results that the formal series $\widetilde{u}_{*,n}(t)$ with $n \geqslant 2$ satisfy the identities

$$\partial_t \widetilde{u}_{*,n}(t) - n(n-1)\widetilde{u}_{*,n}(t) = 0 \quad \text{and} \quad \widetilde{u}_{*,n}(0) = n!.$$

Consequently, $\widetilde{u}_{*,n}(t) = n!e^{n(n-1)t}$ for all $n \geqslant 2$.

Let us now suppose that $\widetilde{u}(t,x)$ is 1-summable in a given direction θ with sum $u(t,x)$. Since

$$\widetilde{u}_{*,n}(t) = \partial_x^n \widetilde{u}(t,x)_{|x=0},$$

we first deduce from Proposition 5.4 that all the formal series $\widetilde{u}_{*,n}(t)$ with $n \geqslant 2$ are 1-summable in the direction θ with sum $u_{*,n}(t) = \partial_x^n u(t,x)_{|x=0}$. Denoting then by $\Sigma_{\theta,>\pi} \times D_r$ the domain of analyticity of $u(t,x)$ and choosing a proper subsector $\Sigma'_{\theta,>\pi} \Subset \Sigma_{\theta,>\pi}$ and a radius $0 < R < r$, we derive from the Cauchy Integral Formula applied to $\partial_x^n u(t,x)$ at $x = 0$ the following estimates for all $t \in \Sigma'_{\theta,>\pi}$ and all $n \geqslant 2$:

$$\left|u_{*,n}(t)\right| = \left|\frac{n!}{2i\pi}\int_{|\xi|=R}\frac{u(t,\xi)}{\xi^{n+1}}d\xi\right| \leqslant \frac{n!}{2\pi}\frac{C\times 2\pi R}{R^{n+1}} = C\frac{n!}{R^n} \tag{6.14}$$

with a convenient constant $C > 1$ independent of t. Consequently, since we have in our case $\widetilde{u}_{*,n}(t) = u_{*,n}(t) = n!e^{n(n-1)t}$, we get the estimates

$$\left|e^{(n-1)t}\right| \leqslant \frac{C}{R}$$

for all $t \in \Sigma'_{\theta,>\pi}$ and all $n \geqslant 2$, that is the sequences $(e^{(n-1)t})_{n\geqslant 2}$ are all bounded on $\Sigma'_{\theta,>\pi}$, which is absurd (indeed, none of these sequences can be bounded on a sector larger than a half plane). Hence, the estimates (6.14) are impossible and the formal solution $\widetilde{u}(t,x)$ is 1-summable in no direction. ■

When the coefficient $a(t,x)$ and the inhomogeneity $\widetilde{f}(t,x)$ have a convenient form, Proposition 6.5 can be reformulated in terms of the initial data $\varphi(x)$. As an example, we investigate below the special case where $a(t,x) = ax$ with $a \in \mathbb{C}^*$, and where $\widetilde{f}(t,x) \equiv 0$. For other examples, we refer for instance to [10, Section 4].

Proposition 6.8 *Let $\widetilde{u}(t,x)$ be the formal solution in $\mathcal{O}(D_{\rho_1})[[t]]$ of the homogeneous linear heat equation*

$$\begin{cases} \partial_t u - ax\partial_x^2 u = 0 \\ u(0,x) = \varphi(x) \end{cases}, (t,x) \in \mathbb{C}^2, \tag{6.15}$$

with $a \in \mathbb{C}^*$ and $\varphi(x) \in \mathcal{O}(D_{\rho_1})$. Then, $\widetilde{u}(t,x)$ is 1-summable in the direction θ if and only if the initial data $\varphi(x)$ can be analytically continued to a sector neighboring the direction $\theta + \arg(a)$ with a global exponential growth of order at most 1 at infinity.

Proof By calculations, it is easy to see that the formal solution $\widetilde{u}(t,x)$ reads in the form

$$\widetilde{u}(t,x) = \sum_{j\geqslant 0} (ax\partial_x^2)^j \varphi(x) \frac{t^j}{j!}$$

with

$$(ax\partial_x^2)^j \varphi(x) = \sum_{n\geqslant 1} \frac{(n-1+j)!}{(n-1)!} \varphi_{0,n+j} \frac{a^j x^n}{n!}$$

for all $j \geqslant 1$. Hence,

$$\widetilde{u}_{*,0}(t) = \varphi_{0,0} \quad \text{and} \quad \widetilde{u}_{*,1}(t) = \sum_{j\geqslant 0} \varphi_{0,j+1}(at)^j.$$

Let us now define the formal 1-Laplace transform $\widetilde{\mathcal{L}}_1^{[x]}\varphi(x)$ of $\varphi(x)$ by

$$\widetilde{\mathcal{L}}_1^{[x]}\varphi(x) = \sum_{n\geqslant 0} \varphi_{0,n}\frac{x^n}{n!}n! = \sum_{n\geqslant 0} \varphi_{0,n}x^n.$$

Then,

$$\widetilde{\mathcal{L}}_1^{[x]}\varphi(at) = \widetilde{u}_{*,0}(t) + at\widetilde{u}_{*,1}(t) \tag{6.16}$$

and, observing that $\widetilde{u}_{*,0}(t)$ is a constant (hence, automatically 1-summable in the direction θ), the following four equivalent assertions hold:

1. $\widetilde{u}(t,x)$ is 1-summable in the direction θ;
2. $\widetilde{u}_{*,1}(t)$ is 1-summable in the direction θ;
3. $\widetilde{\mathcal{L}}_1^{[x]}\varphi(x)$ is 1-summable in the directions $\theta + \arg(a)$;
4. $\varphi(x)$ can be analytically continued to a sector neighboring the direction $\theta+\arg(a)$ with a global exponential growth of order at most 1 at infinity,

which proves Proposition 6.8. Observe here that the equivalence 1–2 is due to Proposition 6.5 and that the equivalence 3–4 stems from the classical Borel-Laplace method (see Proposition 7.18). As for the equivalence 2–3, it is obvious given the identity (6.16). ∎

Let us now look at more general partial differential equations.

6.2.2 *Some More General Problems*

In this section, we consider the very general partial differential equation

$$\begin{cases} \partial_t^\kappa u - \displaystyle\sum_{i\in\mathcal{K}}\sum_{q\in Q_i} P_{i,q}(t,x,u)\partial_t^i\partial_x^q u = \widetilde{f}(t,x) \\ \partial_t^j u(t,x)_{|t=0} = \varphi_j(x),\ j = 0,\ldots,\kappa-1 \end{cases},\quad (t,x)\in\mathbb{C}^2, \tag{6.17}$$

already investigated in Sect. 3.2.3, where

- $\kappa \geqslant 1$ is a positive integer;
- $\mathcal{K}$ is a nonempty subset of $\{0,\ldots,\kappa-1\}$;
- Q_i is a nonempty finite subset of $\mathbb{N}$ for all $i\in\mathcal{K}$;
- $P_{i,q}(t,x,X)$ is a polynomial in X with analytic coefficients on a polydisc $D_{\rho_0}\times D_{\rho_1}$ centered at the origin of $\mathbb{C}^2$ for all $i\in\mathcal{K}$ and all $q\in Q_i$:

$$P_{i,q}(t,x,X) = \sum_{m\in E_{i,q}} a_{i,q,m}(t,x)X^m,$$

where $E_{i,q}$ stands for a nonempty subset of $\mathbb{N}$, and where all the coefficients $a_{i,q,m}(t,x) \in \mathcal{O}(D_{\rho_0} \times D_{\rho_1})$ are not identically zero and with valuation $v_{i,q,m} \geqslant 0$ at $t = 0$: $a_{i,q,m}(t,x) = t^{v_{i,q,m}} a'_{i,q,m}(t,x,)$ with $a'_{i,q,m}(0,x) \not\equiv 0$;

- the inhomogeneity $\widetilde{f}(t,x)$ belongs to $\mathcal{O}(D_{\rho_1})[[t]]$;
- the initial conditions $\varphi_j(x)$ are analytic on D_{ρ_1} for all $j = 0, \ldots, \kappa - 1$.

Recall (see Theorem 3.21) that the Gevrey regularity of the unique formal power series solution $\widetilde{u}(t,x) \in \mathcal{O}(D_{\rho_1})[[t]]$ of Eq. (6.17) depends both on the structure of the equation, that is on its associated linear operator

$$\mathcal{L}_{\kappa,v} = \partial_t^{\kappa} - \sum_{i \in \mathcal{K}} \sum_{q \in Q_i} t^{v_{i,q}} \partial_t^i \partial_x^q, \quad v_{i,q} = \min_{m \in E_{i,q}} v_{i,q,m},$$

which induces a Gevrey regularity entirely determined by its Newton polygon $\mathcal{N}_t(\mathcal{L}_{\kappa,v})$ at $t = 0$, and on the s-Gevrey regularity of the inhomogeneity $\widetilde{f}(t,x)$. More precisely, denoting by s_c the nonnegative rational number equal to the inverse of the smallest positive slope of $\mathcal{N}_t(\mathcal{L}_{\kappa,v})$ if any exists, and equal to 0 otherwise:

- if $s \leqslant s_c$, then $\widetilde{u}(t,x)$ keeps the s_c-Gevrey regularity defined by the structure of Eq. (6.17);
- if $s > s_c$, then $\widetilde{u}(t,x)$ inherits the s-Gevrey regularity of the inhomogeneity $\widetilde{f}(t,x)$.

In particular, when $\widetilde{f}(t,x) \equiv 0$, that is when Eq. (6.17) is homogeneous, the formal solution $\widetilde{u}(t,x)$ is generically s_c-Gevrey; hence, convergent when $s_c = 0$.

As in the case of the heat equation (see Sect. 6.2.1), we now focus on the critical value $s = s_c$, which we assume to be *positive*. We also assume that $k = 1/s_c$ is the *unique* positive slope of the Newton polygon $\mathcal{N}_t(\mathcal{L}_{\kappa,v})$. Under this assumption, we address the following question: *how to characterize the k-summability of* $\widetilde{u}(t,x)$*?*

In the very general form of Eq. (6.17), it does not appear, to our knowledge, that this question has been studied. On the other hand, with more or less strong conditions on the sets $\mathcal{K}$, Q_i and $E_{i,q}$, and on the coefficients $a_{i,q,m}(t,x)$, many answers have been given over the last two decades, using either the Borel-Laplace method (see for instance [12, 44, 45, 68, 69, 82, 88] and the references therein), to which we will return in detail in Sect. 7.5.2, or the characterization by the successive derivatives [105, 107–110] which interests us here and whose main results we present below. As in Sect. 3.2.4 devoted to the study of the Gevrey regularity, we illustrate these on three classical equations: the (semilinear) heat equation, the (semilinear) Euler-Lagrange equation and the generalized Burgers-Korteweg-de Vries equation (in short, the gBKdV equation).

Let us start by considering the special case of the inhomogeneous semilinear equation

$$\begin{cases} \partial_t^{\kappa} u - a(t,x)\partial_x^p u - u^2 P(t,x,u) = \widetilde{f}(t,x) \\ \partial_t^j u(t,x)_{|t=0} = \varphi_j(x),\ j = 0, \ldots, \kappa - 1 \end{cases}, \quad (t,x) \in \mathbb{C}^2, \tag{6.18}$$

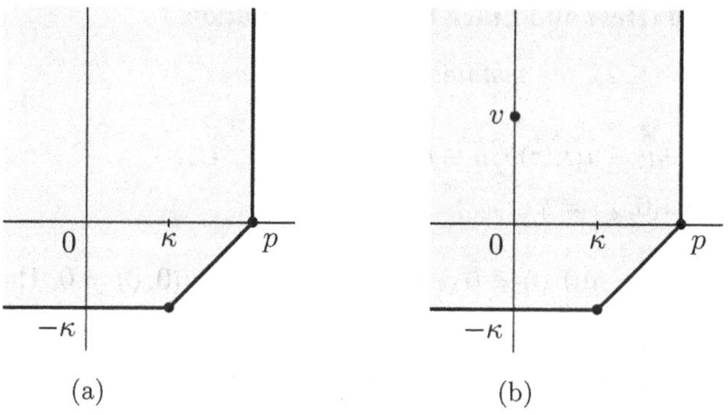

Fig. 6.1 The Newton polygon $\mathcal{N}_t(\mathcal{L}_\kappa)$ of the linear operator $\mathcal{L}_\kappa$ associated with Eq. (6.18). (**a**) Case $P \equiv 0$. (**b**) Case $P \not\equiv 0$

with $p > \kappa \geqslant 1$, $a(0,x) \not\equiv 0$ and $P(t,x,X) \in \mathcal{O}(D_{\rho_0} \times D_{\rho_1})[X]$, possibly zero. This equation naturally extends the case of the heat equation studied earlier and involves in many physical, chemical, biological and ecological problems. For example, for $(\kappa, p) = (1,2)$, Eq. (6.18) is the semilinear heat equation; for $(\kappa, p) = (2,4)$, it is the semilinear Euler-Lagrange equation, etc.

The linear operator $\mathcal{L}_\kappa$ associated with Eq. (6.18) is defined by

$$\mathcal{L}_\kappa = \begin{cases} \partial_t^\kappa - \partial_x^p & \text{if } P \equiv 0 \\ \partial_t^\kappa - \partial_x^p - t^v & \text{if } P \not\equiv 0 \end{cases},$$

where v denotes the smallest valuation at $t = 0$ of the coefficients of P. In particular, as shown in Fig. 6.1, its Newton polygon $\mathcal{N}_t(\mathcal{L}_\kappa)$ has the slope $k = \kappa/(p-\kappa)$ as unique positive slope.

By adapting the proof proposed in Sect. 6.2.1 in the case of the heat equation (6.2), we can prove the following.

Proposition 6.9 ([108]) *Let $\widetilde{u}(t,x)$ be the formal solution in $\mathcal{O}(D_{\rho_1})[[t]]$ of Eq. (6.18). Let* $\arg(t) = \theta \in \mathbb{R}/2\pi\mathbb{Z}$ *be a direction issuing from* 0.

Let us assume that either $a(0,0) \neq 0$, or there exists $q \in \{1, \dots, p-1\}$ such that $\partial_x^n a(t,x)_{|x=0} \equiv 0$ for all $n = 0, \dots, q-1$, and $\partial_x^q a(0,0) \neq 0$. Then,

1. *$\widetilde{u}(t,x)$ is k-summable in the direction θ if and only if the inhomogeneity $\widetilde{f}(t,x)$ and the formal series $\widetilde{u}_{*,n}(t) \in \mathbb{C}[[t]]$ for $n = 0, \dots, p-1$ are k-summable in the direction θ.*
2. *Moreover, the k-sum $u(t,x)$, if any exists, satisfies Eq. (6.18) in which $\widetilde{f}(t,x)$ is replaced by its k-sum $f(t,x)$ in the direction θ.*

Considering the special cases $(\kappa, p) = (1,2)$ and $(\kappa, p) = (2,4)$), we derive in particular the following.

Application 6.10 (Heat and Euler-Lagrange Equations)

- *Case* $(\kappa, p) = (1, 2)$*: the semilinear heat equation*

$$\begin{cases} \partial_t u - a(t,x)\partial_x^2 u - u^2 P(t,x,u) = \widetilde{f}(t,x) \\ u(0,x) = \varphi(x) \end{cases}. \tag{6.19}$$

Assume that either $a(0,0) \neq 0$ or $a(t,0) \equiv 0$ and $\partial_x a(0,0) \neq 0$. Then,

1. $\widetilde{u}(t,x)$ is 1-summable in the direction θ if and only if the inhomogeneity $\widetilde{f}(t,x)$ and the two formal series $\widetilde{u}_{*,0}(t), \widetilde{u}_{*,1}(t) \in \mathbb{C}[[t]]$ are 1-summable in the direction θ.

2. Moreover, the 1-sum $u(t,x)$, if any exists, satisfies Eq. (6.19) in which $\widetilde{f}(t,x)$ is replaced by its 1-sum $f(t,x)$ in the direction θ.

- *Case* $(\kappa, p) = (2, 4)$*: the semilinear Euler-Lagrange equation*

$$\begin{cases} \partial_t^2 u - a(t,x)\partial_x^4 u - u^2 P(t,x,u) = \widetilde{f}(t,x) \\ u(0,x) = \varphi_0(x),\, \partial_t u(t,x)_{|t=0} = \varphi_1(x) \end{cases}. \tag{6.20}$$

Assume that one of the following conditions is satisfied:

- $a(0,0) \neq 0$;
- $a(t,0) \equiv 0$ and $\partial_x a(0,0) \neq 0$;
- $a(t,0) \equiv \partial_x a(t,0) \equiv 0$ and $\partial_x^2 a(0,0) \neq 0$;
- $a(t,0) \equiv \partial_x a(t,0) \equiv \partial_x^2 a(t,0) \equiv 0$ and $\partial_x^3 a(0,0) \neq 0$.

Then,

1. $\widetilde{u}(t,x)$ is 1-summable in the direction θ if and only if the inhomogeneity $\widetilde{f}(t,x)$ and the four formal series $\widetilde{u}_{*,0}(t), \widetilde{u}_{*,1}(t), \widetilde{u}_{*,2}(t), \widetilde{u}_{*,3}(t) \in \mathbb{C}[[t]]$ are 1-summable in the direction θ.

2. Moreover, the 1-sum $u(t,x)$, if any exists, satisfies Eq. (6.20) in which $\widetilde{f}(t,x)$ is replaced by its 1-sum $f(t,x)$ in the direction θ.

Remark 6.11

- When $P \equiv 0$, Eq. (6.19) is reduced to the linear heat equation (6.2) and we find the result already stated in Proposition 6.5.
- As for the linear heat equation (see Proposition 6.8), Proposition 6.9 can be reformulated in terms of the initial datas $\varphi_j(x)$ when the coefficient $a(t,x)$, the semilinearity $P(t,x,u)$ and the inhomogeneity $\widetilde{f}(t,x)$ have a convenient form. We refer to [103] for some examples.

In the case of the general Eq. (6.17), the situation is much more complicated and only the situation where the positive slope of the Newton polygon $\mathcal{N}_t(\mathcal{L}_{\kappa,v})$ is determined by the highest x-derivative order has been studied. More precisely, we have the following.

Theorem 6.12 ([109]) *Let $\widetilde{u}(t,x)$ be the formal solution in $\mathcal{O}(D_{\rho_1})[[t]]$ of Eq. (6.17). Let $\arg(t) = \theta \in \mathbb{R}/2\pi\mathbb{Z}$ be a direction issuing from 0.*

Let us denote by

- *$p^\star$ the maximum of all the x-derivative orders $q \in \bigcup_{i\in\mathcal{K}} Q_i$;*
- *$i^\star$ the maximum of all the t-derivative orders $i \in \mathcal{K}$ such that $p^\star \in Q_i$;*
- *p_i the maximum of all the x-derivative orders $q \in Q_i\backslash\{p^\star\}$ if any exists (that is $Q_i \neq \{p^\star\}$).*

and let us assume the following conditions:

- *$P_{i^\star,p^\star}(0,0,\varphi_0) \neq 0$ (hence, $v_{i^\star,p^\star} = 0$);*
- *$v_{i,p_i} = 0$ for all $i > i^\star$, if some exist;*
- *the unique positive slope k of $\mathcal{N}_t(\mathcal{L}_{\kappa,v})$ is defined by the segment with end points $(\kappa,-\kappa)$ and $(p^\star+i^\star,-i^\star)$; hence $k = (\kappa - i^\star)/(p^\star - \kappa + i^\star)$.*

Then,

1. *$\widetilde{u}(t,x)$ is k-summable in the direction θ if and only if the inhomogeneity $\widetilde{f}(t,x)$ and the formal series $\widetilde{u}_{*,n}(t) \in \mathbb{C}[[t]]$ for $n = 0,\dots,p^\star - 1$ are k-summable in the direction θ.*
2. *Moreover, the k-sum $u(t,x)$, if any exists, satisfies Eq. (6.17) in which $\widetilde{f}(t,x)$ is replaced by its k-sum $f(t,x)$ in the direction θ.*

Application 6.13 (gBKdV Equation) Let us consider the gBKdV equation

$$\begin{cases} \partial_t u - P_{q_1}(t,x,u)\partial_x^{q_1} u - P_{q_2}(t,x,u)\partial_x^{q_2} u = \widetilde{f}(t,x) \\ u(0,x) = \varphi(x) \end{cases}, (t,x) \in \mathbb{C}^2 \qquad (6.21)$$

with $q_1 > \max(1,q_2)$. We have $\kappa = 1$, $\mathcal{K} = \{0\}$ and $Q_0 = \{q_1,q_2\}$; hence, $(i^\star,p^\star) = (0,q_1)$ and $P_{i^\star,p^\star}(t,x,X) = P_{q_1}(t,x,X)$. Assuming $P_{q_1}(0,0,\varphi) \neq 0$, the linear operator $\mathcal{L}_{\kappa,v}$ associated with Eq. (6.21) is defined by $\mathcal{L}_{\kappa,v} = \partial_t - \partial_x^{q_1} - t^{v_2}\partial_x^{q_2}$ with v_2 the smallest valuation at $t = 0$ of the coefficients of P_{q_2}, and its Newton polygon $\mathcal{N}_t(\mathcal{L}_{\kappa,v})$ admits the slope $k = 1/(q_1 - 1)$ as unique positive slope (see Fig. 6.2). We derive then from Theorem 6.12 that the formal solution $\widetilde{u}(t,x)$ of Eq. (6.21) is k-summable in the direction θ if and only if the inhomogeneity $\widetilde{f}(t,x)$ and the q_1 formal series $\widetilde{u}_{*,n}(t)$ with $n = 0,\dots,q_1 - 1$ are k-summable

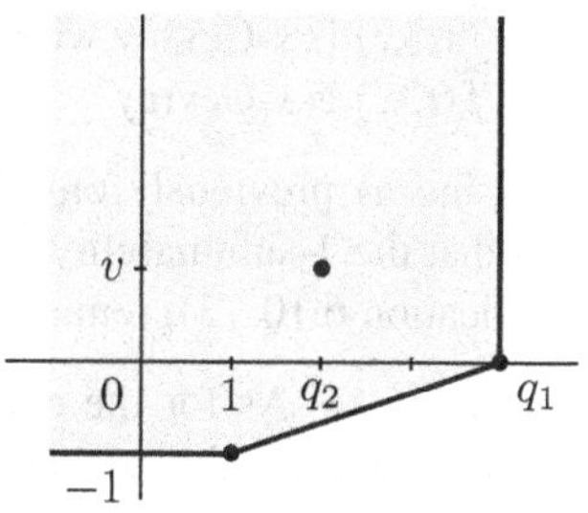

Fig. 6.2 The Newton polygon $\mathcal{N}_t(\mathcal{L}_{\kappa,v})$ of the linear operator $\mathcal{L}_{\kappa,v}$ associated with Eq. (6.21)

in the direction θ. In particular, this result provides a characterization of the $1/2$-summability of the formal solution of the classical Korteweg-de Vries equation

$$\partial_t u + \partial_x^3 u - 6u\partial_x u = 0,$$

and a characterization of the 1-summability of the formal solution of the classical Burgers equation

$$\partial_t u - \partial_x^2 u - 2u\partial_x u = 0.$$

Remark 6.14 When $E_{i^\star,p^\star} = \{0\}$, the polynomial $P_{i^\star,p^\star}(t,x,X)$ is reduced to its analytic coefficient $a_{i^\star,p^\star,0}(t,x)$ and the condition $P_{i^\star,p^\star}(0,0,\varphi_0) \neq 0$ can then be replaced by the condition $a_{i^\star,p^\star,0}(0,0) \neq 0$. In particular, the same applies when Eq. (6.17) is an inhomogeneous semilinear partial differential equation of the form

$$\begin{cases} \partial_t^\kappa u - \displaystyle\sum_{i\in\mathcal{K}}\sum_{q\in Q_i} a_{i,q}(t,x)\partial_t^i\partial_x^q u - u^2 P(t,x,u) = \widetilde{f}(t,x) \\ \partial_t^j u(t,x)_{|t=0} = \varphi_j(x),\ j=0,\dots,\kappa-1 \end{cases},\quad (t,x)\in\mathbb{C}^2. \tag{6.22}$$

Note that when Eq. (6.22) is reduced to Eq. (6.18) (that is when $\mathcal{K}=\{0\}$ and $Q_0 = \{p\}$), we find here one of the conditions which allows to apply Proposition 6.9.

Remark 6.15 The approach developed in Sect. 6.2.1 in the case of the linear heat equation and used in [108, 109] to prove the two results stated in Proposition 6.9 and in Theorem 6.12 can also be applied to other types of partial differential equations. For instance, we saw in Remark 3.28 that the formal power series of the inhomogeneous generalized Boussinesq equation

$$\begin{cases} \partial_t^2 u - a(t,x)\partial_x^4 u - P(t,x,u)\partial_x^2 u - Q(t,x,u)(\partial_x u)^2 = \widetilde{f}(t,x) \\ \partial_t^j u(t,x)_{|t=0} = \varphi_j(x),\ \ j=0,1 \end{cases} \tag{6.23}$$

with $a(0,0)\neq 0$ and with two polynomials $P(t,x,X), Q(t,x,X) \in \mathcal{O}(D_{\rho_0}\times D_{\rho_1})[X]$, possibly zero, admits the same Gevrey regularity that the formal power series solution of the linear Euler-Lagrange equation:

- if $\widetilde{f}(t,x)$ is s-Gevrey with $s\leqslant 1$, then $\widetilde{u}(t,x)$ is generically 1-Gevrey;
- if $\widetilde{f}(t,x)$ is s-Gevrey with $s>1$, then $\widetilde{u}(t,x)$ becomes s-Gevrey.

Focusing as previously on the critical value $s=1$, we can again prove by this way that the 1-summability result established for the Euler-Lagrange equation (see Application 6.10, (2)) remains valid for Eq. (6.23) [110]

Remark 6.16 As for the example of the heat equation (see Remark 6.6), all the results described above show the full value of the summability theory applied to the formal solutions of the partial differential equations of the form (6.17) (with, of course, the additional assumptions given in Proposition 6.9 and Theorem 6.12) or (6.23): when the formal solution $\widetilde{u}(t,x)$ of such a partial differential equation is k-summable in a certain direction θ, then its k-sum $u(t,x)$ defines an analytic solution of this equation on a convenient domain of the form $\Sigma_{\theta,>\pi s} \times D_r$. Furthermore, this solution being s-Gevrey asymptotic to $\widetilde{u}$ at 0 on $\Sigma_{\theta,>\pi s}$, we can deduce some useful information about its behaviour in the neighborhood of the origin $t = 0$.

Chapter 7
Second Characterization of the k-Summability: The Borel-Laplace Method

The method of summation of a formal power series $\widetilde{f}(t) \in \mathbb{C}[[t]]$ by means of Borel and Laplace operators was first introduced by E. Borel in the "simplest case" of level 1 [14, 15], then generalized to any level $k > 0$ by E. Leroy [61], F. Nevanlinna [95] and J.-P. Ramis [99]. We refer for instance to [65, Section 5.3] for more details.

Still considering t as the variable and $x = (x_1, \ldots, x_n)$ as a parameter, this method of summation can be easily extended to the formal power series $\widetilde{u}(t, x) \in \mathcal{O}(D_{\rho_1,\ldots,\rho_n})[[t]]$. The aim of this chapter is to develop this point of view and to present some applications to partial differential equations (Sect. 7.5). We first define the (formal) k-Borel and (formal) k-Laplace operators starting with the case $k = 1$ and we give some important properties (Sects. 7.1 and 7.2). Then, we prove a variant of the classical Nevanlinna's Theorem [65, Theorem 5.3.9, p. 156] and we show how this one provides a characterization of the k-summabiliy in the sense of Definition 5.1 (Sects. 7.3 and 7.4).

7.1 Classical Borel and Laplace Operators

Definition 7.1 (Functional Borel and Laplace Transforms in a Given Direction)

1. *The Borel transform* $\mathcal{B}_{1;\theta}^{[t]}(u)$ *of a function* $u(t, x)$ *with respect to* t *in the direction* θ is defined, when the integral exists, by

$$\mathcal{B}_{1;\theta}^{[t]}(u)(\tau, x) = \frac{1}{2i\pi} \int_{\gamma_{1;\theta}} tu(t, x)e^{\tau/t} d\left(\frac{1}{t}\right)$$

with $\gamma_{1;\theta}$ a Hankel-type contour in the direction θ at the origin (see Fig. 7.1).

P. Remy, *Asymptotic Expansions and Summability*, Lecture Notes in Mathematics 2351, https://doi.org/10.1007/978-3-031-59094-8_7

Fig. 7.1 The contour $\gamma_{1;\theta}$

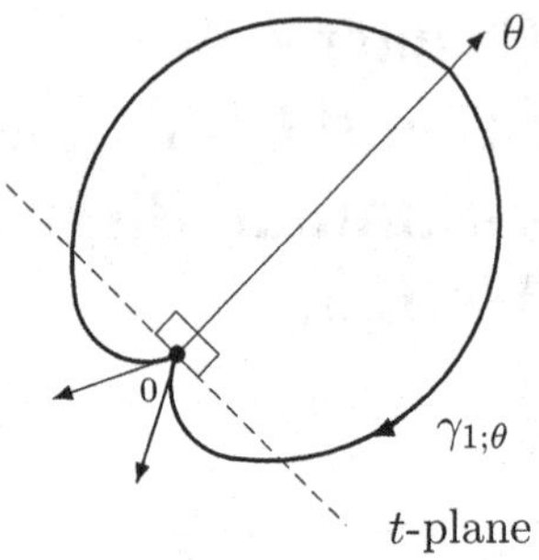

2. *The Laplace transform* $\mathcal{L}^{[\tau]}_{1;\theta}(v)$ *of a function* $v(\tau, x)$ *with respect to* τ *in the direction* θ is defined, when the integral exists, by

$$\mathcal{L}^{[\tau]}_{1;\theta}(v)(t, x) = \frac{1}{t}\int_0^{\infty e^{i\theta}} v(\tau, x)e^{-\tau/t}d\tau.$$

The Hankel-type contour $\gamma_{1;\theta}$ is oriented negatively. Let us observe that we need a contour that ends at 0 since the function is studied near the origin; if we worked at infinity we would take a usual Hankel-type contour at infinity, image of $\gamma_{1;\theta}$ by the change of variable $t = 1/z$. Observe that the tangents at 0 belong to the half-plane directed by $\theta + \pi$ so that $e^{\tau/t}$ is flat on $\gamma_{1;\theta}$ at 0 when $\arg(t)$ is close to θ. Observe also that, thanks to the Cauchy Formula, one can always assume if necessary that $\gamma_{1;\theta}$ is a the boundary, oriented negatively, of a sector bisected by θ, with finite radius and with opening larger than π.

The following proposition is straightforward from the Hankel's Formula for the inverse of the Gamma function.

Proposition 7.2 *Let* $\lambda \in \mathbb{C}$ *such that* $\mathrm{Re}(\lambda) > -1$. *Then,*

$$\mathcal{B}^{[t]}_{1;\theta}(t^\lambda)(\tau) = \frac{\tau^\lambda}{\Gamma(1+\lambda)} \quad \textit{and} \quad \mathcal{L}^{[\tau]}_{1;\theta}(\tau^\lambda)(t) = \Gamma(1+\lambda)t^\lambda$$

for any direction θ.

One can now define the formal Borel (resp. Laplace) transform of a formal power series $\widetilde{u}(t, x) \in \mathcal{O}(D_{\rho_1,\ldots,\rho_n})[[t]]$ (resp. $\widetilde{v}(\tau, x) \in \mathcal{O}(D_{\rho_1,\ldots,\rho_n})[[\tau]]$) as follows.

Definition 7.3 (Formal Borel and Laplace Transforms)

1. *The formal Borel transform* $\widetilde{\mathcal{B}}^{[t]}_1(\widetilde{u})$ *of a formal power series*

$$\widetilde{u}(t, x) = \sum_{j\geqslant 0} u_j(x)t^j \in \mathcal{O}(D_{\rho_1,\ldots,\rho_n})[[t]]$$

with respect to t is the formal power series

$$\widetilde{\mathcal{B}}_1^{[t]}(\widetilde{u})(\tau, x) = \sum_{j\geqslant 0} u_j(x)\frac{\tau^j}{\Gamma(1+j)} = \sum_{j\geqslant 0} \frac{u_j(x)}{j!}\tau^j \in \mathcal{O}(D_{\rho_1,\dots,\rho_n})[[\tau]].$$

2. *The formal Laplace transform* $\widetilde{\mathcal{L}}_1^{[\tau]}(\widetilde{v})$ *of a formal power series*

$$\widetilde{v}(\tau, x) = \sum_{j\geqslant 0} v_j(x)\tau^j \in \mathcal{O}(D_{\rho_1,\dots,\rho_n})[[\tau]]$$

with respect to τ is the formal power series

$$\widetilde{\mathcal{L}}_1^{[\tau]}(\widetilde{v})(t, x) = \sum_{j\geqslant 0} v_j(x)\Gamma(1+j)t^j = \sum_{j\geqslant 0} j!v_j(x)t^j \in \mathcal{O}(D_{\rho_1,\dots,\rho_n})[[t]].$$

Notice that Borel and Laplace operators in a given direction θ on one hand, and formal Borel and formal Laplace operators on the other hand are inverse from each other.

Proposition 7.4 *The following formulæ are valid both for formal power series* $u(t, x)$ *and* $v(t, x)$ ($v(\tau, x)$ *and* $\omega(\tau, x)$ *are then their formal Borel transform) or for functions* ($v(\tau, x)$ *and* $\omega(\tau, x)$ *are then their functional Borel transform in the direction* θ) *when the transformations exist.*

$$\begin{array}{rcl}
 & \widetilde{\mathcal{B}}_1^{[t]}, \mathcal{B}_{1;\theta}^{[t]} & \\
\dfrac{1}{t}u(t,x) & \rightleftarrows & \partial_\tau v(\tau,x) \quad (\textit{assume } u(0,x)\equiv 0) \\
t\partial_t u(t,x) & \rightleftarrows & \tau\partial_\tau v(\tau,x) \\
\partial_t u(t,x) & \rightleftarrows & \partial_\tau\tau\partial_\tau v(\tau,x) = \tau\partial_\tau^2 v(\tau,x) + \partial_\tau v(\tau,x) \\
\partial_t^{-1}u(t,x) & \rightleftarrows & \partial_\tau^{-1}\dfrac{1}{\tau}\partial_\tau^{-1}v(\tau,x) \\
\lambda u(t,x) + \mu v(t,x) & \rightleftarrows & \lambda v(\tau,x) + \mu\omega(\tau,x) \quad (\lambda,\ \mu \textit{ independent of } t) \\
u(t,x)v(t,x) & \rightleftarrows & v * \omega(\tau,x) = \dfrac{d}{d\tau}\displaystyle\int_0^\tau v(\tau-\eta,x)\omega(\eta,x)d\eta \\
 & \widetilde{\mathcal{L}}_1^{[\tau]}, \mathcal{L}_{1;\theta}^{[\tau]} &
\end{array}$$

In the last formula, when $v(\tau, x)$ *and* $\omega(\tau, x)$ *are formal power series in* τ, *the integral is taken term by term.*

Proof These formulæ are easily verified by simple calculations. This is left to the reader. ■

In the following two propositions, we give two important properties on the formal Borel transforms of the 1-Gevrey formal series and of the analytic functions at the origin of $\mathbb{C}^{n+1}$.

Proposition 7.5 *A formal power series $\widetilde{u}(t,x) \in \mathcal{O}(D_{\rho_1,\dots,\rho_n})[[t]]$ is 1-Gevrey if and only if its formal Borel transform $\widehat{u} = \widehat{\mathcal{B}}_1^{[t]}(\widetilde{u})$ with respect to t is convergent at the origin of $\mathbb{C}^{n+1}$.*

Proof Let $\widetilde{u}(t,x) = \sum_{j\geqslant 0} u_j(x)t^j \in \mathcal{O}(D_{\rho_1,\dots,\rho_n})[[t]]$ and

$$\widehat{u}(\tau,x) = \sum_{j\geqslant 0} \frac{u_j(x)}{j!}\tau^j \in \mathcal{O}(D_{\rho_1,\dots,\rho_n})[[\tau]]$$

its formal Borel transform with respect to t.

◁ *The* only if *part.* Let us assume that $\widetilde{u}(t,x)$ is 1-Gevrey. According to Definition 3.1, there exist a radius $0 < r \leqslant \min(\rho_1,\dots,\rho_n)$ and two positive constants $C, K > 0$ such that the inequalities

$$\left|u_j(x)\right| \leqslant CK^j\Gamma(1+j) = CK^j j!$$

hold for all $x \in D_{r,\dots,r}$ and all $j \geqslant 0$. Consequently,

$$\left|\frac{u_j(x)}{j!}\tau^j\right| \leqslant C(K\,|\tau|)^j$$

for all $(\tau,x) \in D_{1/K} \times D_{r,\dots,r}$ and all $j \geqslant 0$, which proves that $\widehat{u}(\tau,x)$ is analytic on $D_{1/K} \times D_{r,\dots,r}$.

◁ *The* if *part.* Let us now assume that $\widehat{u}(\tau,x)$ is convergent at the origin of $\mathbb{C}^{n+1}$. Then, $\widehat{u}(\tau,x)$ is 0-Gevrey and, according to Definition 3.1, there exists a radius $0 < r \leqslant \min(\rho_1,\dots,\rho_n)$ and two positive constants $C, K > 0$ such that the inequalities

$$\left|\frac{u_j(x)}{j!}\right| \leqslant CK^j$$

hold for all $x \in D_{r,\dots,r}$ and all $j \geqslant 0$. Consequently,

$$\left|u_j(x)\right| \leqslant CK^j j!$$

for all $x \in D_{r,\dots,r}$ and all $j \geqslant 0$, which proves that $\widetilde{u}(t,x)$ is 1-Gevrey. ■

Proposition 7.6 *Let $a(t,x)$ be an analytic function on a polydisc $D_{\rho_0,\rho_1,\ldots,\rho_n}$. Then, its formal Borel transform $\widehat{a} = \widetilde{\mathcal{B}}_1^{[t]}(a)$ with respect to t defines an analytic function on $\mathbb{C}\times D_{r,\ldots,r}$ for some $0 < r \leqslant \min(\rho_1,\ldots,\rho_n)$ with global exponential growth of order at most 1 at infinity with respect to τ, that is, there exist two positive constants $A, C > 0$ such that the following estimate holds for all $(\tau,x)\in\mathbb{C}\times D_{r,\ldots,r}$:*

$$|\widehat{a}(\tau,x)| \leqslant C\exp\left(A\,|\tau|\right).$$

Proof Let $a(t,x) = \sum_{j\geqslant 0} a_j(x)t^j \in \mathcal{O}(D_{\rho_0,\rho_1,\ldots,\rho_n})$ and

$$\widehat{a}(\tau,x) = \sum_{j\geqslant 0}\frac{a_j(x)}{j!}\tau^j \in \mathcal{O}(D_{\rho_1,\ldots,\rho_n})[[\tau]]$$

its formal Borel transform with respect to t. By assumption, $a(t,x)$ is 0-Gevrey. Then, according to Definition 3.1, there exist a radius $0 < r \leqslant \min(\rho_1,\ldots,\rho_n)$ and two positive constants $C, A > 0$ such that the inequalities

$$\left|\frac{a_j(x)}{j!}\tau^j\right| \leqslant C\frac{(A\,|\tau|)^j}{j!}$$

hold for all $(\tau,x)\in\mathbb{C}\times D_{r,\ldots,r}$ and all $j\geqslant 0$. Thus, the series $\widehat{a}(\tau,x)$ is normally convergent on $D_R\times D_{r,\ldots,r}$ for any $R > 0$ and, consequently, defines an analytic function on $\mathbb{C}\times D_{r,\ldots,r}$. Moreover, it satisfies the inequality

$$|\widehat{a}(\tau,x)| \leqslant \sum_{j\geqslant 0}\left|\frac{a_j(x)}{j!}\tau^j\right| \leqslant C\exp\left(A\,|\tau|\right)$$

for all $(\tau,x)\in\mathbb{C}\times D_{r,\ldots,r}$, which proves the exponential growth at infinity with respect to τ and ends the proof. ■

For any directions $\alpha < \beta$, let us denote by $\Sigma_{\alpha,\beta}(\infty)$ the infinite open sector

$$\Sigma_{\alpha,\beta}(\infty) = \{t;\, \alpha < \arg(t) < \beta \text{ and } |t| > 0\}.$$

The following proposition provides us an important property of the Laplace transform $\mathcal{L}_{1;\theta}^{[\tau]}$ which will be very useful in the sequel.

Proposition 7.7 *Let $\widetilde{u}(t,x) \in \mathcal{O}(D_{\rho_1,\ldots,\rho_n})[[t]]$ and $\widehat{u}(\tau,x)$ its formal Borel transform with respect to t.*

Assume that $\widehat{u}(\tau,x)$ is convergent at the origin of $\mathbb{C}^{n+1}$ and that its sum $v(\tau,x)$ can be analytically continued to a domain $\Sigma_{\theta_1,\theta_2}(\infty)\times D_{r,\ldots,r}$ for some directions $\theta_1 < \theta_2$ and some radius $0 < r \leqslant \min(\rho_1,\ldots,\rho_n)$, with a global exponential growth of order at most 1 at infinity with respect to τ, that is, there

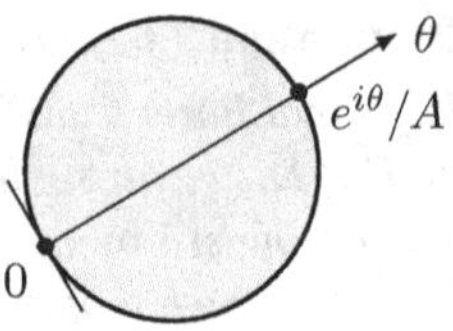

Fig. 7.2 The domain $\mathcal{P}_\theta(A)$

exist two positive constants $A, C > 0$ such that the following estimate holds for all
$(\tau, x) \in \Sigma_{\theta_1, \theta_2}(\infty) \times D_{r,\dots,r}$:

$$|\upsilon(\tau, x)| \leqslant C \exp(A |\tau|).$$

For any direction $\theta \in]\theta_1, \theta_2[$, we denote by

- $\mathcal{P}_\theta(A)$ *the domain* $\mathcal{P}_\theta(A) = \{t; \text{Re}(e^{i\theta}/t) > A\}$;
- u_θ *the Laplace transform of υ with respect to τ in the direction θ:*

$$u_\theta(t, x) = \mathcal{L}_{1;\theta}^{[\tau]}(\upsilon)(t, x) = \frac{1}{t} \int_0^{\infty e^{i\theta}} \upsilon(\tau, x) e^{-\tau/t} d\tau.$$

Then, the functions $u_\theta(t, x)$ glue together into a function $u(t, x)$ defined and analytic on the domain $\bigcup_{\theta \in]\theta_1, \theta_2[} \mathcal{P}_\theta(A) \times D_{r,\dots,r}$.

Proof Let us first observe that $\mathcal{P}_\theta(A)$ is the open disc with center $e^{i\theta}/(2A)$ and radius $1/(2A)$ (see Fig. 7.2). In particular, $\mathcal{P}_\theta(A) \cap \mathcal{P}_{\theta'}(A) \neq \emptyset$ as soon as $|\theta - \theta'| \neq \pi \mod 2\pi$.

According to the assumption on $\upsilon(\tau, x)$, the function $u_\theta(t, x)$ is well-defined and analytic on $\mathcal{P}_\theta(A) \times D_{r,\dots,r}$ for any $\theta \in]\theta_1, \theta_2[$. Indeed, the integrand is analytic and, for any $B > A$, we have

$$\left| \upsilon(\tau, x) e^{-\tau/t} \right| \leqslant C \exp(-(B - A) |\tau|)$$

for all $(\tau, t, x) \in [0, \infty e^{i\theta}[\times \mathcal{P}_\theta(B) \times D_{r,\dots,r}$. Then, Lebesgue's Dominated Convergence Theorem applies and implies the analyticity of $u_\theta(t, x)$ on $\mathcal{P}_\theta(B) \times D_{r,\dots,r}$; hence, on $\mathcal{P}_\theta(A) \times D_{r,\dots,r}$ by making B tend to A.

Let us now prove that all the functions u_θ glue together into an analytic function on $\bigcup_{\theta \in]\theta_1, \theta_2[} \mathcal{P}_\theta(A) \times D_{r,\dots,r}$.

To this end, let us consider two directions $\theta' < \theta''$ in $]\theta_1, \theta_2[$ such that, say, $\theta'' - \theta' < \pi/4$ and let us apply the Cauchy Theorem to $\upsilon(\tau, x) e^{-\tau/t}$ along the boundary γ_R, oriented positively, of a sector $\Sigma_{\theta', \theta''}(R)$ of radius $R > 0$ limited by the lines θ' and θ'' like shown on Fig. 7.3.

Denoting by C_R the arc of circle from $Re^{i\theta'}$ to $Re^{i\theta''}$, we get

$$\int_0^{Re^{i\theta'}} + \int_{C_R} - \int_0^{Re^{i\theta''}} \upsilon(\tau, x) e^{-\tau/t} d\tau = 0;$$

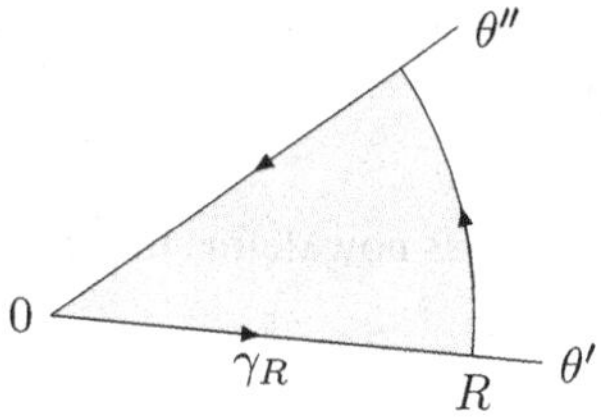

Fig. 7.3 The boundary γ_R of $\Sigma_{\theta',\theta''}(R)$

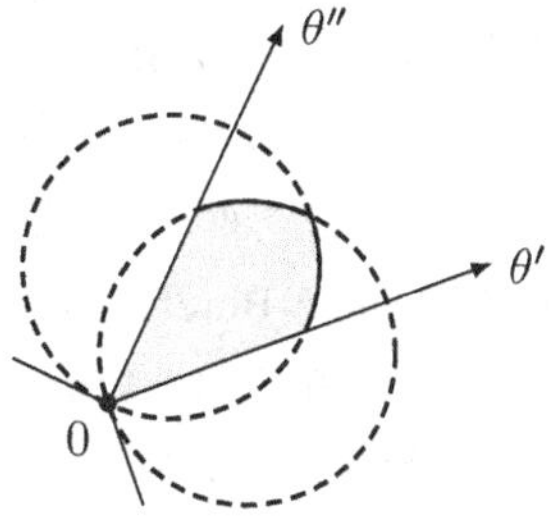

Fig. 7.4 The domain $\mathcal{P}'_{\theta';\theta''}(A)$

hence

$$u_{\theta''}(t,x) - u_{\theta'}(t,x) = \frac{1}{t} \lim_{R\to+\infty} \int_{C_R} \upsilon(\tau,x)e^{-\tau/t}d\tau$$

for any $(t,x) \in (\mathcal{P}_{\theta'}(A) \cap \mathcal{P}_{\theta''}(A)) \times D_{r,\ldots,r}$.

We shall prove that this limit is 0 for any $(t,x) \in \mathcal{P}_{\theta';\theta''}(A) \times D_{r,\ldots,r}$ with $\mathcal{P}_{\theta';\theta''}(A)$ a convenient open subset of $\mathcal{P}_{\theta'}(A) \cap \mathcal{P}_{\theta''}(A)$. Denoting $t = |t|\, e^{i\omega}$, we can write

$$\begin{aligned}\left|\int_{C_R} \upsilon(\tau,x)e^{-\tau/t}d\tau\right| &\leqslant CR\int_{\theta'}^{\theta''} e^{AR}e^{-R\mathrm{Re}(e^{i\theta}/t)}d\theta\\ &= CR\int_{\theta'}^{\theta''} e^{-(\cos(\theta-\omega)/|t|-A)R}d\theta.\end{aligned}$$

Now, let us consider the open domain $\mathcal{P}'_{\theta';\theta''}(A)$ of all the $t \in \mathcal{P}_{\theta'}(A)\cap\mathcal{P}_{\theta''}(A)$ such that $\theta' < \omega < \theta''$ (see Fig. 7.4). Then, from the inequality $|\theta-\omega| < \pi/4$ for all $\theta \in]\theta',\theta''[$, we deduce the estimate

$$\left|\int_{C_R} \upsilon(\tau,x)e^{-\tau/t}d\tau\right| \leqslant CR(\theta''-\theta')\exp\left(-\left(\frac{1}{|t|\sqrt{2}}-A\right)R\right)$$

and the integral tends to 0 as R tends to infinity as soon as $|t| < 1/(A\sqrt{2})$.

Thereby, it results that the functions $u_{\theta'}(t,x)$ and $u_{\theta''}(t,x)$ coincide on $\mathcal{P}_{\theta';\theta''}(A) \times D_{r,\ldots,r}$ with $\mathcal{P}_{\theta';\theta''}(A) = \mathcal{P}'_{\theta';\theta''}(A) \cap \{t;\, |t| < 1/(A\sqrt{2})\}$ and, consequently, are analytic continuations of each other.

The arbitrary choice of the two directions θ' and θ'' in $]\theta_1, \theta_2[$ completes the proof.

■

Let us now define the (formal) k-Borel and (formal) k-Laplace operators for any $k > 0$.

7.2 k-Borel and k-Laplace Operators

Definitions 7.1 and 7.3 can be extended to any $k > 0$ as follows. Let us denote by

- $\mathcal{B}_{k;\theta}^{[t]}$ the k-Borel operator with respect to t in the direction θ;
- $\widetilde{\mathcal{B}}_k^{[t]}$ the formal k-Borel operator with respect to t;
- $\mathcal{L}_{k;\theta}^{[\tau]}$ the k-Laplace operator with respect to τ in the direction θ;
- $\widetilde{\mathcal{L}}_k^{[\tau]}$ the formal k-Laplace operator with respect to τ.

The classical (formal) Borel and Laplace operators introduced previously correspond to the case $k = 1$ and allow to define all the other operators by means of ramifications according to the two following diagrams. Let the ramification ϱ_k be defined by $\varrho_k(t) = z^{1/k}$ and its inverse $\varrho_{1/k}$ by $\varrho_{1/k}(\zeta) = \tau^k$.

The k-Borel operator $\mathcal{B}_{k;\theta}^{[t]}$ with respect to t in the direction θ is defined by the following commutative diagram:

$$\begin{array}{ccc} u(t,x) & \xrightarrow{\ \mathcal{B}_{k;\theta}^{[t]}\ } & \mathcal{B}_{k;\theta}^{[t]}(u)(\tau,x) = v(\tau^k,x) \\ \downarrow \varrho_k & & \uparrow \varrho_{1/k} \\ u(z^{1/k},x) & \xrightarrow{\ \mathcal{B}_{1;k\theta}^{[z]}\ } & v(\zeta,x) \end{array}$$

where

$$v(\zeta, x) = \frac{1}{2i\pi} \int_{\gamma_{1;k\theta}} z u(z^{1/k}, x) e^{\zeta/z} d\left(\frac{1}{z}\right) = \frac{1}{2i\pi} \int_{\gamma_{k;\theta}} t^k u(t, x) e^{\zeta/t^k} d\left(\frac{1}{t^k}\right),$$

the path $\gamma_{k;\theta}$ being deduced from the Hankel-type contour $\gamma_{1;k\theta}$ (see Fig. 7.1) by the ramification $z = t^k$. Again, thanks to the Cauchy Formula, one can always assume if necessary that $\gamma_{k;\theta}$ is a the boundary, oriented negatively, of a sector bisected by θ, with finite radius and with opening larger than π/k.

The k-Laplace operator $\mathcal{L}_{k;\theta}^{[\tau]}$ with respect to τ in the direction θ is defined by the following commutative diagram:

$$\begin{array}{ccc}
v(\tau, x) & \xrightarrow{\mathcal{L}_{k;\theta}^{[\tau]}} & \mathcal{L}_{k;\theta}^{[\tau]}(v)(t,x) = u(t^k, x) \\
\Big\downarrow \varrho_k & & \Big\uparrow \varrho_{1/k} \\
v(\zeta^{1/k}, x) & \xrightarrow{\mathcal{L}_{1;k\theta}^{[\zeta]}} & u(z, x)
\end{array}$$

where

$$u(z,x) = \frac{1}{z}\int_0^{\infty e^{ik\theta}} v(\zeta^{1/k}, x) e^{-\zeta/z} d\zeta = \frac{1}{z}\int_{\tau=0}^{\infty e^{i\theta}} v(\tau, x) e^{-\tau^k/z} d(\tau^k).$$

The formal k-Borel operator $\widetilde{\mathcal{B}}_k^{[t]}$ (resp. formal k-Laplace operator $\widetilde{\mathcal{L}}_k^{[\tau]}$) of a formal power series in t (resp. in τ) is defined by applying $\mathcal{B}_{k;\theta}^{[t]}$ (resp. $\mathcal{L}_{k;\theta}^{[\tau]}$) to each monomial t^j (resp. τ^j) in any direction θ.

We can then state the following definitions generalizing for any $k > 0$ the classical definitions of Borel and Laplace transforms stated in the previous section with $k = 1$.

Definition 7.8 (Functional k-Borel and k-Laplace Transforms in a Given Direction)

1. *The k-Borel transform $\mathcal{B}_{k;\theta}^{[t]}(u)$ of a function $u(t,x)$ with respect to t in the direction θ* is defined, when the integral exists, by

$$\mathcal{B}_{k;\theta}^{[t]}(u)(\tau, x) = \frac{1}{2i\pi}\int_{\gamma_{k;\theta}} t^k u(t,x) e^{(\tau/t)^k} d\left(\frac{1}{t^k}\right)$$

   with $\gamma_{k;\theta}$ a Hankel-type contour in the direction θ at the origin deduced from the Hankel-type contour $\gamma_{1;k\theta}$ (see Fig. 7.1) by the ramification $z = t^k$.
2. *The k-Laplace transform $\mathcal{L}_{k;\theta}^{[\tau]}(v)$ of a function $v(\tau, x)$ with respect to τ in the direction θ* is defined, when the integral exists, by

$$\mathcal{L}_{k;\theta}^{[\tau]}(v)(t,x) = \frac{1}{t^k}\int_{\tau=0}^{\infty e^{i\theta}} v(\tau, x) e^{-(\tau/t)^k} d(\tau^k).$$

Using the formulæ $\mathcal{B}_{k;\theta}^{[t]} = \varrho_{1/k}\mathcal{B}_{1;k\theta}^{[z]}\varrho_k$ and $\mathcal{L}_{k;\theta}^{[\tau]} = \varrho_{1/k}\mathcal{L}_{1;k\theta}^{[\zeta]}\varrho_k$, we easily derive from Proposition 7.2 the following.

Proposition 7.9 *Let* $\lambda \in \mathbb{C}$ *such that* $\mathrm{Re}(\lambda) > -k$. *Let* $s = 1/k$. *Then,*

$$\mathcal{B}_{k;\theta}^{[t]}(t^\lambda)(\tau) = \frac{\tau^\lambda}{\Gamma(1+s\lambda)} \quad \text{and} \quad \mathcal{L}_{k;\theta}^{[\tau]}(\tau^\lambda)(t) = \Gamma(1+s\lambda)t^\lambda$$

for any direction θ.

The formal k-Borel (resp. k-Laplace) transform of a formal power series $\widetilde{u}(t,x) \in \mathcal{O}(D_{\rho_1,\ldots,\rho_n})[[t]]$ (resp. $\widetilde{v}(\tau,x) \in \mathcal{O}(D_{\rho_1,\ldots,\rho_n})[[\tau]]$) is then defined as follows.

Definition 7.10 (Formal k-Borel and k-Laplace Transforms) Let $s = 1/k$.

1. *The formal k-Borel transform* $\widetilde{\mathcal{B}}_k^{[t]}(\widetilde{u})$ *of a formal power series*

$$\widetilde{u}(t,x) = \sum_{j\geqslant 0} u_j(x)t^j \in \mathcal{O}(D_{\rho_1,\ldots,\rho_n})[[t]]$$

with respect to t is the formal power series

$$\widetilde{\mathcal{B}}_k^{[t]}(\widetilde{u})(\tau,x) = \sum_{j\geqslant 0} \frac{u_j(x)}{\Gamma(1+sj)}\tau^j \in \mathcal{O}(D_{\rho_1,\ldots,\rho_n})[[\tau]].$$

2. *The formal k-Laplace transform* $\widetilde{\mathcal{L}}_k^{[\tau]}(\widetilde{v})$ *of a formal power series*

$$\widetilde{v}(\tau,x) = \sum_{j\geqslant 0} v_j(x)\tau^j \in \mathcal{O}(D_{\rho_1,\ldots,\rho_n})[[\tau]]$$

with respect to τ is the formal power series

$$\widetilde{\mathcal{L}}_k^{[\tau]}(\widetilde{v})(t,x) = \sum_{j\geqslant 0} \Gamma(1+sj)v_j(x)t^j \in \mathcal{O}(D_{\rho_1,\ldots,\rho_n})[[t]].$$

Like in the case $k = 1$, the k-Borel and k-Laplace operators in a given direction θ on one hand, and the formal k-Borel and formal k-Laplace operators on the other hand are inverse from each other.

Contrary to the case $k = 1$ (see Proposition 7.4), there are no simple formulæ to easily describe the action of the (formal) k-Borel and k-Laplace operators on the operators ∂_t and ∂_t^{-1} (such formulæ can however be written using the (formal) k-moment-Borel and k-moment-Laplace operators, see Proposition 10.12). On the other hand, concerning their action on the usual algebraic operations, we have the following result which is immediately deduced from Proposition 7.4 and from the relations $\mathcal{B}_{k;\theta}^{[t]} = \varrho_{1/k}\mathcal{B}_{1;k\theta}^{[z]}\varrho_k$ and $\mathcal{L}_{k;\theta}^{[\tau]} = \varrho_{1/k}\mathcal{L}_{1;k\theta}^{[\zeta]}\varrho_k$.

Proposition 7.11 *The following formulæ are valid both for formal power series $u(t,x)$ and $v(t,x)$ ($\upsilon(\tau,x)$ and $\omega(\tau,x)$ are then their formal k-Borel transform) or for functions ($\upsilon(\tau,x)$ and $\omega(\tau,x)$ are then their functional k-Borel transform in the direction θ) when the transformations exist.*

$$\lambda u(t,x)+\mu v(t,x) \quad \overset{\widetilde{\mathcal{B}}_k^{[t]},\,\mathcal{B}_{k;\theta}^{[t]}}{\rightleftarrows} \quad \lambda \upsilon(\tau,x)+\mu\omega(\tau,x) \quad (\lambda,\ \mu \text{ independent of } t)$$

$$u(t,x)v(t,x) \quad \underset{\widetilde{\mathcal{L}}_k^{[\tau]},\,\mathcal{L}_{k;\theta}^{[\tau]}}{\rightleftarrows} \quad \frac{1}{k\tau^{k-1}}\frac{d}{d\tau}\int_0^{\tau^k} \upsilon((\tau^k-\eta)^{1/k},x)\omega(\eta^{1/k},x)d\eta$$

In the last formula, when $\upsilon(\tau,x)$ and $\omega(\tau,x)$ are formal power series in τ, the integral is taken term by term.

The two propositions below extend the results of Propositions 7.5 and 7.6 by giving two important properties on the formal k-Borel transform of a s_1-Gevrey formal series with $s_1 \leqslant s\ (= 1/k)$.

Proposition 7.12 *Let $s = 1/k$. A formal power series $\widetilde{u}(t,x) \in \mathcal{O}(D_{\rho_1,\dots,\rho_n})[[t]]$ is s-Gevrey if and only if its formal k-Borel transform $\widehat{u}_k = \widetilde{\mathcal{B}}_k^{[t]}(\widetilde{u})$ with respect to t is convergent at the origin of $\mathbb{C}^{n+1}$.*

Proof The proof is similar to the one of Proposition 7.5.

Let $\widetilde{u}(t,x) = \sum_{j\geqslant 0} u_j(x)t^j \in \mathcal{O}(D_{\rho_1,\dots,\rho_n})[[t]]$ and

$$\widehat{u}_k(\tau,x) = \sum_{j\geqslant 0} \frac{u_j(x)}{\Gamma(1+sj)}\tau^j \in \mathcal{O}(D_{\rho_1,\dots,\rho_n})[[\tau]]$$

its formal k-Borel transform with respect to t.

◁ *The* only if *part.* Let us assume that $\widetilde{u}(t,x)$ is s-Gevrey. According to Definition 3.1, there exist a radius $0 < r \leqslant \min(\rho_1,\dots,\rho_n)$ and two positive constants $C, K > 0$ such that the inequalities

$$\left|u_j(x)\right| \leqslant CK^j\Gamma(1+sj)$$

hold for all $x \in D_{r,\dots,r}$ and all $j \geqslant 0$. Consequently,

$$\left|\frac{u_j(x)}{\Gamma(1+sj)}\tau^j\right| \leqslant C(K\,|\tau|)^j$$

for all $(\tau,x) \in D_{1/K} \times D_{r,\dots,r}$ and all $j \geqslant 0$, which proves that $\widehat{u}_k(\tau,x)$ is analytic on $D_{1/K} \times D_{r,\dots,r}$.

$\triangleleft$ *The* if *part.* Let us now assume that $\widehat{u}_k(\tau, x)$ is convergent at the origin of $\mathbb{C}^{n+1}$. Then, $\widehat{u}_k(\tau, x)$ is 0-Gevrey and, according to Definition 3.1, there exists a radius $0 < r \leqslant \min(\rho_1, \dots, \rho_n)$ and two positive constants $C, K > 0$ such that the inequalities

$$\left| \frac{u_j(x)}{\Gamma(1+sj)} \right| \leqslant CK^j$$

hold for all $x \in D_{r,\dots,r}$ and all $j \geqslant 0$. Consequently,

$$\left| u_j(x) \right| \leqslant CK^j \Gamma(1+sj)$$

for all $x \in D_{r,\dots,r}$ and all $j \geqslant 0$, which proves that $\widetilde{u}(t, x)$ is s-Gevrey. ∎

Proposition 7.13 *Let $s = 1/k$. Let $\widetilde{u}(t, x) \in \mathcal{O}(D_{\rho_1,\dots,\rho_n})[[t]]$ be a s_1-Gevrey formal series with $s_1 < s$. Then, its formal k-Borel transform $\widehat{u}_k = \widetilde{\mathcal{B}}_k^{[t]}(u)$ with respect to t defines an analytic function on $\mathbb{C} \times D_{r,\dots,r}$ for some $0 < r \leqslant \min(\rho_1, \dots, \rho_n)$ with global exponential growth of order at most $\kappa = (s - s_1)^{-1}$ at infinity with respect to τ, that is, there exist two positive constants $A, C > 0$ such that the following estimate holds for all $(\tau, x) \in \mathbb{C} \times D_{r,\dots,r}$:*

$$|\widehat{u}_k(\tau, x)| \leqslant C \exp\left(A |\tau|^{\kappa}\right).$$

In particular, when $s_1 = 0$, we can state the following important result.

Corollary 7.14 *Let $a(t, x)$ be an analytic function on a polydisc $D_{\rho_0,\rho_1,\dots,\rho_n}$. Then, its formal k-Borel transform $\widehat{a}_k = \widetilde{\mathcal{B}}_k^{[t]}(a)$ with respect to t defines an analytic function on $\mathbb{C} \times D_{r,\dots,r}$ for some $0 < r \leqslant \min(\rho_1, \dots, \rho_n)$ with global exponential growth of order at most k at infinity with respect to τ, that is, there exist two positive constants $A, C > 0$ such that the following estimate holds for all $(\tau, x) \in \mathbb{C} \times D_{r,\dots,r}$:*

$$|\widehat{a}_k(\tau, x)| \leqslant C \exp\left(A |\tau|^{k}\right).$$

Proof of Proposition 7.13 Let $\widetilde{u}(t, x) = \sum_{j \geqslant 0} u_j(x) t^j \in \mathcal{O}(D_{\rho_1,\dots,\rho_n})[[t]]$ a s_1-Gevrey formal series and

$$\widehat{u}_k(\tau, x) = \sum_{j \geqslant 0} \frac{u_j(x)}{\Gamma(1+sj)} \tau^j \in \mathcal{O}(D_{\rho_1,\dots,\rho_n})[[\tau]]$$

its formal k-Borel transform with respect to t. By assumption (see Definition 3.1), there exist a radius $0 < r \leqslant \min(\rho_1, \dots, \rho_n)$ and two positive constants $C', A' > 0$

such that the inequalities

$$\left|\frac{u_j(x)}{\Gamma(1+sj)}\tau^j\right| \leqslant C'\frac{\Gamma(1+s_1 j)\left(A'\,|\tau|\right)^j}{\Gamma(1+sj)}$$

hold for all $x \in D_{r,\dots,r}$ and all $j \geqslant 0$.

Let $\alpha_j = C'\dfrac{\Gamma(1+s_1 j)A'^j}{\Gamma(1+sj)}$. From the Stirling Formula, we first deduce the estimate

$$\alpha_j \underset{j\to+\infty}{\sim} C'\sqrt{\frac{s_1}{s}}\left(A'\frac{s_1^{s_1}}{s^s}e^{s-s_1}\right)^j\left(\frac{1}{j^{s-s_1}}\right)^j. \tag{7.1}$$

Then, since $s > s_1$, the sequence $(\alpha_j^{1/j})$ tends to 0 and, thereby, the Cauchy-Hadamard Theorem tells us that the series $\sum_{j\geqslant 0}\alpha_j T^j$ converges on all $\mathbb{C}$. Consequently, the formal series $\widehat{u}_k(\tau, x)$ is normally convergent on $D_R \times D_{r,\dots,r}$ for any $R > 0$; hence, defines an analytic function on $\mathbb{C} \times D_{r,\dots,r}$.

To prove now that $\widehat{u}_k(\tau, x)$ admits an exponential growth of order at most $\kappa = (s - s_1)^{-1}$ at infinity with respect to τ, it is sufficient to verify that the same exponential growth occurs for the series $\sum_{j\geqslant 0}\alpha_j T^j$, which stems from the relation

$$\kappa = \limsup_{j\to+\infty}\frac{j\ln j}{\ln(1/\alpha_j)} \tag{7.2}$$

(see for instance [62, Thm. 2 page 6]) and from the estimate

$$\frac{j\ln j}{\ln(1/\alpha_j)} \underset{j\to+\infty}{\sim} \left(s - s_1 - \frac{\ln\left(A'\dfrac{s_1^{s_1}}{s^s}e^{s-s_1}\right)}{\ln j} - \frac{\ln\left(C'\sqrt{\dfrac{s_1}{s}}\right)}{j\ln j}\right)^{-1}$$

we derive from (7.1). This ends the proof of Proposition 7.13 ∎

We end this section by the generalization of Proposition 7.7 to the k-Laplace transform $\mathcal{L}^{[\tau]}_{k;\theta}$ for any $k > 0$.

Proposition 7.15 *Let $\widetilde{u}(t, x) \in \mathcal{O}(D_{\rho_1,\dots,\rho_n})[[t]]$ and $\widehat{u}_k(\tau, x)$ its formal k-Borel transform with respect to t.*

Assume that $\widehat{u}_k(\tau, x)$ is convergent at the origin of $\mathbb{C}^{n+1}$ and that its sum $v_k(\tau, x)$ can be analytically continued to a domain $\Sigma_{\theta_1,\theta_2}(\infty) \times D_{r,\dots,r}$ for some directions $\theta_1 < \theta_2$ and some radius $0 < r \leqslant \min(\rho_1, \dots, \rho_n)$, with a global *exponential growth of order at most k at infinity with respect to τ, that is, there exist two positive constants $A, C > 0$ such that the following estimate holds for all $(\tau, x) \in \Sigma_{\theta_1,\theta_2}(\infty) \times D_{r,\dots,r}$:*

$$|v_k(\tau, x)| \leqslant C \exp\left(A\,|\tau|^k\right).$$

For any direction $\theta \in]\theta_1, \theta_2[$, we denote by

- *$\mathcal{P}_{k;\theta}(A)$ the domain $\mathcal{P}_{k;\theta}(A) = \{t;\ \mathrm{Re}(e^{i\theta}/t^k) > A\}$;*
- *$u_{k;\theta}$ the k-Laplace transform of v_k with respect to τ in the direction θ:*

$$u_{k;\theta}(t, x) = \mathcal{L}_{k;\theta}^{[\tau]}(v_k)(t, x) = \frac{1}{t^k}\int_{\tau=0}^{\infty e^{i\theta}} v_k(\tau, x)e^{-(\tau/t)^k}\, d(\tau^k).$$

Then, the functions $u_{k;\theta}(t, x)$ glue together into a function $u_k(t, x)$ defined and analytic on the domain $\bigcup_{\theta\in]\theta_1,\theta_2[} \mathcal{P}_{k;\theta}(A) \times D_{r,\dots,r}$.

Proof Setting $\zeta = \tau^k$ and $z = t^k$, the integral reduces to an usual 1-Laplace transform in the direction $k\theta$ and we conclude by Proposition 7.7. ■

Let us now turn to the characterization of the k-summability in terms of the k-Borel and k-Laplace operators. As a first step, we start by proving a variant of the classical Nevanlinna's Theorem.

7.3 Nevanlinna's Theorem

Throughout this section, we fix a positive real number $k > 0$ and a direction θ_0 issuing from 0 which we assume, even if it means rotating the variable t, to be the direction $\theta_0 = 0$.

Let us first describe the curves and domains we will be concerned with. To this end, we consider four copies of $\mathbb{C}$:

- one with coordinate t (referred as the *Laplace plane*) and one with coordinate τ (referred as the *Borel plane*);
- one with coordinate $Z = 1/t^k$ and one with coordinate $\zeta = \tau^k$.

◁ In the t-plane, we consider for any $\ell > 0$ the domain $\mathfrak{s}_{k;\ell}$ (called *Borel disc* when $k = 1$ and *Fatou petal* when $k \neq 1$, see Fig. 7.5) defined by

$$\mathfrak{s}_{k;\ell} = \left\{t;\ \mathrm{Re}\left(\frac{1}{t^k}\right) > \ell^k \text{ and } |\arg(t)| < \frac{\pi}{2k}\right\}.$$

Observe that when $k < 1/2$, the domain $\mathfrak{s}_{k;\ell}$ has an angle larger than 2π at 0. It lives then on the Riemann surface $\widetilde{\mathbb{C}}$ of the logarithm with similar shape.

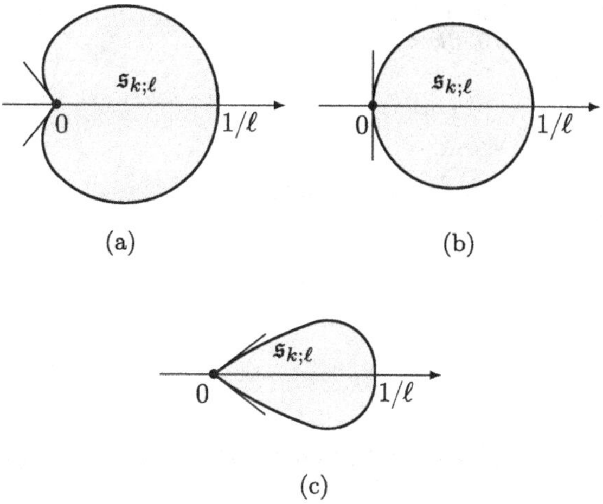

Fig. 7.5 Fatou petals and Borel disc $\mathfrak{s}_{k;\ell}$ in the t-plane. **(a)** $0 < k < 1$. **(b)** $k = 1$. **(c)** $k > 1$

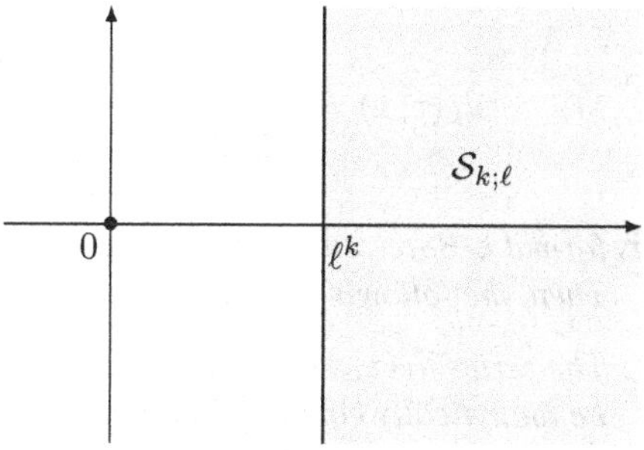

Fig. 7.6 Half-plane $\mathcal{S}_{k;\ell}$ in the Z-plane

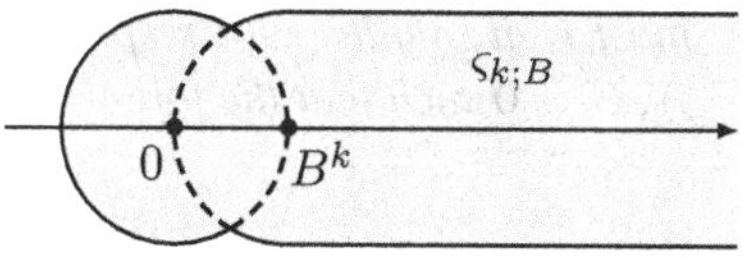

Fig. 7.7 Domain $\varsigma_{k;B}$ in the ζ-plane

◁ In the Z-plane, we consider the image $\mathcal{S}_{k;\ell}$ of $\mathfrak{s}_{k;\ell}$ by the map $Z = 1/t^k$. This is the half-plane $\{Z;\, \mathrm{Re}(Z) > \ell^k\}$ (see Fig. 7.6).

◁ In the ζ-plane, we consider for any $B > 0$ the domain $\varsigma_{k;B} = D(0, B^k) \cup \varsigma'_{k;B}$ union of the open disc $D(0, B^k)$ with center 0 and radius B^k, and of the set $\varsigma'_{k;B}$ of the points of $\mathbb{C}$ at a distance less than B^k of the line $[B^k, +\infty[$ (see Fig. 7.7).

◁ In the τ-plane, we consider for any $B > 0$ the domain $\sigma_{k;B} = D(0, B) \cup \sigma'_{k;B}$ union of the open disc $D(0, B)$ with center 0 and radius B, and of the image $\sigma'_{k;B}$ of $\varsigma'_{k;B}$ by the map $\tau = \zeta^{1/k}$ for the choice of the principal determination of the kth-root (see Fig. 7.8).

Observe that all these domains depend on k.

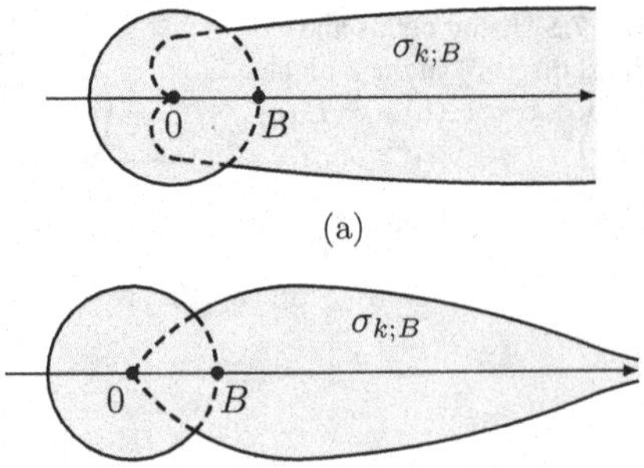

Fig. 7.8 Domain $\sigma_{k;B}$ in the τ-plane. (**a**) $0 < k < 1$. (**b**) $k > 1$

Theorem 7.16 (Nevanlinna's Theorem) *Let* $\widetilde{u}(t,x) = \sum_{j\geqslant 0} u_j(x)t^j \in \mathcal{O}(D_{\rho_1,\dots,\rho_n})[[t]]$ *be a formal power series and*

$$\widehat{u}_k(\tau,x) = \sum_{j\geqslant 0} \frac{u_j(x)}{\Gamma(1+sj)}\tau^j \in \mathcal{O}(D_{\rho_1,\dots,\rho_n})[[\tau]] \text{ with } s = \frac{1}{k}$$

its formal k-Borel transform with respect to t.

Then, the following two assertions are equivalent:

1. *The series $\widehat{u}_k(\tau,x)$ is convergent at the origin of $\mathbb{C}^{n+1}$ and its sum $v_k(\tau,x)$ can be analytically continued to a domain $\sigma_{k;B} \times D_{r,\dots,r}$ for some $B > 0$ and some radius $0 < r < \min(\rho_1,\dots,\rho_n)$ with a global exponential growth of order at most k at infinity with respect to τ, that is, there exist two positive constants $A, C > 0$ such that the following estimate holds for all $(\tau,x) \in \sigma_{k;B} \times D_{r,\dots,r}$:*

$$|v_k(\tau,x)| \leqslant C\exp\left(A\,|\tau|^k\right). \tag{7.3}$$

2. *There exist a function $u(t,x)$ holomorphic on a domain $\mathfrak{s}_{k;\ell_0} \times D_{r',\dots,r'}$ for some $\ell_0 > 0$ and some radius $0 < r' < \min(\rho_1,\dots,\rho_n)$, and two positive constants $D, K > 0$ such that the following estimate holds for all $(t,x) \in \mathfrak{s}_{k;\ell_0} \times D_{r',\dots,r'}$ and all $J \geqslant 1$:*

$$\left|u(t,x) - \sum_{j=0}^{J-1} u_j(x)t^j\right| \leqslant DK^J\Gamma(1+sJ)\,|t|^J. \tag{7.4}$$

Moreover, the functions $u(t,x)$ and $v_k(\tau,x)$ are k-Laplace and k-Borel transforms of each other: given $\ell > \ell_0$ and a radius $0 < r'' \leqslant \min(r, r')$, they satisfy the identities

- $u(t,x) = \dfrac{1}{t^k}\displaystyle\int_0^{+\infty} v_k(\tau,x)e^{-(\tau/t)^k}\,d(\tau^k)$ *for all* $(t,x) \in \mathfrak{s}_{k;\ell} \times D_{r'',\dots,r''}$*;*
- $v_k(\tau,x) = \dfrac{1}{2i\pi}\displaystyle\int_{\mathrm{Re}(1/t^k)=\ell^k} t^k u(t,x)e^{(\tau/t)^k}\,d\left(\dfrac{1}{t^k}\right)$ *for all* $(\tau,x) \in \mathbb{R}_+^* \times D_{r'',\dots,r''}$.

Observe that condition (2) is not a mere asymptotic condition on $\mathfrak{s}_{k;\ell_0} \times D_{r',\dots,r'}$, but a uniform one.

Proof ◁ (1) *implies* (2). Assume that the formal power series $\widehat{u}_k(\tau,x)$ is convergent at the origin of $\mathbb{C}^{n+1}$ and that its sum $v_k(\tau,x)$ can be analytically continued to a domain $\sigma_{k;B} \times D_{r,\dots,r}$ with the global exponential growth (7.3). Setting

$$u(t,x) = \frac{1}{t^k}\int_0^{+\infty} v_k(\tau,x)e^{-(\tau/t)^k}\,d(\tau^k),$$

we shall prove that there exists a convenient $\ell_0 > 0$ such that $u(t,x)$ is well-defined, analytic and satisfies condition (7.4) on the domain $\mathfrak{s}_{k;\ell_0} \times D_{r,\dots,r}$.

Applying first the condition (7.3), it is clear that $u(t,x)$ is defined and analytic for any (t,x) satisfying $\mathrm{Re}(1/t^k) > A$ and $x \in D_{r,\dots,r}$; hence, adding the condition $|\arg(t)| < \pi/(2k)$, on any domain $\mathfrak{s}_{k;\ell} \times D_{r,\dots,r}$ with $\ell^k > A$. We choose in the sequel such a ℓ we denote by ℓ_0.

Let us now set $\zeta = \tau^k$ and let us denote by ζ^s the principal k^{th} root of ζ (recall that $s = 1/k$). Then,

$$u(t,x) = \frac{1}{t^k}\int_0^{+\infty} v_k(\zeta^s,x)e^{-\zeta/t^k}\,d\zeta \tag{7.5}$$

and, from the Cauchy's Theorem, we can integrate as well on a path γ_b homotopic to $[0,+\infty[$ in $\varsigma_{k;B}$ and defined as follows: the path γ_b follows a straight line from 0 to b^k and continues along a horizontal line from b^k to $+\infty$. Besides, we choose b with argument, say, $\beta = \pi/(4k)$ and $\mathrm{Re}(b) > 0$ so small that the whole path γ_b be included in $\varsigma_{k;B}$ (see Fig. 7.9). The integral (7.5) can be then decomposed into

$$u(t,x) = u^b(t,x) + v^b(t,x)$$

with

$$u^b(t,x) = \frac{1}{t^k}\int_0^{b^k} v_k(\zeta^s,x)e^{-\zeta/t^k}\,d\zeta \quad \text{and}$$

$$v^b(t,x) = \frac{1}{t^k}\int_{b^k}^{+\infty} v_k(\zeta^s,x)e^{-\zeta/t^k}\,d\zeta.$$

Fig. 7.9 The path γ_b in the ζ-plane

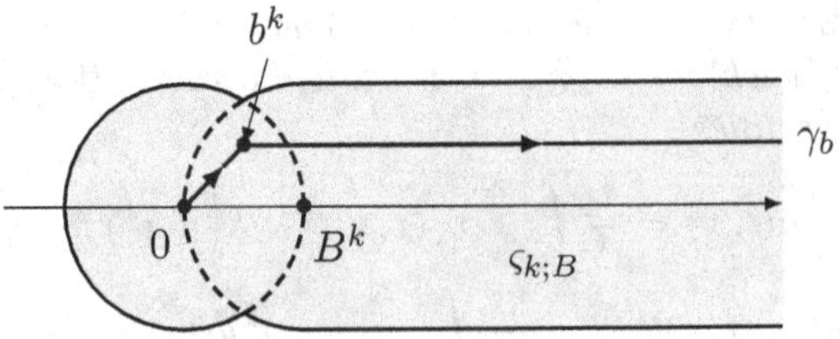

Fig. 7.10 Sector Σ_δ and domain $\mathfrak{s}_{k;\ell_0} \cap \Sigma_\delta$ (hachured) in the case $k = 1$

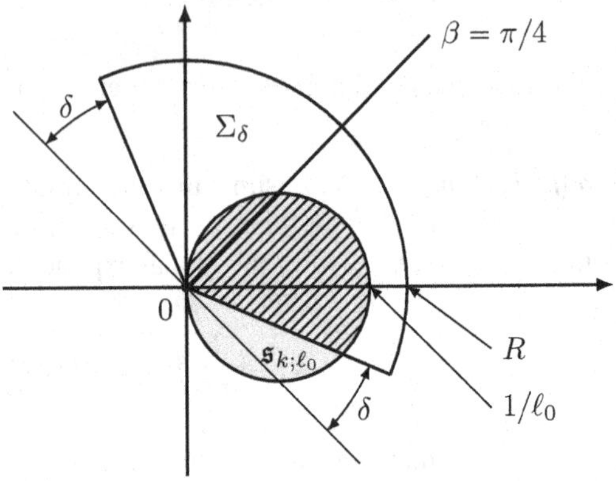

Next, let us choose in the Laplace t-plane a sector Σ_δ bisected by β, with radius $R > 1/\ell_0$, and with opening $\pi/k - 2\delta$, where δ is assumed in $]0, \pi/(4k)[$ so that Σ_δ contains the upper part $\mathfrak{s}_{k;\ell_0} \cap \{t;\, \mathrm{Im}(t) \geqslant 0\}$ of $\mathfrak{s}_{k;\ell_0}$ (see Fig. 7.10 for an example of sector Σ_δ in the case $k = 1$).

From the proof of Theorem 4.20 (the sector Σ_δ can in fact be seen as a proper subsector of a convenient sector with opening less than π/k), we know that $u^b(t, x)$ satisfies a global s-Gevrey asymptotic condition like (7.4) on Σ_δ: there exist two positive constants $D', K' > 0$ such that the following estimate holds for all $(t, x) \in \Sigma_\delta \times D_{r,\dots,r}$ and all $J \geqslant 1$:

$$\left| u^b(t, x) - \sum_{j=0}^{J-1} u_j(x) t^j \right| \leqslant D' K'^J \Gamma(1 + sJ)\, |t|^J .$$

On the other hand, we can prove that $v^b(t, x)$ is k-exponentially flat on $\Sigma_\delta \cap \mathfrak{s}_{k;\ell_0}$ as follows. Parameterizing the second part of γ_b by $\zeta = b^k + \eta$, we first get

$$v^b(t, x) = \frac{e^{-b^k/t^k}}{t^k} \int_0^{+\infty} \upsilon_k((b^k + \eta)^s, x) e^{-\eta/t^k} d\eta;$$

hence, using the assumption (7.3),

$$\left|v^b(t,x)\right| \leqslant Ce^{A|b|^k}\frac{e^{-\mathrm{Re}(b^k/t^k)}}{|t|^k}\int_0^{+\infty} e^{-(\mathrm{Re}(1/t^k)-A)\eta}d\eta$$

for all $(t,x)\in\mathfrak{s}_{k;\ell_0}\times D_{r,\dots,r}$. Now, let us observe that

- if $t\in\Sigma_\delta$, then $\left|\arg(b^k/t^k)\right| < \pi/2-k\delta$ and, consequently, there exists a constant $c>0$ such that $-\mathrm{Re}(b^k/t^k)\leqslant -c/|t|^k$,
- if $t\in\mathfrak{s}_{k;\ell_0}$, then

$$\int_0^{+\infty} e^{-(\mathrm{Re}(1/t^k)-A)\eta}d\eta \leqslant \int_0^{+\infty} e^{-(\ell_0^k-A)\eta}d\eta = \frac{1}{\ell_0^k-A}.$$

Thereby, the following estimate holds for all $(t,x)\in(\Sigma_\delta\cap\mathfrak{s}_{k;\ell_0})\times D_{r,\dots,r}$:

$$\left|v^b(t,x)\right| \leqslant \frac{Ce^{A|b|^k}}{\ell_0^k-A}e^{-c/(2|t|^k)},$$

which proves the k-exponentially flat of $v^b(t,x)$ on $\Sigma_\delta\cap\mathfrak{s}_{k;\ell_0}$. Then, using the proof of Proposition 4.19, we deduce there exist two positive constants D'', $K''>0$ such that the following estimate holds for all $(t,x)\in(\Sigma_\delta\cap\mathfrak{s}_{k;\ell_0})\times D_{r,\dots,r}$ and all $J\geqslant 1$:

$$\left|v^b(t,x)\right| \leqslant D''K''^J\Gamma(1+sJ)\,|t|^J.$$

In conclusion, taking $D=\max(D',D'')$ and $K=\max(K',K'')$, we have proved that $u(t,x)$ satisfies condition (7.4) on $(\Sigma_\delta\cap\mathfrak{s}_{k;\ell_0})\times D_{r,\dots,r}$.

Symmetrically, choosing $\overline{b}$ instead of b and the path $\gamma_{\overline{b}}$ symmetric from γ_b with respect to the real axis, we can also prove that $u(t,x)$ satisfies condition (7.4) on $(\Sigma'_\delta\cap\mathfrak{s}_{k;\ell_0})\times D_{r,\dots,r}$, where Σ'_δ stands for the symmetric sector of Σ_δ with respect to the real axis. Consequently, since $\mathfrak{s}_{k;\ell_0}$ is included in $\Sigma_\delta\cup\Sigma'_\delta$, we conclude that $u(t,x)$ satisfies condition (7.4) on $\mathfrak{s}_{k;\ell_0}\times D_{r,\dots,r}$, which achieves the proof of this part.

$\lhd$ (2) *implies* (1). Let us first observe that Theorem 7.16 being satisfied for any monomial $u_j(x)t^j$, we can suppose in this part that $u_0(x)\equiv 0$, that is the formal series $\widetilde{u}(t,x)$ has valuation 1 with respect to t.

Our assumption is the following: there exist a function $u(t,x)$ holomorphic on a domain $\mathfrak{s}_{k;\ell_0}\times D_{r',\dots,r'}$ and two positive constants D, $K>0$ such that the following estimate holds for all $(t,x)\in\mathfrak{s}_{k;\ell_0}\times D_{r',\dots,r'}$ and all $J\geqslant 1$:

$$\left|u(t,x)-\sum_{j=0}^{J-1}u_j(x)t^j\right| \leqslant DK^J\Gamma(1+sJ)\,|t|^J. \tag{7.6}$$

In particular, reasoning as in Proposition 4.3, we get

$$|u_j(x)| \leqslant DK^j\Gamma(1+sj) \tag{7.7}$$

for all $x \in D_{r',\dots,r'}$ and all $j \geqslant 1$ and, consequently, the formal k-Borel transform $\widehat{u}_k(\tau,x)$ of $\widetilde{u}(t,x)$ defines an analytic function on $D_{1/K} \times D_{r',\dots,r'}$. Observe that we can always suppose $K \geqslant 1$ in the estimates (7.6) and (7.7).

Again, let us set $Z = 1/t^k$ and $\zeta = \tau^k$, and let us denote by Z^s and ζ^s the principal k^{th} roots of Z and ζ (recall that $s = 1/k$). By assumption, the function $U(Z,x) = u(1/Z^s, x)$ is analytic on the domain $\mathcal{S}_{k;\ell_0} \times D_{r',\dots,r'}$.

Let us choose $\ell > \ell_0$ and let us set

$$\Upsilon(\zeta,x) = \frac{1}{2i\pi}\int_{\ell^k+i\mathbb{R}} \frac{1}{Z}U(Z,x)e^{Z\zeta}dZ \tag{7.8}$$

for all $\zeta > 0$ and $x \in D_{r',\dots,r'}$. Observe that this formula makes sense. Indeed, condition (7.6) implying the inequalities

$$\left|U(Z,x) - \sum_{j=0}^{J-1}\frac{u_j(x)}{Z^{sj}}\right| \leqslant \frac{DK^J\Gamma(1+sJ)}{|Z|^{sJ}} \tag{7.9}$$

for all $(Z,x) \in \mathcal{S}_{k;\ell_0} \times D_{r',\dots,r'}$ and all $J \geqslant 1$, we have in particular for $J = 1$ the estimate

$$\left|\frac{1}{Z}U(Z,x)\right| \leqslant \frac{DK\Gamma(1+s)}{|Z|^{1+s}} \tag{7.10}$$

for all $(Z,x) \in \mathcal{S}_{k;\ell_0} \times D_{r',\dots,r'}$ (recall that $u_0(x) \equiv 0$).

To conclude this part, we must now prove the following four points:

(i) The function $U(Z,x)$ can be written in the form

$$U(Z,x) = Z\int_0^{+\infty} \Upsilon(\zeta,x)e^{-Z\zeta}d\zeta.$$

(ii) The series $\widehat{u}_k(\tau,x)$ converges to $\Upsilon(\tau^k,x)$ for all $0 < \tau < 1/K$ and all $x \in D_{r',\dots,r'}$; hence, the analytic continuation of $\Upsilon(\tau^k,x)$ by $v_k(\tau,x)$ to the domain $D_{1/K} \times D_{r',\dots,r'}$ ($\zeta = 0$ might be a branch point for the series $\Upsilon(\zeta,x)$ itself).

(iii) The function $\Upsilon(\zeta,x)$ can be analytically continued to $\varsigma'_{k;1/K} \times D_{r',\dots,r'}$.

(iv) The function $\Upsilon(\zeta,x)$ has a global exponential growth on $\varsigma'_{k;1/B} \times D_{r,\dots,r}$ of order at most 1 at infinity with respect to ζ for any $B > K$ and any radius $0 < r < r'$.

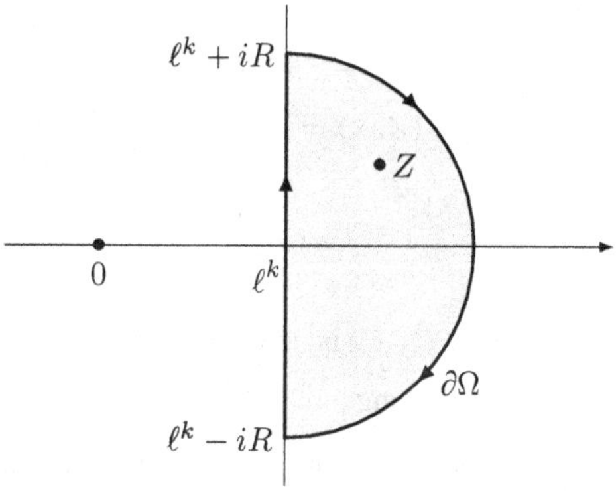

Fig. 7.11 The domain Ω and its boundary $\partial\Omega$

Proof of Point (i) Let $(Z, x) \in \mathcal{S}_{k;\ell} \times D_{r',\ldots,r'}$ and let us enclose Z in a domain Ω limited by the vertical line at ℓ^k and an arc of a circle centered at 0 with radius R as shown on Fig. 7.11.

Denoting by $\partial\Omega$ the boundary of Ω oriented clockwise and applying the Cauchy Integral Formula, we can write

$$\frac{1}{Z}U(Z, x) = \frac{1}{2i\pi}\int_{\partial\Omega}\frac{1}{\eta}U(\eta, x)\frac{d\eta}{Z-\eta}.$$

Let us now observe that, due to the estimate (7.10), the integral along the half-circle tends to zero as R tends to infinity. Consequently,

$$U(Z, x) = \frac{Z}{2i\pi}\int_{\ell^k+i\mathbb{R}}\frac{1}{\eta}U(\eta, x)\frac{d\eta}{Z-\eta}.$$

Wrting then

$$\frac{1}{Z-\eta} = \int_0^{+\infty} e^{-(Z-\eta)\zeta}d\zeta$$

so that

$$U(Z, x) = \frac{Z}{2i\pi}\int_{\ell^k+i\mathbb{R}}\frac{1}{\eta}U(\eta, x)\left(\int_0^{+\infty} e^{-(Z-\eta)\zeta}d\zeta\right)d\eta$$

and applying the Fubini Theorem to the iterated integral (the estimate (7.10) implies indeed the inequality

$$\left|\frac{1}{\eta}U(\eta, x)e^{-(Z-\eta)\zeta}\right| \leqslant \frac{DK\Gamma(1+s)}{|\eta|^{1+s}}e^{-(\mathrm{Re}(Z)-\ell^k)\zeta}$$

with $\mathrm{Re}(Z) - \ell^k > 0$), we get

$$\begin{aligned} U(Z,x) &= \frac{Z}{2i\pi}\int_0^{+\infty}\left(\int_{\ell^k+i\mathbb{R}}\frac{1}{\eta}U(\eta,x)e^{\eta\zeta}d\eta\right)e^{-Z\zeta}d\zeta \\ &= Z\int_0^{+\infty}\Upsilon(\zeta,x)e^{-Z\zeta}d\zeta. \end{aligned}$$

Moreover, $\Upsilon(\zeta,x)$ is independant of $\ell > \ell_0$ (apply the Cauchy Theorem to $\frac{1}{\eta}U(\eta,x)e^{\eta\zeta}$ along a rectangle with vertical sides at $\mathrm{Re}(Z) = \ell^k$ and $\mathrm{Re}(Z) = \ell'^k$, $\ell^k \neq \ell'^k$, and let the horizontal sides go to infinity).

Proof of Point (ii) Let us set

$$R_J(Z,x) = U(Z,x) - \sum_{j=0}^{J-1}\frac{u_j(x)}{Z^{sj}}$$

for all $(Z,x) \in S_{k;\ell_0} \times D_{r',\dots,r'}$ and all $J \geqslant 1$. According to the Laplace Formula

$$\frac{1}{\Gamma(\alpha)} = \frac{1}{2\pi}\int_{-\infty}^{+\infty}\frac{e^{a+ib}}{(a+ib)^{\alpha}}db \quad, \mathrm{Re}(\alpha) > 0, a > 0$$

(see for instance [54]) and the identity (7.8), we can first write

$$\Upsilon(\zeta,x) = \sum_{j=0}^{J-1}\frac{u_j(x)}{\Gamma(1+sj)}\zeta^{sj} + \Upsilon_J(\zeta,x) \tag{7.11}$$

with

$$\Upsilon_J(\zeta,x) = \frac{1}{2i\pi}\int_{\ell^k+i\mathbb{R}}\frac{1}{Z}R_J(Z,x)e^{Z\zeta}dZ$$

for all $\zeta > 0$, all $x \in D_{r',\dots,r'}$ and all $J \geqslant 1$.

Let us now set $\zeta = \ell^k + iv$ and let us apply the inequalities (7.9). Then, the following estimate holds for all $\zeta > 0$, all $x \in D_{r',\dots,r'}$ and all $J \geqslant 1$:

$$\begin{aligned} |\Upsilon_J(\zeta,x)| &\leqslant \frac{DK^J\Gamma(1+sJ)e^{\ell^k\zeta}}{2\pi\ell^{k(1+sJ)}}\int_{-\infty}^{+\infty}\frac{dv}{\sqrt{1+(v/\ell^k)^2}^{1+sJ}} \\ &\leqslant D'\left(\frac{K}{\ell}\right)^J\Gamma(1+sJ)e^{\ell^k\zeta} \end{aligned}$$

with (recall that $\ell > \ell_0$)

$$D' = \frac{D}{2\pi \ell_0^k} \int_{-\infty}^{+\infty} \frac{dv}{\sqrt{1+(v/\ell^k)^2}^{1+sJ}} < +\infty.$$

Let us take $0 < \zeta < (1/K)^k$, $x \in D_{r',\dots,r'}$ and $J \geqslant 1$, and let us consider the right-hand side of the second inequality above as a function of ℓ. Since the function $y(\ell) = e^{\ell^k \zeta}/\ell^J$ for $\ell > 0$ reaches its minimal value at $\ell_1(J) = (sJ)^s/\zeta^s$ with $y(\ell_1(J)) = e^{sJ}(sJ)^{-sJ}\zeta^{sJ}$, we choose $j_0 = j_0(\zeta)$ so large that $\ell_1(j_0) > \ell_0$. Thereby, for $J \geqslant j_0(\zeta)$, we have also $\ell_1(J) > \ell_0$ and, since $\Upsilon_J(\zeta, x)$ does not depend on $\ell > \ell_0$, we can take $\ell = \ell_1(J)$. Then, $\Upsilon_J(\zeta, x)$ satisfies

$$|\Upsilon_J(\zeta, x)| \leqslant D' K^J \Gamma(1+sJ) y(\ell_1(J)) = D'(\zeta^s K)^J \Gamma(1+sJ) e^{sJ}(sJ)^{-sJ}$$

and tends to 0 as J tends to infinity. Indeed, $0 < \zeta^s K < 1$ and

$$\Gamma(1+sJ)e^{sJ}(sJ)^{-sJ} \underset{J\to+\infty}{\sim} \sqrt{2\pi s J}$$

from the Stirling's Formula.

Hence,

$$\Upsilon(\tau^k, x) = \sum_{j\geqslant 0} \frac{u_j(x)}{\Gamma(1+sj)} \tau^j = \widehat{u}_k(\tau, x)$$

for all $0 < \tau < 1/K$ and all $x \in D_{r',\dots,r'}$, and, consequently, $\Upsilon(\tau^k, x)$ can be analytically continued by $v_k(\tau, x)$ to the domain $D_{1/K} \times D_{r',\dots,r'}$.

From now on, we denote $v_k(\tau, x) = \Upsilon(\tau^k, x)$.

Proof of Point (iii) Given $\zeta_0 \geqslant (1/K)^k$, we shall prove that the Taylor Series

$$\widehat{\Upsilon}_{\zeta_0}(\zeta, x) = \sum_{J\geqslant 0} \frac{1}{J!} \frac{\partial^J \Upsilon}{\partial \zeta^J}(\zeta_0, x)(\zeta - \zeta_0)^J$$

with respect to ζ of $\Upsilon(\zeta, x)$ at ζ_0 is well-defined and holomorphic on $D(\zeta_0, (1/K)^k) \times D_{r',\dots,r'}$ ($D(\zeta_0, (1/K)^k)$ stands for the open disc with center ζ_0 and radius $(1/K)^k$) and converges to $\Upsilon(\zeta, x)$ for ζ real and x with real components.

To this end, let us first prove that

- for any $x \in D_{r',\dots,r'}$, the functions $\zeta \longmapsto \Upsilon(\zeta, x)$ is infinitely derivable for $\zeta > 0$;
- for any $\zeta_0 > 0$ and any $J \geqslant 0$, the functions $x \longmapsto \dfrac{\partial^J \Upsilon}{\partial \zeta^J}(\zeta_0, x)$ are holomorphic on $D_{r',\dots,r'}$.

For $J = 0$, the holomorphy of the maps $x \longmapsto \Upsilon(\zeta_0, x)$ stems from the identity (7.8) and the estimate (7.10). Let us consider $J \geqslant 1$ and let us choose a nonnegative integer $m \in \mathbb{N}$ such that $k(J+1) < m \leqslant k(J+1)+1$. From identity (7.11), we can write

$$\Upsilon(\zeta, x) = \sum_{j=0}^{m} \frac{u_j(x)}{\Gamma(1+sj)}\zeta^{sj} + \frac{1}{2i\pi}\int_{\ell^k+i\mathbb{R}} \frac{1}{Z} R_{m+1}(Z, x) e^{Z\zeta} dZ \tag{7.12}$$

for all $\zeta > 0$ and all $x \in D_{r',\dots,r'}$, and we look at the μth derivative with respect to ζ of the integrand for $1 \leqslant \mu \leqslant J$. From the estimate (7.9), we have

$$\left|\frac{1}{Z} R_{m+1}(Z, x) Z^{\mu} e^{Z\zeta}\right| \leqslant \frac{DK^{m+1}\Gamma(1+s(m+1))e^{\ell^k\zeta}}{|Z|^{1-\mu+s(m+1)}} \leqslant \frac{D''}{|Z|^{1+s}}, \tag{7.13}$$

where the constant D'' is independent of ζ as long as ζ stays bounded and also independent of any $x \in D_{r',\dots,r'}$. Then, applying Lebesgue's Theorem, we can conclude that $\zeta \in]0, +\infty[\longmapsto \Upsilon(\zeta, x)$ can be derivated J times for any $x \in D_{r',\dots,r'}$ and that $x \longmapsto \dfrac{\partial^J \Upsilon}{\partial \zeta^J}(\zeta_0, x)$ is holomorphic on $D_{r',\dots,r'}$ for any $\zeta_0 > 0$.

Now, let us choose $B > K \geqslant 1$ and $J \geqslant 1$, and let us estimate $\dfrac{\partial^J \Upsilon}{\partial \zeta^J}(\zeta, x)$ for any $\zeta \geqslant (1/B)^k$ and any $x \in D_{r',\dots,r'}$. From (7.12), we can write

$$\frac{\partial^J \Upsilon}{\partial \zeta^J}(\zeta, x) = I_J(\zeta, x) + I'_J(\zeta, x),$$

where

$$I_J(\zeta, x) = \sum_{j=1}^{m} \frac{u_j(x)}{\Gamma(1+sj)} sj(sj-1)\dots(sj-J+1)\zeta^{sj-J} \quad \text{and}$$

$$I'_J(\zeta, x) = \frac{1}{2i\pi}\int_{\ell^k+i\mathbb{R}} \frac{1}{Z} R_{m+1}(Z, x) Z^J e^{Z\zeta} dZ.$$

Applying the Gevrey estimates (7.7) and the assumption on m, we first get

$$\frac{|u_j(x)|}{\Gamma(1+sj)} \leqslant DK^j \leqslant DB^j$$

for all $j \geqslant 1$ and all $x \in D_{r',\dots,r'}$; hence

$$
\begin{aligned}
|I_J(\zeta,x)| &\leqslant Dsm(sm-1)\dots(sm-J+1)\zeta^{-J}\sum_{j=1}^{m}(B\zeta^s)^j \\
&\leqslant Dsm(sm-1)\dots(sm-J+1)\zeta^{-J}\times m(B\zeta^s)^m \\
&\leqslant kD(J+2+s)(J+1+s)\dots(2+s)B^m\zeta^{sm-J} \\
&\leqslant kDB^{k+1}\Gamma(J+3+s)B^{kJ}\max(\zeta,\zeta^{1+s})
\end{aligned}
$$

for all $\zeta \geqslant (1/B)^k$ and all $x \in D_{r',\dots,r'}$. On the other hand, setting $Z = \ell^k + iv$ and applying the estimates (7.13) and the assumption on m, we obtain

$$
\begin{aligned}
\left|I'_J(\zeta,x)\right| &\leqslant \frac{DB^{m+1}\Gamma(1+s(m+1))e^{\ell^k\zeta}}{2\pi\,\ell^{k(1-J+s(m+1))}}\int_{-\infty}^{+\infty}\frac{dv}{\sqrt{1+(v/\ell^k)^2}^{\,1-J+s(m+1)}} \\
&\leqslant D'''\Gamma(J+2+2s)B^{kJ}e^{\ell^k\zeta}
\end{aligned}
$$

for all $\zeta \geqslant (1/B)^k$ and all $x \in D_{r',\dots,r'}$, where D''' is the positive constant defined by

$$
D''' = \frac{DB^{k+2}}{2\pi}\max\left(\frac{1}{\ell^{2k+1}},\frac{1}{\ell^{2k+2}}\right)\int_{-\infty}^{+\infty}\frac{dv}{\sqrt{1+(v/\ell^k)^2}^{\,2+s}} < +\infty.
$$

Adding these two estimates, we see that there exists a positive constant $\alpha_B > 0$ such that

$$
\left|\frac{\partial^J\Upsilon}{\partial\zeta^J}(\zeta,x)\right| \leqslant \alpha_B\Gamma(J+3+2s)B^{kJ}e^{\ell^k\zeta} \tag{7.14}
$$

for all $J \geqslant 1$, all $\zeta \geqslant (1/B)^k$ and all $x \in D_{r',\dots,r'}$.

Let us now fix $\zeta_0 \geqslant (1/K)^k$. Then, for any $B > K$, the function $\widehat{\Upsilon}_{\zeta_0}(\zeta,x)$ defined by

$$
\widehat{\Upsilon}_{\zeta_0}(\zeta,x) = \sum_{J\geqslant 0}\frac{1}{J!}\frac{\partial^J\Upsilon}{\partial\zeta^J}(\zeta_0,x)(\zeta-\zeta_0)^J
$$

is analytic with respect to ζ on $D(\zeta_0,(1/B)^k)$ for any $x \in D_{r',\dots,r'}$, and analytic with respect to x on $D_{r',\dots,r'}$ for any $\zeta \in D(\zeta_0,(1/B)^k)$. Consequently, Hartogs' Theorem applies and implies that $\widehat{\Upsilon}_{\zeta_0}(\zeta,x)$ is analytic on $D(\zeta_0,(1/B)^k)\times D_{r',\dots,r'}$, and thereby on $D(\zeta_0,(1/K)^k)\times D_{r',\dots,r'}$ by making B tend to K.

We are left to prove that $\widehat{\Upsilon}_{\zeta_0}(\zeta, x)$ converges to $\Upsilon(\zeta, x)$ for ζ real and x with real components. To do that, let us fix $x \in]0, r'[^n$ and let us write the Taylor-Lagrange Formula

$$\Upsilon(\zeta, x) = \sum_{j=0}^{J-1} \frac{1}{j!} \frac{\partial^j \Upsilon}{\partial \zeta^j}(\zeta_0, x)(\zeta - \zeta_0)^j + \Psi_J(\zeta, x)$$

with

$$\Psi_J(x) = \frac{1}{J!} \frac{\partial^J \Upsilon}{\partial \zeta^J}(\zeta_0 + \theta(\zeta - \zeta_0), x)(\zeta - \zeta_0)^J, \quad 0 < \theta < 1.$$

Next, let us choose $\zeta_0 \geqslant (1/K)^k$ and $\zeta \geqslant (1/B)^k$ with $B > K$. Then, $\zeta_0 + \theta(\zeta - \zeta_0) > (1/B)^k$ and we can apply the estimate (7.14) to $\dfrac{\partial^J \Upsilon}{\partial \zeta^J}(\zeta_0 + \theta(\zeta - \zeta_0), x)$ so that

$$|\Psi_J(\zeta, x)| \leqslant \frac{\alpha_B \Gamma(J + 3 + 2s) B^{kJ} \, |\zeta - \zeta_0| \, J}{J!} e^{\ell^k \max(\zeta_0, \zeta)}$$

and $\Psi_J(\zeta, x)$ tends to 0 as J tends to infinity as soon as $\max((1/B)^k, \zeta_0 - (1/B)^k) < \zeta < \zeta_0 + (1/B)^k$. Therefore, $\widehat{\Upsilon}_{\zeta_0}(\zeta, x)$ and $\Upsilon(\zeta, x)$ coincide on the domain $]\max((1/B)^k, \zeta_0 - (1/B)^k), \zeta_0 + (1/B)^k[\times]0, r'[^n$.

This proves that $\Upsilon(\zeta, x)$ admits an analytic continuation to the domain $\varsigma'_{k;1/B} \cap \{\mathrm{Re}(\zeta) > (1/B)^k\} \times D_{r',\ldots,r'}$. Since the two intervals $]0, (1/K)^k[$ and $]\max((1/B)^k, \zeta_0 - (1/B)^k), \zeta_0 + (1/B)^k[$ overlap for instance for $\zeta_0 = (1/K)^k$, this analytic continuation fit the analytic continuation by $v_k(\zeta^s, x)$ on the domain $D_{(1/K)^k} \cap \varsigma'_{k;1/B} \cap \{\mathrm{Re}(\zeta) > (1/B)^k\} \times D_{r',\ldots,r'}$.

Finally, letting B tend to K allows us to extend the analytic continuation of $\Upsilon(\zeta, x)$ up to $\varsigma'_{k;1/K} \times D_{r',\ldots,r'}$ and, consequently, to extend the analytic continuation of $v_k(\tau, x)$ to the full domain $\sigma_{k;1/K} \times D_{r',\ldots,r'}$.

Proof of Point (iv) Let $B > K \geqslant 1$ and $0 < r < r'$. Since $v_k(\tau, x)$ is analytic on the domain $D_{1/K} \times D_{r',\ldots,r'}$ (see the beginning of the proof *(2) implies (1)* page 110), it is bounded on the smaller domain $D_{1/B} \times D_{r,\ldots,r}$. Then, $\Upsilon(\zeta, x)$ is bounded on $D_{(1/B)^k} \times D_{r,\ldots,r}$ and we are left to prove the exponential estimate in the domains $D(\zeta_0, (1/B)^k) \times D_{r,\ldots,r}$ for any $\zeta_0 \geqslant (1/K)^k$.

The analytic continuation of $\Upsilon(\zeta, x)$ to $D(\zeta_0, (1/B)^k) \times D_{r,\ldots,r}$ is given by the Taylor series $\widehat{\Upsilon}_{\zeta_0}(\zeta, x)$:

$$\Upsilon(\zeta, x) = \sum_{J \geqslant 0} \frac{1}{J!} \frac{\partial^J \Upsilon}{\partial \zeta^J}(\zeta_0, x)(\zeta - \zeta_0)^J.$$

Applying then the estimates (7.14) with $B' \in]K, B[$, we get

$$|\Upsilon(\zeta, x)| \leqslant \alpha_{B'} \sum_{J \geqslant 0} \frac{\Gamma(J+3+2s)}{J!} B'^{kJ} \, |\zeta - \zeta_0|^J \, e^{\ell^k \zeta_0}$$

$$\leqslant \alpha_{B'} \sum_{J \geqslant 0} \frac{\Gamma(J+3+2s)}{J!} \left(\frac{B'}{B}\right)^{kJ} e^{\ell^k/B^k} e^{\ell^k \mathrm{Re}(\zeta)} < +\infty$$

for all $(\zeta, x) \in D(\zeta_0, (1/B)^k) \times D_{r,\dots,r}$ (write $\zeta = (\zeta_0 - \mathrm{Re}(\zeta)) + \mathrm{Re}(\zeta)$ and observe that $|\zeta_0 - \mathrm{Re}(\zeta)| \leqslant (1/B)^k$). Consequently, there exist two positive constants $C, A > 0$ such that

$$|\Upsilon(\zeta, x)| \leqslant C \exp(A\,|\zeta|)$$

for all $(\zeta, x) \in \varsigma'_{k;1/B} \times D_{r,\dots,r}$. Hence, the result which achieves the proof of the second part of Theorem 7.16 and the rest of the statement. ■

Remark 7.17

- The proof "(2) implies (1)" shows that one can choose any $B > K$ and any radius $0 < r < r'$ and that $v_k(\tau, x)$ is actually analytic on $\sigma_{k;1/K} \times D_{r',\dots,r'}$. Thus, condition (2) implies that $v_k(\tau, x)$ has an exponential growth of order at most k on $\sigma_{k;1/K} \times D_{r',\dots,r'}$ with respect to τ (meaning that there exist global exponential estimates with respect to τ in restriction to any proper sub-domain $\sigma_{k;1/B} \times D_{r,\dots,r}$ of $\sigma_{k;1/K} \times D_{r',\dots,r'}$).
- On the contrary, starting from condition (1) with a constant B and a radius r, there is no similar control on K, but the radius r' can always be chosen equal to the initial radius r.

7.4 Borel-Laplace Method and k-Summability

We are now able to state the characterization of the k-summability in terms of the k-Borel and k-Laplace transforms.

Proposition 7.18 (Borel-Laplace Method) *A formal power series $\widetilde{u}(t,x) \in \mathcal{O}(D_{\rho_1,\dots,\rho_n})[[t]]$ is k-summable in the direction $\arg(t) = \theta$ if and only if the following two conditions are satisfied:*

1. *the formal k-Borel transform $\widehat{u}_k(\tau, x)$ of $\widetilde{u}(t,x)$ with respect to t is convergent at the origin of $\mathbb{C}^{n+1}$;*
2. *the sum $v_k(\tau, x)$ of $\widehat{u}_k(\tau, x)$ can be analytically continued to a domain $\Sigma_{\theta_1,\theta_2}(\infty) \times D_{r,\dots,r}$ for some directions $\theta_1 < \theta < \theta_2$ and some radius $0 < r \leqslant \min(\rho_1, \dots, \rho_n)$, with a global exponential growth of order at most k at infinity with respect to τ, that is, there exist two positive constants $A, C > 0$*

such that the following estimate holds for all $(\tau, x) \in \Sigma_{\theta_1,\theta_2}(\infty) \times D_{r,\dots,r}$*:*

$$|\upsilon_k(\tau, x)| \leqslant C \exp\left(A\,|\tau|^k\right).$$

Moreover, the k-sum $u(t, x)$ *of* $\widetilde{u}(t, x)$ *in the direction* θ*, if any exists, is given by the k-Laplace transform of* $\upsilon_k(\tau, x)$ *with respect to* τ *in the direction* θ*:*

$$u(t, x) = \mathcal{L}^{[\tau]}_{k;\theta}(\upsilon_k)(t, x) = \frac{1}{t^k}\int_{\tau=0}^{\infty e^{i\theta}} \upsilon_k(\tau, x)e^{-(\tau/t)^k}\,d(\tau^k).$$

Proof Recall that $s = 1/k$.

◁ *The* if *part.* By assumption, there exists a radius $R > 0$ and a radius $0 < r' \leqslant r$ such that $\widehat{u}_k(\tau, x)$ converges on $D_R \times D_{r',\dots,r'}$. Let us set $\mathcal{D} = D_R \times \Sigma_{\theta_1,\theta_2}(\infty)$ and let us choose two directions α and β satisfying the conditions $\theta_1 < \alpha < \theta < \beta < \theta_2$ and $\beta - \alpha < \pi/(4k)$.

For each of the directions $d \in \{\alpha, \theta, \beta\}$, there exists a tubular neighborhood $\sigma_{k;B_d}$ of the line $[0, \infty e^{id}[$ which is contained in $\mathcal{D}$ so that condition (1) of Nevanlinna's Theorem holds on $\sigma_{k;B_d} \times D_{r',\dots,r'}$. Thereby, denoting by $u_d(t, x) = \mathcal{L}^{[\tau]}_{k;d}(\upsilon_k)(t, x)$ the k-Laplace transform of $\upsilon_k(\tau, x)$ with respect to τ in the direction d, Nevanlinna's Theorem implies that there exist $\ell > 0$ and two positive constants $D, K > 0$ such that, for all directions $d \in \{\alpha, \theta, \beta\}$:

- $u_d(t, x)$ is holomorphic on the domain $\mathfrak{s}_{k;\ell}(d) \times D_{r',\dots,r'}$ with $\mathfrak{s}_{k;\ell}(d)$ the Fatou petal

$$\mathfrak{s}_{k;\ell}(d) = \left\{t;\ \mathrm{Re}\left(\frac{1}{t^k}\right) > \ell^k \text{ and } |\arg(t) - d| < \frac{\pi}{2k}\right\}$$

  bisected by the direction d and opening πs;
- the following estimate holds for all $(t, x) \in \mathfrak{s}_{k;\ell}(d) \times D_{r',\dots,r'}$ and all $J \geqslant 1$:

$$\left|u_d(t, x) - \sum_{j=0}^{J-1} u_j(x)t^j\right| \leqslant DK^J\Gamma(1 + sJ)\,|t|^J.$$

From the choice of α and β, the intersection $\mathfrak{s}_{k;\ell}(\alpha) \cap \mathfrak{s}_{k;\ell}(\theta) \cap \mathfrak{s}_{k;\ell}(\beta)$ is non-empty and, from Proposition 7.15, the functions $u_\alpha(t, x)$, $u_\theta(t, x)$ and $u_\beta(t, x)$ glue together into an analytic function $u(t, x)$ on $(\mathfrak{s}_{k;\ell}(\alpha) \cup \mathfrak{s}_{k;\ell}(\theta) \cup \mathfrak{s}_{k;\ell}(\beta)) \times D_{r',\dots,r'}$ satisfying the inequalities

$$\left|u(t, x) - \sum_{j=0}^{J-1} u_j(x)t^j\right| \leqslant DK^J\Gamma(1 + sJ)\,|t|^J \tag{7.15}$$

for all $(t, x) \in (\mathfrak{s}_{k;\ell}(\alpha) \cup \mathfrak{s}_{k;\ell}(\theta) \cup \mathfrak{s}_{k;\ell}(\beta)) \times D_{r',\dots,r'}$ and all $J \geqslant 1$.

The union $\mathfrak{s}_{k;\ell}(\alpha) \cup \mathfrak{s}_{k;\ell}(\theta) \cup \mathfrak{s}_{k;\ell}(\beta)$ containing a sector $\Sigma_{\theta,>\pi s}$ bisected by θ and with opening larger than πs, inequalities (7.15) show that $u(t,x)$ is the k-sum of $\widetilde{u}(t,x)$ in the direction θ in the sense of Definition 5.1. Hence, $\widetilde{u}(t,x)$ is k-summable in the direction θ and its sum in this direction is given, for instance, by the k-Laplace transform $\mathcal{L}^{[\tau]}_{k;\theta}(v_k)$.

$\triangleleft$ *The* only if *part.* Let us now assume that $\widetilde{u}(t,x)$ is k-summable in the direction θ. From Definition 5.1, there exist a sector $\Sigma_{\theta,>\pi s}$ bisected by θ and with opening larger than πs, a radius $0 < r' \leqslant \min(\rho_1,\ldots,\rho_n)$, a function $u(t,x)$ analytic on $\Sigma_{\theta,>\pi s} \times D_{r',\ldots,r'}$, and two positive constants $D, K > 0$ such that the following estimate holds for all $(t,x) \in \Sigma_{\theta,>\pi s} \times D_{r',\ldots,r'}$ and all $J \geqslant 1$:

$$\left| u(t,x) - \sum_{j=0}^{J-1} u_j(x) t^j \right| \leqslant D K^J \Gamma(1+sJ)\, |t|^J . \tag{7.16}$$

Let us write $\Sigma_{\theta,>\pi s}$ in the form

$$\Sigma_{\theta,>\pi s} = \left\{ t;\, 0 < |t| < R \text{ and } |\arg(t) - \theta| < \frac{\pi}{2k} + \delta \right\}$$

with a convenient radius $R > 0$ and a convenient $\delta > 0$.

By construction, there exists $\ell > 0$ so that, for any direction $d \in]\theta - \delta, \theta + \delta[$, we can draw a Fatou petal $\mathfrak{s}_{k;\ell}(d)$ bisected by d, with opening π/k and contained in $\Sigma_{\theta,>\pi s}$. In particular, inequalities (7.16) show that the function $u(t,x)$ satisfies condition (2) of Nevanlinna's Theorem on each of these Fatou petals with same constants D and K. Then, Nevanlinna's Theorem implies that the formal k-Borel transform $\widehat{u}_k(\tau,x)$ is convergent at the origin of $\mathbb{C}^{n+1}$ and that its sum $v_k(\tau,x)$ can be analytically continued to any tubular neighborhood of the lines $[0, \infty e^{id}[$ with thickness $1/K$ times $D_{r,\ldots,r}$ (with any radius $0 < r < r'$). Moreover, it satisfies on each of these neighborhoods an exponential estimate of order at most k with respect to τ with same constants (see the proof "(2) implies (1)" of Theorem 7.16 and Remark 7.17).

The result follows by observing that the union of these neighborhoods jointly with the disc of convergence of $\widehat{u}_k(\tau,x)$ contains a domain $\Sigma_{\theta_1,\theta_2}(\infty) \times D_{r,\ldots,r}$ with two convenient directions θ_1 and θ_2 satisfying $\theta_1 < \theta < \theta_2$, and with a convenient radius $0 < r < r'$ on which $v_k(\tau,x)$ is analytic and admits a global exponential growth of order at most k with respect to τ. This ends the proof of Proposition 7.18. ■

Remark 7.19 In addition to providing another characterization of the k-sumability, the Borel-Laplace method is also of practical interest, since it provides an explicit construction of the k-sum. In particular, this can be used to obtain a numerical approximation of this function by adapting the algorithms already written in the framework of the summable series in $\mathbb{C}[[t]]$ (see, for instance, [27, 51, 93, 127–129], and [28] for an adaptation to the case of several variables).

As a direct consequence of Proposition 7.18, we can state the following result which proves the converse of Proposition 6.2.

Corollary 7.20 *Let $\widetilde{u}(t,x) \in \mathcal{O}(D_{\rho_1,\dots,\rho_n})[[t]]$ and $k > 0$.*
Assume that $\widetilde{u}(t,x)$ is k-summable in any direction.
Then, $\widetilde{u}(t,x)$ defines an analytic function at the origin of $\mathbb{C}^{n+1}$.

Proof According to the assumption on $\widetilde{u}(t,x)$, it results from Proposition 7.18 that the formal k-Borel transform $\widehat{u}_k(\tau,x)$ defines an analytic function on $\mathbb{C} \times D_{r,\dots,r}$ for some radius $r > 0$ with an exponential growth of order at most k at infinity with respect to τ. Then, choosing a direction θ, the k-Laplace transform $\mathcal{L}_{k;\theta}^{[\tau]}(\widehat{u}_k)$ is well-defined and, from Proposition 7.15, defines an analytic function $u(t,x)$ on a domain $D_R\backslash\{0\} \times D_{r,\dots,r}$ with a convenient radius $R > 0$.

Since $u(t,x)$ is by construction the k-sum of $\widetilde{u}(t,x)$ in the direction θ (see Proposition 7.18), it is bounded when $|t|$ tends to 0 and, therefore, is analytic on $D_R \times D_{r,\dots,r}$ by the Removable Singularities Theorem. Consequently, for any $x \in D_{r,\dots,r}$, $\widetilde{u}(t,x)$ is the Taylor expansion at $t = 0$ of an analytic function and, thereby, coincides with $u(t,x)$ for all $(t,x) \in D_R \times D_{r,\dots,r}$, which ends the proof. ■

7.5 Some Applications to Partial Differential Equations

As in Chap. 6 devoted to the characterization of the summability in terms of the successive derivatives (see Sect. 6.2), we shall now illustrate our characterization of the summability by the Borel-Laplace method (Proposition 7.18) on some partial differential equations. Contrary to what we had previously obtained, this new approach will allow us to obtain a characterization of the summability of the formal series solutions, not in terms of the series itself, but in terms of the initial data.

Before stating various general results, let us start by detailing, as in Sects. 3.2 and 6.2, the case of the heat equation in order to introduce some theoretical and technical tools that can be used in this kind of study.

7.5.1 Example: The Heat Equation

Let us consider the homogeneous linear heat equation with constant coefficients

$$\begin{cases} \partial_t u - a\partial_x^2 u = 0 \\ u(0,x) = \varphi(x) \end{cases}, (t,x) \in \mathbb{C}^2, \tag{7.17}$$

where $a \in \mathbb{C}^*$ is a nonzero complex constant, and where $\varphi(x) \in \mathcal{O}(D_{\rho_1})$. The unique formal power series $\widetilde{u}(t,x) \in \mathcal{O}(D_{\rho_1})[[t]]$ solution of Eq. (7.17) is given by

$$\widetilde{u}(t,x) = \sum_{j\geqslant 0} \varphi^{(2j)}(x)\frac{(at)^j}{j!}, \tag{7.18}$$

where $\varphi^{(k)}$ stands for the kth derivative of φ.

In Sect. 6.2.1, Proposition 6.5, we proved a characterization of the 1-summability of $\widetilde{u}(t,x)$ in a given direction $\arg(t) = \theta$ by means of the two formal power series

$$\widetilde{u}_{*,0}(t) = \sum_{j\geqslant 0} \varphi^{(2j)}(0)\frac{(at)^j}{j!} \quad \text{and} \quad \widetilde{u}_{*,1}(t) = \sum_{j\geqslant 0} \varphi^{(2j+1)}(0)\frac{(at)^j}{j!}.$$

In the present section, we propose to prove another characterization of the 1-summability, this time in terms of the initial data $\varphi(x)$, by using the Borel-Laplace method. More precisely, we shall prove the following.

Proposition 7.21 *The formal solution $\widetilde{u}(t,x) \in \mathcal{O}(D_{\rho_1})[[t]]$ of the heat equation (7.17) is 1-summable in the direction θ if and only if the initial data $\varphi(x)$ can be analytically continued to sectors neighboring the directions $\frac{1}{2}(\theta + \arg(a)) \mod \pi$ with a global exponential growth of order at most 2 at infinity.*

Let us immediately observe that Proposition 7.21 allows us to recover the result of S. Kowalevskaya stating that there can not exist any locally analytic solutions of the heat equation (7.17) in the variable x.

Corollary 7.22 ([53]) *The formal solution $\widetilde{u}(t,x) \in \mathcal{O}(D_{\rho_1})[[t]]$ of the heat equation (7.17) is convergent if and only if the initial data $\varphi(x)$ is an entire function with an exponential growth of order at most 2 at infinity.*

Moreover, in this case, $\widetilde{u}(t,x)$ defines an analytic function on $D_R \times \mathbb{C}$ for a convenient radius $R > 0$.

Proof The sufficient condition and the domain of holomorphy are proved in Remark 3.11 and the necessary condition is straightforward from Propositions 6.2 and 7.21. ∎

Let us now turm to the proof of Proposition 7.21.

Proof of Proposition 7.21 The case $a = 1$ was proved by D. A. Lutz, M. Miyake and R. Schäfke in [67]. We extend below their proof to any nonzero constant $a \in \mathbb{C}^*$.

Let us set

$$\theta_a = \frac{1}{2}(\theta + \arg(a))$$

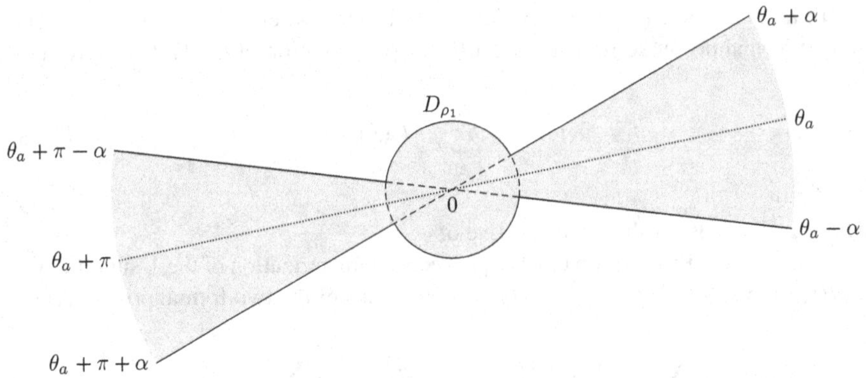

Fig. 7.12 The holomorphy domain $D_{\rho_1} \cup \mathcal{C}_{\theta_a-\alpha,\theta_a+\alpha}$ of $\varphi(x)$

and, for any directions $\theta_1 < \theta_2$, let us denote by $\mathcal{C}_{\theta_1,\theta_2}$ the double cone

$$\mathcal{C}_{\theta_1,\theta_2} = \Sigma_{\theta_1,\theta_2}(\infty) \cup \Sigma_{\theta_1+\pi,\theta_2+\pi}(\infty).$$

$\triangleleft$ *The* if *part.* By assumption, there exists $\alpha > 0$ such that $\varphi(x)$ is analytic on the domain $D_{\rho_1} \cup \mathcal{C}_{\theta_a-\alpha,\theta_a+\alpha}$ (see Fig. 7.12) with a global exponential growth of order at most 2 at infinity.

To prove that $\widetilde{u}(t,x)$ is 1-summable in the direction θ, we shall use the characterization of the summability given in Proposition 7.18. To do that, we consider the formal Borel transform

$$\widehat{u}(\tau,x) = \sum_{j\geqslant 0} \varphi^{(2j)}(x)\frac{(a\tau)^j}{(j!)^2} \quad , x \in D_{\rho_1}$$

of $\widetilde{u}(t,x)$ and we must prove the following two points:

- $\widehat{u}(\tau,x)$ is convergent at the origin of $\mathbb{C}^2$;
- the sum $v(\tau,x)$ of $\widehat{u}(\tau,x)$ can be analytically continued to a domain $\Sigma_{\theta_1,\theta_2}(\infty) \times D_r$ for some directions $\theta_1 < \theta < \theta_2$ and some radius $0 < r \leqslant \rho_1$ with a global exponential growth of order at most 1 at infinity with respect to τ.

Let us choose a radius $r \in]0, \rho_1/2[$. From the Cauchy Integral Formula, we can first write $\widehat{u}(\tau,x)$ in the form

$$\widehat{u}(\tau,x) = \frac{1}{2i\pi}\sum_{j\geqslant 0}\binom{2j}{j}(a\tau)^j \int_{|\xi|=r} \frac{\varphi(\xi+x)}{\xi^{2j+1}}d\xi \tag{7.19}$$

for all $x \in D_r$. Let us now observe that the following identity holds for all $|s| < 1/4$:

$$\sum_{j \geqslant 0} \binom{2j}{j} s^j = (1 - 4s)^{-1/2}.$$

Then, for all $|\xi| = r$, the series

$$\sum_{j \geqslant 0} \binom{2j}{j} \left(\frac{a\tau}{\xi^2}\right)^j$$

is normally convergent on all the compact sets of $D_{r^2/(4a)}$ and, consequently, we can permute the series and the integral in (7.19). Choosing in particular a determination of $\xi^{1/2}$ so that $(\xi^2)^{1/2} = \xi$, we get

$$\widehat{u}(\tau, x) = \frac{1}{2i\pi} \int_{|\xi|=r} \varphi(\xi + x)(\xi^2 - 4a\tau)^{-1/2} d\xi \quad (= \upsilon(\tau, x)) \qquad (7.20)$$

for all $(\tau, x) \in D_{r^2/(4a)} \times D_r$, which proves the convergence of $\widehat{u}(\tau, x)$ at the origin of $\mathbb{C}^2$.

Let us now prove that the sum $\upsilon(\tau, x)$ of $\widehat{u}(\tau, x)$ can be analytically continued to a domain $\Sigma_{\theta_1, \theta_2}(\infty) \times D_{r'}$ for some directions $\theta_1 < \theta < \theta_2$ and some radius $0 < r' \leqslant r$ with a global exponential growth of order at most 1 at infinity with respect to τ.

To do that, let us first observe that the function $\xi \longmapsto (\xi^2 - 4a\tau)^{1/2}$ with $\tau \in \mathbb{C}^*$ is univalent in the ξ-plane outside the segment joining the two points $\pm 2(a\tau)^{1/2}$. Therefore, thanks to the Cauchy Theorem, we can still deform the integration path $|\xi| = r$ in (7.20) into any simple, piecewise smooth curve Γ_τ surrounding this segment if necessary, provided that all the different translated curves $x + \Gamma_\tau$ with $x \in D_r$ are still contained in the holomorphy domain $D_{\rho_1} \cup \mathcal{C}_{\theta_a - \alpha, \theta_a + \alpha}$ of $\varphi(x)$.

Let us choose two directions θ_1 and θ_2 such that $\theta_1 < \theta < \theta_2$ and close enough to θ so that the sector $\Sigma_{\theta_1, \theta_2}(\infty)$ be contained in the set $\{s^2/(4a)$ such that $s \in \mathcal{C}_{\theta_a - \frac{\alpha}{3}, \theta_a + \frac{\alpha}{3}}\}$. By construction, we have $\pm 2(a\tau)^{1/2} \in \mathcal{C}_{\theta_a - \frac{\alpha}{3}, \theta_a + \frac{\alpha}{3}}$ for any $\tau \in \Sigma_{\theta_1, \theta_2}(\infty)$ and we can deform the integration path $|\xi| = r$ in (7.20) by the path Γ_τ consisting of the boundary of $D_r \cup \mathcal{C}_{\theta_a - \frac{\alpha}{2}, \theta_a + \frac{\alpha}{2}}$ cutting off at the points $\pm(2 |a\tau|^{1/2} + 1)e^{i\theta_a}$ as shown in Fig. 7.13.

Observe that all the translated curves $x + \Gamma_\tau$ are still contained in $D_{\rho_1} \cup \mathcal{C}_{\theta_a - \alpha, \theta_a + \alpha}$ for any $x \in D_r$ as soon as the radius $r \in]0, \rho_1/2[$ is small enough. Then, by these choices, the function $\upsilon(\tau, x)$ given by (7.20) can be analytically continued on the domain $\Sigma_{\theta_1, \theta_2}(\infty) \times D_r$.

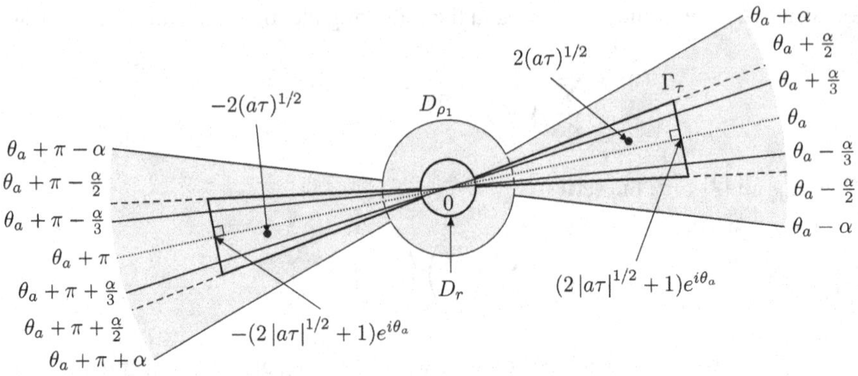

Fig. 7.13 The holomorphy domain of $\varphi(x)$ (gray area), the integration path Γ_τ (thick black line) and the double cone $C_{\theta_a - \frac{\alpha}{3}, \theta_a + \frac{\alpha}{3}}$

As for the exponential growth of $v(\tau, x)$, it stems both from the exponential growth of $\varphi(x)$ and from the fact that for all $x \in D_r$, all $\tau \in \Sigma_{\theta_1,\theta_2}(\infty)$ and all $\xi \in \Gamma_\tau$, we have

$$|\xi + x| \leqslant |\xi| + |x| \leqslant \max\left(r, (2\sqrt{|a\tau|} + 1)\left(1 + \tan\frac{\alpha}{2}\right)\right) + r$$

and

$$\left|\xi^2 - 4a\tau\right| \geqslant \delta$$

for some convenient constant $\delta > 0$. This completes the proof of the *if* part.

◁ *The* only if *part.* Let us now suppose that $\widetilde{u}(t, x)$ is 1-summable in the direction θ. From Proposition 7.18, we deduce then that there exist $\alpha > 0$ and two radii $R, r > 0$ such that the formal Borel transform $\widehat{u}(\tau, x)$ of $\widetilde{u}(t, x)$ with respect to t defines an analytic function in the domain $(D_R \cup \Sigma_{\theta-\alpha,\theta+\alpha}(\infty)) \times D_r$ with a global exponential growth of order at most 1 at infinity with respect to τ. Under this assumption, we must prove that the initial data $\varphi(x)$ can be analytically continued to sectors neighboring the directions θ_a and $\theta_a + \pi$ with a global exponential growth of order at most 2 at infinity.

First, we observe, due to Proposition 7.4, that $\widehat{u}(\tau, x)$ satisfies the Fuchs type partial differential equation

$$(\partial_\tau \tau \partial_\tau)\widehat{u}(\tau, x) = a\partial_x^2 \widehat{u}(\tau, x). \tag{7.21}$$

Therefore, we can see $\widehat{u}(\tau, x)$ as a solution of the Cauchy problem (7.21) together with the initial data

$$\widehat{u}(\tau, 0) = \psi_0(\tau) \quad \text{and} \quad \partial_x \widehat{u}(\tau, x)_{|x=0} = \psi_1(\tau)$$

with two convenient functions $\psi_0(\tau)$ and $\psi_1(\tau)$ which can always be assumed, thanks to our assumption on $\widehat{u}(\tau, x)$, analytic on $D_R \cup \Sigma_{\theta-\alpha,\theta+\alpha}(\infty)$ with a global exponential growth of order at most 1 at infinity (replacing if necessary R and α by $R' < R$ and $\alpha' < \alpha$). Solving this problem, we easily get

$$\widehat{u}(\tau, x) = \sum_{n\geqslant 0}(\partial_\tau \tau \partial_\tau)^n \psi_0(\tau)\frac{x^{2n}}{(2n)!a^n} + \sum_{n\geqslant 0}(\partial_\tau \tau \partial_\tau)^n \psi_1(\tau)\frac{x^{2n+1}}{(2n+1)!a^n}.$$

Let us now observe that $\varphi(x) = \widehat{u}(0, x)$ and

$$(\partial_\tau \tau \partial_\tau)^n \psi_i(\tau)_{|\tau=0} = n!\psi_i^{(n)}(0)$$

for all $i = 0, 1$. Then, choosing a radius $w \in]0, R[$ and a radius $0 < r' < \min(\rho_1, 2\sqrt{w\,|a|})$, and applying the Cauchy Integral Formula, we get

$$\begin{aligned}
\varphi(x) &= \sum_{n\geqslant 0} n!\psi_0^{(n)}(0)\frac{x^{2n}}{(2n)!a^n} + \sum_{n\geqslant 0} n!\psi_1^{(n)}(0)\frac{x^{2n+1}}{(2n+1)!a^n}\\
&= \frac{1}{2i\pi}\sum_{n\geqslant 0}\int_{|\xi|=w}\frac{n!n!}{(2n)!}\frac{\psi_0(\xi)x^{2n}}{a^n\xi^{n+1}}d\xi\\
&\qquad + \frac{1}{2i\pi}\sum_{n\geqslant 0}\int_{|\xi|=w}\frac{n!n!}{(2n+1)!}\frac{\psi_1(\xi)x^{2n+1}}{a^n\xi^{n+1}}d\xi;
\end{aligned}$$

hence,

$$\begin{aligned}
\varphi(x) &= \frac{1}{2i\pi}\int_{|\xi|=w} {}_2F_1\left(1, 1; \frac{1}{2}; \frac{x^2}{4a\xi}\right)\frac{\psi_0(\xi)}{\xi}d\xi\\
&\qquad + \frac{x}{2i\pi}\int_{|\xi|=w} {}_2F_1\left(1, 1; \frac{3}{2}; \frac{x^2}{4a\xi}\right)\frac{\psi_1(\xi)}{\xi}d\xi
\end{aligned} \tag{7.22}$$

for all $x \in D_{r'}$, where ${}_2F_1$ stands for the Gauss hypergeometric function which is given for all $|s| < 1$ by

$${}_2F_1(a_1, a_2; a_3; s) = \sum_{n\geqslant 0}\frac{(a_1)_n(a_2)_n}{(a_3)_n}\frac{s^n}{n!}$$

with $(a_i)_n = \dfrac{\Gamma(a_i+n)}{\Gamma(a_i)}$ for all $n \geqslant 0$. Observe that the permutation of the series and the integral is licit since the series is normally convergent on $D_{r'}$ for all $|\xi| = w$. Observe also that the functions ${}_2F_1\left(1, 1; \frac{1}{2}; s\right)$ and ${}_2F_1\left(1, 1; \frac{3}{2}; s\right)$ can be analytically continued on $\mathbb{C}\backslash[1, +\infty[$ and have possible singular points at

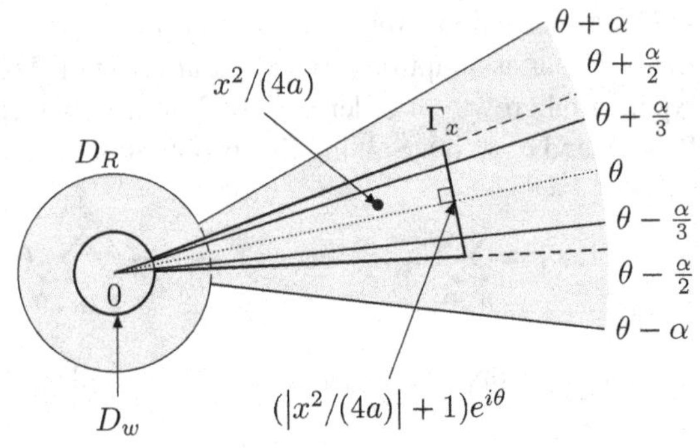

Fig. 7.14 The holomorphy domain of $\psi_0(\tau)$ and $\psi_1(\tau)$ (gray area), the integration path Γ_x (thick black line) and the sector $\Sigma_{\theta-\frac{\alpha}{3},\theta+\frac{\alpha}{3}}(\infty)$

$s = 1$ and $s = \infty$ which are both of regular type [121]. In particular, they satisfy a growth condition of polynomial type when s tends to 1 and s tends to ∞ in any sector with finite opening angle.

Let us now consider the double cone $\mathcal{C}_{\theta_a-\frac{\alpha}{6},\theta_a+\frac{\alpha}{6}}$. Since $x^2/(4a) \in \Sigma_{\theta-\frac{\alpha}{3},\theta+\frac{\alpha}{3}}(\infty)$ for any $x \in \mathcal{C}_{\theta_a-\frac{\alpha}{6},\theta_a+\frac{\alpha}{6}}$, we can deform the integration path $|\xi| = w$ in (7.22) by the path Γ_x consisting of the boundary of $D_w \cup \Sigma_{\theta-\frac{\alpha}{2},\theta+\frac{\alpha}{2}}(\infty)$ cutting off at the point $(|x^2/(4a)| + 1)e^{i\theta}$ as shown in Fig. 7.14, so that Γ_x be contained in the holomorphy domain of the ψ_i's and $x^2/(4a\xi) \notin [1, +\infty[$ for any $\xi \in \Gamma_x$.

Then, by this choice, the function $\varphi(x)$ given by (7.22) can be analytically continued to the domain $\mathcal{C}_{\theta_a-\frac{\alpha}{6},\theta_a+\frac{\alpha}{6}}$. Moreover, thanks to the growth condition of the ψ_i's and the inequality

$$|\xi| \leqslant \max\left(w, \left(\left|x^2/(4a)\right| + 1\right)\left(1 + \tan\frac{\alpha}{2}\right)\right),$$

it is clear that $\varphi(x)$ admits a global exponential growth of order at most 2 at infinity on $\mathcal{C}_{\theta_a-\frac{\alpha}{6},\theta_a+\frac{\alpha}{6}}$, which ends the proof of Proposition 7.21. ∎

Let us now conclude this section on the Borel-Laplace method by stating some more general results that can be obtained by adapting the above proof.

7.5.2 *Some More General Problems*

We start with a natural extension of the heat equation (7.17). Given two positive integers $p > \kappa \geqslant 1$ and a nonzero constant $a \in \mathbb{C}^*$, we consider the homogeneous linear equation

$$\begin{cases} \partial_t^\kappa u - a\partial_x^p u = 0 \\ u(0, x) = \varphi(x) \in \mathcal{O}(D_{\rho_1}) \\ \partial_t^j u(t, x)_{|t=0} = 0,\ j = 1, \ldots, \kappa - 1 \end{cases}, (t, x) \in \mathbb{C}^2. \tag{7.23}$$

Recall (see Eq. (6.18) page 84) that this equation involves in many physical, chemical, biological and ecological problems; for example: the heat equation (case $(\kappa, p) = (1, 2)$), the Euler-Lagrange equation (case $(\kappa, p) = (2, 4)$), etc.

The unique formal power series $\widetilde{u}(t, x) \in \mathcal{O}(D_{\rho_1})[[t]]$ solution of Eq. (7.23) is given by

$$\widetilde{u}(t, x) = \sum_{j \geqslant 0} \varphi^{(pj)}(x) \frac{a^j t^{\kappa j}}{(\kappa j)!}$$

and, from Theorem 3.16, we know that it is generically s-Gevrey with $s = p/\kappa - 1$ (indeed, the Newton polygon of the linear operator $\partial_t^\kappa - \partial_x^p$ admits a unique positive slope, the two end points being $(\kappa, -\kappa)$ and $(p, 0)$).

Then, by adapting the proof of Proposition 7.21, one can prove the following.

Proposition 7.23 *Let $\widetilde{u}(t, x) \in \mathcal{O}(D_{\rho_1})$ be the formal solution of Eq. (7.23) and $k = 1/s = \kappa/(p - \kappa)$. Then,*

1. *$\widetilde{u}(t, x)$ is k-summable in the direction θ if and only if the initial data $\varphi(x)$ can be analytically continued to sectors neighboring the directions $\dfrac{1}{p}(\kappa\theta + \arg(a))$ mod $(2\pi/p)$ with a global exponential growth of order at most $k + 1$ at infinity.*
2. *$\widetilde{u}(t, x)$ is convergent if and only if the initial data $\varphi(x)$ is an entire function with an exponential growth of order at most $k + 1$ at infinity. Moreover, in this case, $\widetilde{u}(t, x)$ defines an analytic function on $D_R \times \mathbb{C}$ for a convenient radius $R > 0$. In other words, Eq. (7.23) does not admit any locally analytic solutions in the variable x.*

The case $a = 1$ is due to M. Miyake and a proof can be found in [88]. Notice that Proposition 7.23 can also be proved by using the characterization of the k-summability via the successive derivatives (see for instance [103]).

As a second example of results that can be obtained using the Borel-Laplace method, we now consider the equation

$$\begin{cases} L(u) = 0, \quad L = \partial_t^\kappa - \displaystyle\sum_{i \in \mathcal{K}} \sum_{q \in Q_i} a_{i,q} t^{rq-i} \partial_t^{\kappa-i} \partial_x^{pq} \\ \partial_t^j u(t, x)_{|t=0} = 0, \; j = 0, \dots, \kappa - 2 \\ \partial_t^{\kappa-1} u(t, x)_{|t=0} = \varphi(x) \in \mathcal{O}(D_{\rho_1}) \end{cases}, \; (t, x) \in \mathbb{C}^2 \qquad (7.24)$$

studied by K. Ichinobe in [44], where

- (p, r) are two relatively prime positive integers;
- $\kappa \geqslant 1$ is a positive integer;
- $\mathcal{K}$ is a nonempty subset of $\{1, \dots, \kappa\}$;
- Q_i is a nonempty finite subset of $\{q \in \mathbb{N}\backslash\{0\}$ such that $rq \geqslant i\}$ for all $i \in \mathcal{K}$;
- $a_{i,q} \in \mathbb{C}^*$ is a nonzero constant for all $i \in \mathcal{K}$ and all $q \in Q_i$.

Besides, we assume that the Newton polygon

$$\mathcal{N}_t(L) = C(\kappa, -\kappa) \cup \bigcup_{i\in\mathcal{K}} \bigcup_{q\in Q_i} C(pq+\kappa-i, rq-\kappa)$$

of the operator L (see Definition 3.14) admits a unique positive slope $k > 0$, the two end points being $(\kappa, -\kappa)$ and $(pq^*+\kappa-i^*, rq^*-\kappa)$ with a convenient (unique) pair $(i^*, q^*) \in \mathcal{K} \times Q_{i^*}$. Thus, $k = rq^*/(pq^* - i^*)$ and, from Theorem 3.16, we know that the unique formal power series $\widetilde{u}(t,x) \in \mathcal{O}(D_{\rho_1})[[t]]$ solution of Eq. (7.24) is generically $1/k$-Gevrey.

As in the case of the ordinary differential equations (see for instance [65, Section 3.3.3.3]), we attach to the slope k of the Newton polygon $\mathcal{N}_t(L)$ the characteristic equation defined as the equation

$$\sum_{(i,q)\in\mathcal{S}_k} a_{i,q}\alpha^q = 1 \tag{7.25}$$

of unknown $\alpha \in \mathbb{C}$ and degree q^*, where $\mathcal{S}_k$ is the set

$$\mathcal{S}_k = \{(i,q) \in \mathcal{K} \times Q_i;\ 1 \leqslant q \leqslant q^* \text{ and } i/q = i^*/q^*\}$$

of all the pairs (i,q) whose the associated points $(pq+\kappa-i, rq-\kappa)$ belong to the edge of slope k.

Adapting then the proof of Proposition 7.21, one can prove the following.

Proposition 7.24 ([44]) *Let $\widetilde{u}(t,x) \in \mathcal{O}(D_{\rho_1})[[t]]$ be the formal solution of Eq. (7.24) and $k = rq^*/(pq^* - i^*)$.*

Let $\alpha_1, \ldots, \alpha_{q^*} \in \mathbb{C}$ be the roots of the characteristic equation (7.25).

Assume that, for a fixed direction θ, the initial data $\varphi(x)$ can be analytically continued to sectors neighboring the directions

$$\theta_{n,m} = \frac{1}{p}(r\theta - (\arg(\alpha_n) + 2\pi m)) \mod (2\pi)$$

for all $n = 1, \ldots, q^*$ and all $m = 0, \ldots, p-1$, with a global exponential growth of order at most kp/r at infinity.

Then, $\widetilde{u}(t,x)$ is k summable in all the direction $\theta + 2\pi\ell/r$ with $\ell = 0, \ldots, r-1$.

Other similar results on linear partial differential equations with variable coefficients can be found, for example, in [45] or as special cases of the papers [96, 125, 134] dealing more generally with multisummability. Note that for these last three references, the definitions of the Newton polygon and of the Borel-Laplace operators are significantly different from those we use here.

Let us end this section with a last example of using the Borel-Laplace method by considering the Burgers equation

$$\begin{cases} \partial_t u - \partial_x^2 u - 2u\partial_x u = 0 \\ u(0,x) = \varphi(x) \in \mathcal{O}(D_{\rho_1}) \end{cases} . \tag{7.26}$$

In Application 3.27, we saw that its unique formal solution $\widetilde{u}(t,x) \in \mathcal{O}(D_{\rho_1})[[t]]$ is generically 1-Gevrey and, in Application 6.13, we gave a characterization of the 1-summability of this solution in terms of the two formal power series $\widetilde{u}_{*,0}(t)$ and $\widetilde{u}_{*,1}(t)$.

Using the Borel-Laplace method and observing that the formal solutions of the heat and Burgers equations are closely related by the Cole-Hopf transformation[1] [43], one can deduce from Proposition 7.21 a new characterization of the 1-summability of the formal solution of Eq. (7.26) in terms of the initial data $\varphi(x)$. We will observe in particular in the statement below, due to G. Lysik, that the nonlinear character of the Burgers equation deeply modifies the nature of the analytic continuation of $\varphi(x)$ which is no longer holomorphic, but becomes meromorphic.

Proposition 7.25 ([68]) *Let $\widetilde{u}(t,x) \in \mathcal{O}(D_{\rho_1})[[t]]$ be the formal solution of Eq. (7.26) and $\theta \in \mathbb{R}/2\pi\mathbb{Z}$ a direction.*

1. *Assume that $\widetilde{u}(t,x)$ is* 1*-summable in the direction θ. Then, the initial data $\varphi(x)$ can be meromorphically continued to sectors neighboring the directions $\theta/2$ and $\theta/2+\pi$ with at most simple poles with positive integer residues.*
2. *Assume that the initial data $\varphi(x)$ can be extended on sectors neighboring the directions $\theta/2$ and $\theta/2+\pi$ to a meromorphic function of the form*

$$\varphi(x) = \frac{m}{x} + \sum_{n=1}^{+\infty}\left(\frac{1}{x-x_n} + \frac{1}{x_n} + \frac{x}{x_n^2}\right) + v(x)$$

where

- *$m \in \mathbb{N}$ is a nonnegative integer;*
- *the poles $x_n \in \mathbb{C}^*$ are nonzero and satisfy the following condition:*

$$\sum_{n\geqslant 1} \frac{1}{|x_n|^{2+\varepsilon}} < +\infty \quad \textit{for any } \varepsilon > 0;$$

- *$v(x)$ is analytic and satisfies the estimate $|v(x)| \leqslant a\,|x| + b$ for all x with some convenient $a, b < +\infty$.*

Then, $\widetilde{u}(t,x)$ is 1*-summable in the direction θ.*

---

[1] This transformation tells us that $u(t,x)$ is a solution of the heat equation if and only if its logarithmic derivative with respect to x is a solution of the Burgers equation.

Part III
Moment Summability

The characterization of the k-summability by the Borel-Laplace method introduced in Chap. 7 can be generalized by using, instead of the usual k-Borel and k-Laplace operators, integral operators whose kernels are functions more general than exponentials.

This new characterization (see Proposition 9.1), referred in the literature as *k-moment summability* and described by W. Balser in [5, 6, 8], is based on the key notion of *moment function* (see Definition 8.1), which allows to define two new families of integral operators: the *k-moment-Borel* and *k-moment-Laplace* operators (see Sect. 8.2). Thus, in the classical Borel-Laplace method, the two k-Borel and k-Laplace operators can be replaced by their moment analogues, which, through the choice of the moment function, gives great flexibility to the study of the k-summability of formal power series in $\mathcal{O}(D_{\rho_1,\ldots,\rho_n})[[t]]$. In particular, when these series are the formal solutions of certain partial differential equations, this approach can make it possible, with a suitable choice of moment function, to simplify the initial problem by making the calculations easier (see for instance the case of the heat equation developed in Sect. 9.2).

In addition to this great flexibility, the introduction of these moment functions also extends the range of possible applications by considering, instead of the standard derivation operators ∂_t and ∂_x, their moment analogues $\partial_{m;t}$ and $\partial_{m;x}$ (see Definition 10.2 for the exact definition of these operators), thus defining a new and wide class of functional equations, referred as *moment partial differential equations* which includes, for suitable choices of moment functions, a large number of classical equations such as, for example, the partial differential equations, the fractional partial differential equations, the q-difference equations, etc. Naturally, this new class of equations is highly suitable for applying the k-moment-summability method. However, given the very nature of the moment functions, the methods usually used in the classical partial differential equations (see the previous chapters) are largely modified and become, in general, much more technical. We refer to Chap. 10 below for some examples.

All along of this third part, we denote by $k > 0$ a positive real number and by $s = 1/k$ its inverse. For any direction $\theta \in \mathbb{R}/2\pi\mathbb{Z}$ and any $\alpha > 0$, we also denote by $\Sigma_{\theta,=\alpha}(\infty)$ the infinite open sector

$$\Sigma_{\theta,=\alpha}(\infty) = \Big\{t;\, \theta - \frac{\alpha}{2} < \arg(t) < \theta + \frac{\alpha}{2} \text{ and } |t| > 0\Big\}$$

bisected by θ and with opening α.

Chapter 8
Moment Functions and Moment Operators

In this section, we define some generalizations of the k-Borel and k-Laplace operators and we investigate some of their properties. To do that, let us first introduce pairs of functions serving as kernels for these operators, as well as their associated moment functions.

8.1 Kernel Functions and Moment Functions

Let us start with the case $s < 2$ (or, equivalently, $k > 1/2$).

Definition 8.1 (Kernel and Moment Functions of Small Order $s < 2$) A pair (e, E) of $\mathbb{C}$-valued functions is called *kernel functions of order $s < 2$* if the three following conditions hold:

1. The function e satisfies the following points:
   (a) e is holomorphic on the sector $\Sigma_{0,=\pi s}(\infty)$;
   (b) $e(t) > 0$ for all $t > 0$;
   (c) there exists $\alpha > 0$ such that, for any proper subsector $\Sigma \Subset \Sigma_{0,=\pi s}(\infty)$, there exist two positive constants $K_\Sigma, r_\Sigma > 0$ such that the following estimate holds for all $t \in \Sigma$ with $|t| < r_\Sigma$:

$$|e(t)| \leqslant K_\Sigma \, |t|^\alpha . \tag{8.1}$$

   (d) e is k-exponentially flat at infinity, that is, for any infinite proper subsector $\Sigma \Subset \Sigma_{0,=\pi s}(\infty)$, there exist two positive constants $A_\Sigma, C_\Sigma > 0$ such that the following estimate holds for all $t \in \Sigma$:

$$|e(t)| \leqslant C_\Sigma \exp\left(-A_\Sigma \, |t|^k\right). \tag{8.2}$$

P. Remy, *Asymptotic Expansions and Summability*, Lecture Notes in Mathematics 2351, https://doi.org/10.1007/978-3-031-59094-8_8

2. The function E satisfies the following points:

   (a) E is entire on $\mathbb{C}$;

   (b) E has a global exponential growth of order at most k at infinity, that is, there exist two positive constants $A', C' > 0$ such that the following estimate holds for all $t \in \mathbb{C}$:

$$|E(t)| \leqslant C' \exp\left(A' |t|^k\right); \tag{8.3}$$

   (c) there exists $\beta > 0$ such that, for any proper subsector $\Sigma \Subset \Sigma_{\pi,=\pi(2-s)}(\infty)$, there exist two positive constants $K'_\Sigma, R'_\Sigma > 0$ such that the following estimate holds for all $t \in \Sigma$ with $|t| > R'_\Sigma$:

$$|E(t)| \leqslant K'_\Sigma |t|^{-\beta}. \tag{8.4}$$

3. The functions e and E are connected by a corresponding *moment function m of order s* as follows:

   (a) the function m is defined by the Mellin transform of e:

$$m(\lambda) = \int_0^{+\infty} t^{\lambda-1} e(t)dt \quad \text{for all } \mathrm{Re}(\lambda) \geqslant 0; \tag{8.5}$$

   (b) the function E has the power series expansion

$$E(t) = \sum_{j \geqslant 0} \frac{t^j}{m(j)} \quad \text{for all } t \in \mathbb{C}. \tag{8.6}$$

Remark 8.2 Conditions (1)-(c) and (2)-(c) may be replaced by the following weaker conditions (see [6, 8]):

- the function $t \longmapsto t^{-1}e(t)$ is integrable at the origin on $\Sigma_{0,=\pi s}(\infty)$, meaning that the integral

$$\int_0^{t_0} t^{-1} \left|e(te^{i\theta})\right| dt$$

  exists for arbitrary $t_0 > 0$ and any direction $\theta \in]-\frac{\pi s}{2}, \frac{\pi s}{2}[$;
- the function $t \longmapsto t^{-1}E(1/t)$ is integrable at the origin on $\Sigma_{\pi,=\pi(2-s)}(\infty)$ in the above sense.

However, for what we have seen in this part, the conditions given in Definition 8.1 are sufficient.

The example below presents some kernel functions and their corresponding moment functions.

Example 8.3

1. Let $k > 1/2$ and $e(t) = kt^k \exp(-t^k)$. Then, the corresponding moment function m is given by $m(\lambda) = \Gamma(1 + s\lambda)$ and the function E is the Mittag-Leffler's function of index s:

$$E_s(t) = \sum_{j \geqslant 0} \frac{t^j}{\Gamma(1+sj)}.$$

Using well-known properties of this function (see Chap. 14), it is easy to see that all requirements of above (see Point 2 of Definition 8.1) are satisfied. Observe that for $k = 1$, we have $e(t) = te^{-t}$ and $E(t) = e^t$, and we find the kernel functions used in the definition of the classical Borel and Laplace operators (see Definition 7.1).

2. Slightly more generally, the function $e(t) = kt^{\alpha k} \exp(-t^k)$ with $k > 1/2$ and $\alpha > 0$ is also a kernel function of order $s < 2$. The corresponding moment function m is now given by $m(\lambda) = \Gamma(\alpha + s\lambda)$ and the function E is the generalized Mittag-Leffler's function $E_{s,\alpha}$ of indices (s, α):

$$E_{s,\alpha}(t) = \sum_{j \geqslant 0} \frac{t^j}{\Gamma(\alpha+sj)}.$$

Again, all the requirements of above are satisfied (see Chap. 14).

3. In the two examples above, the kernel function e of order s is obtained from that one of order 1 by the change of variable $t \mapsto t^k$. This generalizes to arbitrary kernels as follows. Let e be a kernel function of order $s < 2$ and m and E its corresponding moment and entire functions. Let $\alpha \in]0, 2/s[$. Then, the function $e(\cdot; \alpha)$ defined by $e(t; \alpha) = e(t^{1/\alpha})/\alpha$ is a kernel function of order $\alpha s < 2$; its corresponding moment function is $m(\alpha\lambda)$ and its corresponding entire function $E(\cdot; \alpha)$ is given by the integral

$$E(t; \alpha) = \frac{1}{2i\pi} \int_\gamma E(\tau) \frac{\tau^{\alpha-1}}{\tau^\alpha - t} d\tau,$$

where γ is a Hankel-type path as in Hankel's Formula for the inverse of the Gamma function, that is a path starting from infinity along the ray $\arg(\tau) = -\pi$, circling the origin in the counterclockwise and backing to infinity along the ray $\arg(\tau) = \pi$. Observe that, applying this result to the function $e(t) = t\exp(-t)$ (see the first example above), we have $E(t) = e^t$ and $e(t; 1/k) = kt^k \exp(-t^k)$, and we find here the integral representation of the Mittag-Leffler's function $E_s(t) = E(t; 1/k)$ (see Proposition 14.2).

4. For $\alpha > 1$, let us denote by C_α the entire function defined by

$$C_\alpha(t) = \frac{\alpha}{2i\pi} \int_\gamma \exp\left(\tau - t\tau^{1/\alpha}\right) d\left(\tau^{1/\alpha}\right),$$

where the integration path γ is as above. Using the identity

$$\exp\left(-t\tau^{1/\alpha}\right) = \sum_{j\geqslant 0} \frac{(-t)^j \tau^{j/\alpha}}{j!}$$

and the Hankel's Formula for the inverse of the Gamma function, we get the following power series expansion

$$C_\alpha(t) = \sum_{j\geqslant 0} \frac{(-t)^j}{j!\Gamma\left(1-(j+1)/\alpha\right)}$$

for all $t \in \mathbb{C}$ and we conclude that C_α has an exponential growth of order at most $\alpha/(\alpha-1)$ at infinity (see identity (7.2) page 103). Then, the function e defined by $e(t) = tC_\alpha(t)$ is a kernel function of order $1-1/\alpha$ and its corresponding moment function m is given by

$$m(\lambda) = \frac{\Gamma(1+\lambda)}{\Gamma(1+\lambda/\alpha)}.$$

More generally, as an application of example (3) above, let us choose $s_1 > s_2 > 0$ with $s_1 - s_2 < 2$, and let us define the function e_{s_1,s_2} by $e_{s_1,s_2}(t) = \frac{1}{s_1} t^{1/s_1} C_{s_1/s_2}(t^{1/s_1})$. Then, e_{s_1,s_2} is a kernel function of order $s_1 - s_2$ and its corresponding moment function m_{s_1,s_2} is given by

$$m_{s_1,s_2}(\lambda) = \frac{\Gamma(1+s_1\lambda)}{\Gamma(1+s_2\lambda)}.$$

In particular, for $s_1 = 2$ and $s_2 = 1$, we get

$$e_{2,1}(t) = \frac{1}{2} t^{1/2} C_{2,1}(t^{1/2}) = \frac{1}{2\sqrt{\pi}} t^{1/2} e^{-t/4},$$

$$m_{2,1}(\lambda) = \frac{\Gamma(1+2\lambda)}{\Gamma(1+\lambda)} \quad \text{and} \quad E_{2,1}(t) = \sum_{j\geqslant 0} \frac{j!}{(2j)!} t^j.$$

Observe that by the inverse Mellin transform and by (8.6), the moment function m uniquely determines the kernel functions e and E.

Observe also that the integral (8.5) being absolutely and locally uniformly convergent, the function m is holomorphic for $\mathrm{Re}(\lambda) > 0$ and continuous up to the imaginary axis, and the values $m(\lambda)$ are positive real numbers for all $\lambda \geqslant 0$. Moreover, thanks to the estimates (8.2) and (8.3) and the identities (8.5) and (8.6), there exist four positive constants $c, C, a, A > 0$ such that the following estimate holds for all $j \geqslant 0$:

$$ca^j \Gamma(1+sj) \leqslant m(j) \leqslant CA^j \Gamma(1+sj). \tag{8.7}$$

Definition 8.4 (s-Gevrey-Type Sequence) A sequence $(m(j))_{j\geqslant 0}$ of positive real numbers is called a *s-Gevrey-type sequence* if it satisfies an inequality of the form (8.7) for all $j \geqslant 0$.

Example 8.5

1. The sequence $(\Gamma(\alpha + sj))_{j\geqslant 0}$ associated with the moment function of Example (8.3), (2), is a s-Gevrey-type sequence.
2. The sequence $(\Gamma(\alpha+s_1 j)/\Gamma(1+s_2 j))_{j\geqslant 0}$ associated with the moment function of Example (8.3), (4), is a $(s_1 - s_2)$-Gevrey-type sequence. In particular, the sequence $((2j)!/j!)_{j\geqslant 0}$ is a 1-Gevrey-type sequence.

When $s \geqslant 2$ (or, equivalently, $k \leqslant 1/2$), the second condition of Definition 8.1 can not be satisfied since the set $\Sigma_{\pi,=\pi(2-s)}(\infty)$ is not defined. It means that we must define the kernel functions of high order $s \geqslant 2$ in another way.

Definition 8.6 (Kernel Function of Order $s \geqslant 2$) A function e is called *kernel function of order $s > 0$* if we can find a kernel function $\widetilde{e}$ of order $\widetilde{s} < 2$ so that

$$e(t) = \frac{\widetilde{s}}{s}\widetilde{e}\left(t^{\widetilde{s}/s}\right) \tag{8.8}$$

for all $t \in \Sigma_{0,=\pi s}(\infty)$. Note that the sector $\Sigma_{0,=\pi s}(\infty)$ may have opening larger than 2π, in which case the function e will have a branch point at the origin.

Observe that, thanks to Example 8.3, (3), we conclude that if a kernel function of *some* order $\widetilde{s} < 2$ exists so that relation (8.8) holds, then there exists one for *any* such $\widetilde{s}$. In particular, if s happens to be smaller than 2, then we can choose $\widetilde{s} = s$; hence, e is a kernel function in the sense of Definition 8.1. Moreover, to verify that e is a kernel function of order s, we may always assume that $\widetilde{s} = s/p$ for a sufficiently large $p \in \mathbb{N}$. This leads then us to the following characterization of a kernel function of order s.

Proposition 8.7 *Let $s > 0$. Then, a function e is a kernel function of order s if and only if the following two conditions hold:*

1. *The function e satisfies condition (1) of Definition 8.1.*
2. *For some $p \in \mathbb{N}$ with $s/p < 2$, the function $\widetilde{E}_p$ defined by*

$$\widetilde{E}_p(t) = \sum_{j\geqslant 0} \frac{t^j}{m(j/p)}$$

with m the corresponding moment function associated with e (see relation (8.5)) is entire on $\mathbb{C}$ and with a global exponential growth of order at most pk at infinity. Moreover, there exists $\beta > 0$ such that, for any proper subsector $\Sigma \Subset \Sigma_{\pi,=\pi(2-s/p)}(\infty)$, there exist two positive constants $K'_\Sigma, r'_\Sigma > 0$ such that the following estimate holds for all $t \in \Sigma$ with $|t| > r'_\Sigma$:

$$\left|\widetilde{E}_p(t)\right| \leqslant K'_\Sigma\, |t|^{-\beta}.$$

Remark 8.8 Thanks to Example 8.3, (3), the pair $(\widetilde{e}_p, \widetilde{E}_p)$ with $\widetilde{e}_p$ as in (8.8) is a pair of kernel functions of order $s/p < 2$ with the corresponding moment function $\widetilde{m}_p(\lambda) = m(\lambda/p)$. In particular, since the sequence $(\widetilde{m}_p(j))$ is a s/p-Gevrey-type sequence (see remark page 137), then the sequence $(m(j))$ is a s-Gevrey-type sequence and the function

$$E(t) = \sum_{j\geqslant 0} \frac{t^j}{m(j)}$$

is an entire function on $\mathbb{C}$ with a global exponential growth of order at most k at infinity. Observe also that, in the case $s < 2$, we can always choose $p = 1$, and we find the conditions given in Definition 8.1.

The example below presents some kernel functions of order $s > 0$ which generalize those given in previous Example 8.3.

Example 8.9

1. For any $k, \alpha > 0$, the function $e(t) = kt^{\alpha k}\exp(-t^k)$ is a kernel function of order s and its corresponding moment function m is given by $m(\lambda) = \Gamma(\alpha + s\lambda)$.
2. For any $s_1 > s_2 > 0$, the function $e_{s_1,s_2}(t) = \frac{1}{s_1} t^{1/s_1} C_{s_1/s_2}(t^{1/s_1})$ is a kernel function of order $s_1 - s_2$ and its corresponding moment function m_{s_1,s_2} is given by

$$m_{s_1,s_2}(\lambda) = \frac{\Gamma(1 + s_1\lambda)}{\Gamma(1 + s_2\lambda)}.$$

Let us now turn to the k-moment operators which generalize the k-Borel and k-Laplace operators.

8.2 k-Moment Operators

8.2.1 k-Moment-Laplace Operators

In this section, we consider a kernel function e of order $s > 0$ and its corresponding moment function m.

Definition 8.10 (Functional k-Moment-Laplace Transform in a Given Direction) The k-moment-Laplace transform $\mathcal{L}_{k;\theta}^{[m;\tau]}(v)$ of a function $v(\tau, x)$ with respect to τ in the direction θ is defined, when the integral exists, by

$$\mathcal{L}_{k;\theta}^{[m;\tau]}(v)(t, x) = \int_0^{\infty e^{i\theta}} v(\tau, x) e\left(\frac{\tau}{t}\right) \frac{d\tau}{\tau}.$$

Remark 8.11

- When e is the kernel function defined by $e(t) = kt^k \exp(-t^k)$, then the k-moment-Laplace transform coincides with the classical k-Laplace transform (see Definition 7.8, (2) page 99).
- When e is the kernel function $e_{s_1,s_2}(t) = \frac{1}{s_1} t^{1/s_1} C_{s_1/s_2}(t^{1/s_1})$ with $s_1 > s_2 > 0$ (see Example 8.3, (4)), then the operator $\mathcal{A}_{s_1,s_2;\theta}$ defined by

$$\begin{aligned}\mathcal{A}_{s_1,s_2;\theta}(v)(t, x) &= \mathcal{L}_{(s_1-s_2)^{-1};\theta}^{[m;\tau]}(v)(t, x)\\ &= t^{-1/s_1} \int_{\tau=0}^{\infty e^{i\theta}} v(\tau, x) C_{s_1/s_2}\left(\left(\frac{\tau}{t}\right)^{1/s_1}\right) d\left(\tau^{1/s_1}\right)\end{aligned}$$

is called *acceleration operator of indices* (s_1, s_2) *with respect to* τ *in the direction* θ. This operator, or to be precise one of a slightly modified form, has been introduced by J. Ecalle in [31–33] in a very general setting and plays a central role in the multisummability theory of the formal power series (see for instance [75], [6, Section 11] and [65, Section 7.6]).

The two following propositions extend to the k-moment-Laplace operators the results of Propositions 7.9 and 7.15 obtained for the k-Laplace operators.

Proposition 8.12 *Let* $\lambda \in \mathbb{C}$. *Assume that* $\lambda = 0$ *or* $\mathrm{Re}(\lambda) > 0$.
Then, $\mathcal{L}_{k;\theta}^{[m;\tau]}(\tau^\lambda)(t) = m(\lambda)t^\lambda$ *for any direction* θ.

Proof Use the change of variable $u = \tau/t$, then apply identity (8.5). ∎

In particular, this allows us to define the formal k-moment-Laplace transform of a formal power series $\widetilde{v}(\tau, x) \in \mathcal{O}(D_{\rho_1,\ldots,\rho_n})[[\tau]]$.

Definition 8.13 (Formal k-Moment-Laplace Transform) The formal k-moment-Laplace transform $\widetilde{\mathcal{L}}_k^{[m;\tau]}(\widetilde{v})$ of a formal power series

$$\widetilde{v}(\tau,x)=\sum_{j\geqslant 0} v_j(x)\tau^j \in \mathcal{O}(D_{\rho_1,\dots,\rho_n})[[\tau]]$$

with respect to τ is the formal power series

$$\widetilde{\mathcal{L}}_k^{[m;\tau]}(\widetilde{v})(t,x)=\sum_{j\geqslant 0} m(j)v_j(x)t^j \in \mathcal{O}(D_{\rho_1,\dots,\rho_n})[[t]].$$

Observe that, in the case where the kernel function e is given by $e(t) = kt^k\exp(-t^k)$, then $m(j) = \Gamma(1+sj)$ (see Example 8.3, (1)) and Definition 8.13 coincides with the classical definition of the formal k-Laplace transform $\widetilde{\mathcal{L}}_k^{[\tau]}$ (see Definition 7.10, (2)).

Proposition 8.14 *Let $v(\tau,x)$ be an analytic function at the origin of $\mathbb{C}^{n+1}$. Assume that $v(\tau,x)$ can be analytically continued to a domain $\Sigma_{\theta_1,\theta_2}(\infty)\times D_{r,\dots,r}$ for some directions $\theta_1<\theta_2$ and some radius $r>0$, with a global exponential growth of order at most k at infinity with respect to τ, that is, there exist two positive constants $A, C>0$ such that the following estimate holds for all $(\tau,x)\in\Sigma_{\theta_1,\theta_2}(\infty)\times D_{r,\dots,r}$:*

$$|v(\tau,x)|\leqslant C\exp\left(A\,|\tau|^k\right). \tag{8.9}$$

Let $\varepsilon\in]0,\pi s/2[$ and $A_\varepsilon, C_\varepsilon>0$ such that

$$|e(t)|\leqslant C_\varepsilon\exp\left(-A_\varepsilon\,|t|^k\right) \tag{8.10}$$

for all $\arg(t)\in]-\pi s/2+\varepsilon,\pi s/2-\varepsilon[$ (see condition (1)-(d) of Definition 8.1). For any direction $\theta\in]\theta_1,\theta_2[$, we denote by

- *$\mathcal{P}_{s;\theta}(A,\varepsilon,A_\varepsilon)$ the open sector*

$$\mathcal{P}_{s;\theta}(A,\varepsilon,A_\varepsilon)=\left\{t;\,0<|t|<\left(\frac{A_\varepsilon}{A}\right)^s \text{ and } |\theta-\arg(t)|<\frac{\pi s}{2}-\varepsilon\right\}$$

- *$u_{k;\theta}^{[m]}$ the k-moment-Laplace transform of v with respect to τ in the direction θ:*

$$u_{k;\theta}^{[m]}(t,x)=\mathcal{L}_{k;\theta}^{[m;\tau]}(v)(t,x)=\int_0^{\infty e^{i\theta}} v(\tau,x)e\left(\frac{\tau}{t}\right)\frac{d\tau}{\tau}.$$

Then, the functions $u_{k;\theta}^{[m]}(t,x)$ glue together into a function $u^{[m]}(t,x)$ defined and analytic on the domain $\bigcup_{\theta\in]\theta_1,\theta_2[}\mathcal{P}_{s;\theta}(A,\varepsilon,A_\varepsilon)\times D_{r,\dots,r}$.

Proof ◁ Let us first start by proving the holomorphy of $u_{k;\theta}^{[m]}(t,x)$ on $\mathcal{P}_{s;\theta}(A,\varepsilon,A_\varepsilon)\times D_{r,\dots,r}$. To do that, let us consider, for any small enough $b,\eta>0$, the subdomain $\mathcal{P}_{s;\theta}^{[b,\eta]}(A,\varepsilon,A_\varepsilon)$ of $\mathcal{P}_{s;\theta}(A,\varepsilon,A_\varepsilon)$ defined by

$$\mathcal{P}_{s;\theta}^{[b,\eta]}(A,\varepsilon,A_\varepsilon)=\left\{t;\,b<|t|<\left(\frac{A_\varepsilon}{A+\eta}\right)^s \text{ and } |\theta-\arg(t)|<\frac{\pi s}{2}-\varepsilon\right\}.$$

By assumption,

- the function $\upsilon(\tau,x)$ is analytic on $D_K\times D_{r,\dots,r}$ for some convenient radius $K>0$;
- there exists $\alpha, K_\varepsilon>0$ and a radius $r_\varepsilon\in]0,K/b[$ such that

$$|e(z)|\leqslant K_\varepsilon\,|z|^\alpha$$

  for all z such that $|z|<r_\varepsilon$ and $|\arg(z)|<\pi s/2-\varepsilon$ (see condition (1)-(c) of Definition 8.1).

Let $r'\in]0,r[$. From conditions above, the function $\upsilon(\tau,x)$ is bounded by a convenient positive constant $M_{\varepsilon,b,r'}$ on the closed polydisc $\overline{D}_{r_\varepsilon b}\times\overline{D}_{r',\dots,r'}$ and the following estimate holds for all $(\tau,t,x)\in]0,r_\varepsilon be^{i\theta}]\times\mathcal{P}_{s;\theta}^{[b,\eta]}(A,\varepsilon,A_\varepsilon)\times D_{r',\dots,r'}$:

$$\left|\frac{1}{\tau}\upsilon(\tau,x)e\left(\frac{\tau}{t}\right)\right|\leqslant\frac{K_\varepsilon M_{\varepsilon,b,r'}}{b}\,|\tau|^{\alpha-1}$$

with $\alpha-1>-1$. Then, Lebesgue's Dominated Convergence Theorem applies and implies that the integral

$$\int_0^{r_\varepsilon be^{i\theta}}\upsilon(\tau,x)e\left(\frac{\tau}{t}\right)\frac{d\tau}{\tau}$$

defined an analytic function on $\mathcal{P}_{s;\theta}^{[b,\eta]}(A,\varepsilon,A_\varepsilon)\times D_{r',\dots,r'}$.

On the other hand, from the estimates (8.9) and (8.10), we get also the estimate

$$\left|\frac{1}{\tau}\upsilon(\tau,x)e\left(\frac{\tau}{t}\right)\right|\leqslant\frac{CC_\varepsilon}{|\tau|}\exp\left(\left(A-\frac{A_\varepsilon}{|t|^k}\right)|\tau|^k\right)\leqslant\frac{CC_\varepsilon}{|\tau|}\exp\left(-\eta\,|\tau|^k\right)$$

for all $(\tau,t,x)\in[r_\varepsilon be^{i\theta},\infty e^{i\theta}[\times\mathcal{P}_{s;\theta}^{[b,\eta]}(A,\varepsilon,A_\varepsilon)\times D_{r',\dots,r'}$. Hence, applying again Lebesgue's Dominated Convergence Theorem, we deduce that the integral

$$\int_{r_\varepsilon be^{i\theta}}^{\infty e^{i\theta}}\upsilon(\tau,x)e\left(\frac{\tau}{t}\right)\frac{d\tau}{\tau}$$

defined also an analytic function on $\mathcal{P}_{s;\theta}^{[b,\eta]}(A,\varepsilon,A_\varepsilon)\times D_{r',\dots,r'}$.

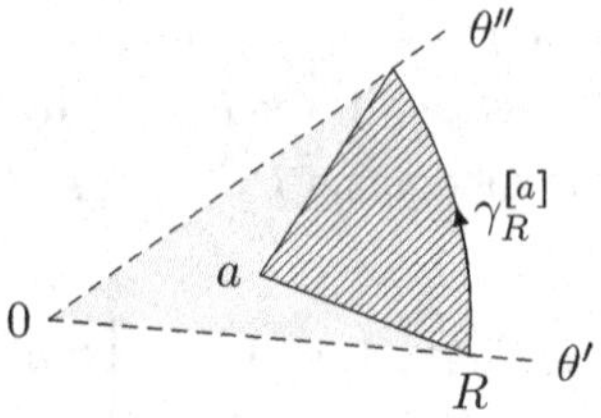

Fig. 8.1 The sector $\Sigma_{\theta',\theta''}(R)$ (gray area), the sector $\Sigma^{[a]}_{\theta',\theta''}(R)$ (hatched area) and its boundary $\gamma_R^{[a]}$

Consequently, the function $u^{[m]}_{k;\theta}(t,x)$ is analytic on $\mathcal{P}^{[b,\eta]}_{s;\theta}(A,\varepsilon,A_\varepsilon)\times D_{r',\dots,r'}$ for any small enough $b,\eta>0$ and any $r'\in]0,r[$; hence, analytic on $\mathcal{P}_{s;\theta}(A,\varepsilon,A_\varepsilon)\times D_{r,\dots,r}$ by making b, η and r' tend to 0, 0 and r respectively.

◁ Let us now prove that all the functions $u^{[m]}_{k;\theta}$ glue together into an analytic function on $\bigcup_{\theta\in]\theta_1,\theta_2[}\mathcal{P}_{s;\theta}(A,\varepsilon,A_\varepsilon)\times D_{r,\dots,r}$. To do that, let us consider two directions $\theta'<\theta''$ in $]\theta_1,\theta_2[$ with $\theta''-\theta'<\pi s-2\varepsilon$ so that the sectors $\mathcal{P}_{s;\theta'}(A,\varepsilon,A_\varepsilon)$ and $\mathcal{P}_{s;\theta''}(A,\varepsilon,A_\varepsilon)$ overlap. We denote by $\mathcal{P}_{s;\theta';\theta''}(A,\varepsilon,A_\varepsilon)$ their intersection:

$$\begin{aligned}&\mathcal{P}_{s;\theta';\theta''}(A,\varepsilon,A_\varepsilon)\\&\quad=\left\{t;\,0<|t|<\left(\frac{A_\varepsilon}{A}\right)^s\text{ and }\theta''-\frac{\pi s}{2}+\varepsilon<\arg(t)<\theta'+\frac{\pi s}{2}-\varepsilon\right\}.\end{aligned}$$

Let us also consider a sector $\Sigma_{\theta',\theta''}(R)$ of radius $R>0$ limited by the lines θ' and θ'', and let us define, for any $a\in\Sigma_{\theta',\theta''}(R)$ the sector $\Sigma^{[a]}_{\theta',\theta''}(R)$ with vertex a like shown on Fig. 8.1.

Applying the Cauchy Theorem to $v(\tau,x)e(\tau/t)/\tau$ along the boundary $\gamma_R^{[a]}$, oriented positively, of $\Sigma^{[a]}_{\theta',\theta''}(R)$ and making a tend to 0, we get

$$\int_0^{Re^{i\theta'}}+\int_{C_R}-\int_0^{Re^{i\theta''}}v(\tau,x)e\left(\frac{\tau}{t}\right)\frac{d\tau}{\tau}=0;$$

hence,

$$u^{[m]}_{k;\theta''}(t,x)-u^{[m]}_{k;\theta'}(t,x)=\lim_{R\to+\infty}\int_{C_R}v(\tau,x)e\left(\frac{\tau}{t}\right)\frac{d\tau}{\tau}$$

for any $(t,x)\in\mathcal{P}_{s;\theta';\theta''}(A,\varepsilon,A_\varepsilon)\times D_{r,\dots,r}$, where C_R stands for the arc of circle from $Re^{i\theta'}$ to $Re^{i\theta''}$.

We shall now prove that this limit is 0. Applying the estimates (8.9) and (8.10), we get

$$\begin{aligned}\left|\int_{C_R} v(\tau,x)e\left(\frac{\tau}{t}\right)\frac{d\tau}{\tau}\right| &\leqslant \int_{\theta'}^{\theta''}\left|v(Re^{i\theta},x)\right|\left|e\left(\frac{Re^{i\theta}}{t}\right)\right|d\theta\\ &\leqslant \int_{\theta'}^{\theta''} Ce^{AR^k}C_\varepsilon e^{-A_\varepsilon R^k/|t|^k}d\theta\\ &= (\theta''-\theta')CC_\varepsilon \exp\left(\left(A-\frac{A_\varepsilon}{|t|^k}\right)R^k\right)\end{aligned}$$

for all $(t,x)\in\mathcal{P}_{s;\theta';\theta''}(A,\varepsilon,A_\varepsilon)\times D_{r,\dots,r}$, and the integral tends to 0 as R tends to infinity since $|t|^k < A_\varepsilon/A$.

Consequently, the functions $u^{[m]}_{k;\theta'}(t,x)$ and $u^{[m]}_{k;\theta''}(t,x)$ coincide on $\mathcal{P}_{s;\theta';\theta''}(A,\varepsilon,A_\varepsilon)\times D_{r,\dots,r}$ and are thus analytic continuation of each other. The arbitrary choice of the two directions θ' and θ'' in $]\theta_1,\theta_2[$ completes the proof. ■

As direct consequences of Proposition 8.14, we can state in particular the two following results which will be useful in the sequel.

Corollary 8.15 *Let $v(\tau,x)$ be an analytic function at the origin of $\mathbb{C}^{n+1}$. Assume that $v(\tau,x)$ can be analytically continued to a domain $\Sigma_{\theta_1,\theta_2}(\infty)\times D_{r,\dots,r}$ for some directions $\theta_1<\theta_2$ and some radius $r>0$, with a global exponential growth of order at most k at infinity with respect to τ.*

Let $\theta\in]\theta_1,\theta_2[$. Then, the k-moment-Laplace transform $\mathcal{L}^{[m;\tau]}_{k;\theta}$ of v with respect to τ in the direction θ defines an analytic function on a domain of the form $\Sigma\times D_{r,\dots,r}$, where Σ is a sector bisected by θ and with opening larger than πs.

Proof Since $\theta\in]\theta_1,\theta_2[$, there exists $\eta>0$ such that the infinite sector $\Sigma_{\theta-\eta,\theta+\eta}(\infty)$ bisected by θ and with opening 2η be contained in $\Sigma_{\theta_1,\theta_2}(\infty)$. Applying then Proposition 8.14 to this sector, we deduce that $\mathcal{L}^{[m;\tau]}_{k;\theta}$ defines, for any $\varepsilon\in]0,\pi s/2[$, an analytic function on a domain of the form $\Sigma_\varepsilon\times D_{r,\dots,r}$, where Σ_ε is a sector bisected by θ and with opening $\pi s+2(\eta-\varepsilon)$. Corollary 8.15 follows by choosing ε in $]0,\min(\eta,\pi s/2)[$. ■

Corollary 8.16 *Let $v(\tau,x)$ be an analytic function on $\mathbb{C}\times D_{r,\dots,r}$ for some radius $r>0$, with a global exponential growth of order at most k at infinity with respect to τ.*

Let $\theta\in\mathbb{R}/2\pi\mathbb{Z}$. Then, the k-moment-Laplace transform $\mathcal{L}^{[m;\tau]}_{k;\theta}$ of v with respect to τ in the direction θ defines an analytic function at the origin of $\mathbb{C}^{n+1}$. More precisely, writing $v(\tau,x)$ in the form

$$v(\tau,x)=\sum_{j\geqslant 0} v_j(x)\tau^j$$

with $v_j(x) \in \mathcal{O}(D_{r,\dots,r})$ and $\tau \in \mathbb{C}$, we have

$$\mathcal{L}_{k;\theta}^{[m;\tau]}(v)(t,x) = \sum_{j\geqslant 0} m(j) v_j(x) t^j \in \mathcal{O}(D_\rho \times D_{r,\dots,r})$$

for some small enough radius $\rho > 0$.

Proof Corollary 8.16 is straightforward from Propositions 8.12 and 8.14. ∎

8.2.2 *k*-Moment-Borel Operators

◁ The Case $s \in]0, 2[$

Let us first consider a kernel function e of small order $s \in]0, 2[$ together with its corresponding moment function m and entire function E.

Definition 8.17 (Functional k-Moment-Borel Transform in a Given Direction) The k-moment-Borel transform $\mathcal{B}_{k;\theta}^{[m;t]}(u)$ of a function $u(t,x)$ with respect to t in the direction θ is defined, when the integral exists, by

$$\mathcal{B}_{k;\theta}^{[m;t]}(u)(\tau,x) = -\frac{1}{2i\pi}\int_{\gamma_{k;\theta}} u(t,x) E\left(\frac{\tau}{t}\right)\frac{dt}{t},$$

where the integration path $\gamma_{k;\theta}$ is as in the definition of the classical k-Borel transform (see Definition 7.8, (1) page 99).

When e is the kernel function defined by $e(t) = kt^k \exp(-t^k)$ with $k > 1/2$, we saw in the previous section that the k-Laplace and the k-moment-Laplace transforms coincide. Proposition 8.18 below tells us that the same occurs for the k-(moment)-Borel transforms. Recall that in this case, the corresponding entire function E associated with e is the Mittag-Leffler's function E_s of index s (see Example 8.3, (1)).

Proposition 8.18 *Let $\alpha > \pi s$ and $r > 0$. Let $u(t,x)$ be an analytic function on $\Sigma_{\theta,=\alpha} \times D_{r,\dots,r}$ and continuous at the origin of $\Sigma_{\theta,=\alpha}$ for all $x \in D_{r,\dots,r}$. Then, the k-Borel transform*

$$\mathcal{B}_{k;\theta}^{[t]}(u)(\tau,x) = \frac{1}{2i\pi}\int_{\gamma_{k;\theta}} t^k u(t,x) e^{(\tau/t)^k} d\left(\frac{1}{t^k}\right)$$

and the k-moment-Borel transform

$$\mathcal{B}_{k;\theta}^{[m;t]}(u)(\tau,x) = -\frac{1}{2i\pi}\int_{\gamma_{k;\theta}} u(t,x) E_s\left(\frac{\tau}{t}\right)\frac{dt}{t}$$

are well-defined, analytic on the domain $\Sigma_{\theta,=\alpha-\pi s}(\infty) \times D_{r,\dots,r}$, *and coincide on this domain. In other words,*

$$\mathcal{B}^{[t]}_{k;\theta}(u)(\tau, x) = -\frac{1}{2i\pi}\int_{\gamma_{k;\theta}} u(t, x)E_s\left(\frac{\tau}{t}\right)\frac{dt}{t}$$

for all $(\tau, x) \in \Sigma_{\theta,=\alpha-\pi s}(\infty) \times D_{r,\dots,r}$.

Proof ◁ Let us start by proving the analyticity of $\mathcal{B}^{[t]}_{k;\theta}(u)$. Fist of all, let us observe that, for any direction $\theta' \in]\theta-(\alpha-\pi s)/2, \theta+(\alpha-\pi s)/2[$, the sector $\Sigma_{\theta,=\alpha}$ contains a proper subsector $\Sigma_{\theta',=\pi s+\varepsilon}$ bisected by θ' and with opening $\pi s+\varepsilon > \pi s$ for some convenient $\varepsilon \in]0, \pi s/2[$ (depending, of course, of the direction θ').

Let us now fix such a direction θ' and let us choose for the path $\gamma_{k;\theta'}$ the boundary, oriented negatively, of the associated sector $\Sigma_{\theta',=\pi s+\varepsilon}$. Then, for all $\tau \in \Sigma_{\theta',=\varepsilon}(\infty)$ and all $t \in \gamma_{k;\theta'}$, we have

$$-\frac{\pi}{2} - k\varepsilon < \arg\left(\left(\frac{\tau}{t}\right)^k\right) < \frac{\pi}{2} + k\varepsilon,$$

and, therefore, the exponential function $\exp((\tau/t)^k)$ decreases along the two radial parts of $\gamma_{k;\theta'}$. Consequently, the integral

$$\mathcal{B}^{[t]}_{k;\theta'}(u)(\tau, x) = \frac{1}{2i\pi}\int_{\gamma_{k;\theta'}} t^k u(t, x)e^{(\tau/t)^k} d\left(\frac{1}{t^k}\right)$$

converges absolutely and compactly; hence, defines an analytic function on $\Sigma_{\theta',=\varepsilon}(\infty) \times D_{r,\dots,r}$.

We conclude with Cauchy's Theorem and the Analytic Continuation Theorem by observing that a change of the direction θ' in $]\theta - (\alpha - \pi s)/2, \theta + (\alpha - \pi s)/2[$ results in analytic continuation of $\mathcal{B}^{[t]}_{k;\theta'}(u)$.

In the same way, one can prove, thanks to the properties of the Mittag-Leffler's function E_s recalled in Chap. 14, that the k-moment-Borel transform $\mathcal{B}^{[m;t]}_{k;\theta}(u)$ also defines an analytic function on $\Sigma_{\theta,=\alpha-\pi s}(\infty) \times D_{r,\dots,r}$.

◁ Let us now prove that $\mathcal{B}^{[t]}_{k;\theta}(u)$ and $\mathcal{B}^{[m;t]}_{k;\theta}(u)$ coincide on $\Sigma_{\theta,=\alpha-\pi s}(\infty) \times D_{r,\dots,r}$. To do that, let us fix $\beta \in]\pi s, \alpha[$ and let us choose for the path $\gamma_{k;\theta}$ the boundary, oriented negatively, of a proper subsector of $\Sigma_{\theta,=\alpha}$ of the form $\Sigma_{\theta,=\pi s+\varepsilon}$ with some convenient small enough $\varepsilon > 0$ chosen so that $\beta + \varepsilon < \min(\alpha, 2\pi, 2\pi s)$. Denoting by R the radius of this proper subsector, we also define the subdomain $\Sigma_{\theta,\beta,R}$ of $\Sigma_{\theta,=\alpha-\pi s}(\infty)$ by

$$\Sigma_{\theta,\beta,R} = \left\{\tau;\, |\tau| > R \text{ and } |\theta - \arg(\tau)| < \frac{\beta - \pi s}{2}\right\}.$$

Doing that, the following estimates

$$\left|\frac{\tau}{t}\right| > 1 \quad \text{and} \quad \left|\arg\left(\frac{\tau}{t}\right)\right| < \frac{\beta+\varepsilon}{2}$$

hold for all $(\tau, t) \in \Sigma_{\theta,\beta,R} \times \gamma_{k;\theta}$, and we derive from Proposition 14.3 the identity

$$E_s(\tau/t) = k\exp((\tau/t)^k) + \widetilde{E}_s(\tau/t)$$

with $(\tau/t)\widetilde{E}_s(\tau/t)$ bounded for all $(\tau, t) \in \Sigma_{\theta,\beta,R} \times \gamma_{k;\theta}$.

Then, we can write

$$\mathcal{B}^{[t]}_{k;\theta}(u)(\tau, x) - \mathcal{B}^{[m;t]}_{k;\theta}(u)(\tau, x) = -\frac{1}{2i\pi}\int_{\gamma_{k;\theta}} u(t, x)\widetilde{E}_s\left(\frac{\tau}{t}\right)\frac{dt}{t}$$

for all $(\tau, x) \in \Sigma_{\theta,\beta,R} \times D_{r,\dots,r}$, where the integral is zero accordingly the Cauchy Integral Theorem. Consequently, $\mathcal{B}^{[t]}_{k;\theta}(u)$ and $\mathcal{B}^{[m;t]}_{k;\theta}(u)$ coincide on $\Sigma_{\theta,\beta,R} \times D_{r,\dots,r}$ and we conclude by the Analytic Continuation Theorem. ■

The following proposition extends to the k-moment-Borel operators the result of Proposition 7.9 obtained for the k-Borel operators.

Proposition 8.19 *Let $\lambda \in \mathbb{C}$. Assume that $\lambda = 0$ or $\mathrm{Re}(\lambda) > 0$. Then*

$$\mathcal{B}^{[m;t]}_{k;\theta}(t^\lambda)(\tau) = \frac{\tau^\lambda}{m(\lambda)}$$

for any direction θ. Moreover,

$$\frac{1}{m(\lambda)} = \frac{1}{2i\pi}\int_\gamma E(z)z^{-\lambda-1}dz,$$

where γ is a Hankel-type path as in Hankel's Formula for the inverse of the Gamma function, that is a path starting from infinity along the ray $\arg(\tau) = -\pi$, circling the origin in the counterclockwise and backing to infinity along the ray $\arg(\tau) = \pi$.

Proof Using the change of variable $z = \tau/t$ and Cauchy's Theorem to deform the path of integration, it is clear that

$$\mathcal{B}^{[m;t]}_{k;\theta}(t^\lambda)(\tau) = \tau^\lambda \times \frac{1}{2i\pi}\int_\gamma E(z)z^{-\lambda-1}dz, \tag{8.11}$$

that is $\mathcal{B}^{[m;t]}_{k;\theta}(t^\lambda)(\tau)$ is equal to τ^λ times a constant. In particular, we can apply to $\mathcal{B}^{[m;t]}_{k;\theta}(t^\lambda)$ the k-moment-Laplace transform $\mathcal{L}^{[m;\tau]}_{k;\theta}$. Interchanging the order of integration, we get

$$\mathcal{L}_{k;\theta}^{[m;\tau]}(\mathcal{B}_{k;\theta}^{[m;t]}(t^\lambda))(t) = -\frac{1}{2i\pi}\int_{\gamma_{k;\theta}}\left(\int_0^{\infty e^{i\theta}} E\left(\frac{\tau}{z}\right)e\left(\frac{\tau}{t}\right)\frac{d\tau}{\tau}\right)z^\lambda\frac{dz}{z}.$$

Let us now observe that, for all $(z,t) \neq (0,0)$ such that $|z/t|$ is small enough and $\arg(z) = \arg(t) = \theta$, the condition (3)-(b) of Definition 8.1 and Corollary 8.16 imply

$$\int_0^{\infty e^{i\theta}} E\left(\frac{\tau}{z}\right)e\left(\frac{\tau}{t}\right)\frac{d\tau}{\tau} = \sum_{j\geqslant 0} m(j)\frac{(t/z)^j}{m(j)} = \frac{z}{z-t}.$$

Moreover, this formula extends to any values $(z,t) \neq (0,0)$ such that $\arg(z) \neq \arg(t) \mod 2\pi$ since its both sides are well-defined. Indeed, we can always choose θ so that $|\theta - \arg(z)| < \pi s/2$ and $|\pi - \theta + \arg(t)| < \pi(2-s)$, implying the absolute convergence of the integral accordingly the properties of kernel functions.

Consequently,

$$\mathcal{L}_{k;\theta}^{[m;\tau]}(\mathcal{B}_{k;\theta}^{[m;t]}(t^\lambda))(t) = -\frac{1}{2i\pi}\int_{\gamma_{k;\theta}}\frac{z^\lambda}{z-t}dz$$

and, since $\gamma_{k;\theta}$ has a nonnegative orientation, we conclude that

$$\mathcal{L}_{k;\theta}^{[m;\tau]}(\mathcal{B}_{k;\theta}^{[m;t]}(t^\lambda))(t) = t^\lambda$$

from the Cauchy Integral Formula. Proposition 8.19 follows then from Proposition 8.12 by applying $\mathcal{L}_{k;\theta}^{[m;\tau]}$ to the both sides of (8.11). ∎

As in the previous section, Proposition 8.19 allows us in particular to define the formal k-moment-Borel transform of a formal power series $\widetilde{u}(t,x) \in \mathcal{O}(D_{\rho_1,\ldots,\rho_n})[[t]]$.

Definition 8.20 (Formal k-Moment-Borel Transform) The formal *k-moment-Borel transform* $\widetilde{\mathcal{B}}_k^{[m;t]}(\widetilde{u})$ of a formal power series

$$\widetilde{u}(t,x) = \sum_{j\geqslant 0} u_j(x)t^j \in \mathcal{O}(D_{\rho_1,\ldots,\rho_n})[[t]]$$

with respect to t is the formal power series

$$\widetilde{\mathcal{B}}_k^{[m;t]}(\widetilde{u})(\tau,x) = \sum_{j\geqslant 0}\frac{u_j(x)}{m(j)}\tau^j \in \mathcal{O}(D_{\rho_1,\ldots,\rho_n})[[\tau]].$$

Observe that, in the case where the kernel function e is given by $e(t) = kt^k \exp(-t^k)$, then $m(j) = \Gamma(1+sj)$ (see Example 8.3, (1)) and Definition 8.20 coincides with the classical definition of the formal k-Borel transform $\widetilde{\mathcal{B}}_k^{[t]}$ (see Definition 7.10, (1)).

Observe also that, the sequence $(m(j))$ being a s-Gevrey-type sequence (see remark page 137), all the results stated for the formal k-Borel transform $\widetilde{\mathcal{B}}_k^{[t]}$ in Sect. 7.2 can be obviously extended to the formal k-moment-Borel transform $\widetilde{\mathcal{B}}_k^{[m;t]}$. More precisely, Propositions 7.12 and 7.13, and Corollary 7.14 can be reformulated as follows.

Proposition 8.21 *A formal power series $\widetilde{u}(t,x) \in \mathcal{O}(D_{\rho_1,\ldots,\rho_n})[[t]]$ is s-Gevrey if and only if its formal k-moment-Borel transform $\widehat{u}_k^{[m]} = \widetilde{\mathcal{B}}_k^{[m;t]}(\widetilde{u})$ with respect to t is convergent at the origin of $\mathbb{C}^{n+1}$.*

Proposition 8.22 *Let $\widetilde{u}(t,x) \in \mathcal{O}(D_{\rho_1,\ldots,\rho_n})[[t]]$ be a s_1-Gevrey formal series with $s_1 < s$. Then, its formal k-moment-Borel transform $\widehat{u}_k^{[m]} = \widetilde{\mathcal{B}}_k^{[m;t]}(u)$ with respect to t defines an analytic function on $\mathbb{C} \times D_{r,\ldots,r}$ for some $0 < r \leqslant \min(\rho_1,\ldots,\rho_n)$ with global exponential growth of order at most $\kappa = (s-s_1)^{-1}$ at infinity with respect to τ, that is, there exist two positive constants $A, C > 0$ such that the following estimate holds for all $(\tau,x) \in \mathbb{C} \times D_{r,\ldots,r}$:*

$$\left|\widehat{u}_k^{[m]}(\tau,x)\right| \leqslant C \exp\left(A\,|\tau|^{\kappa}\right).$$

Corollary 8.23 *Let $a(t,x)$ be an analytic function on a polydisc $D_{\rho_0,\rho_1,\ldots,\rho_n}$. Then, its formal k-moment-Borel transform $\widehat{a}_k^{[m]} = \widetilde{\mathcal{B}}_k^{[m;t]}(a)$ with respect to t defines an analytic function on $\mathbb{C} \times D_{r,\ldots,r}$ for some $0 < r \leqslant \min(\rho_1,\ldots,\rho_n)$ with global exponential growth of order at most k at infinity with respect to τ, that is, there exist two positive constants $A, C > 0$ such that the following estimate holds for all $(\tau,x) \in \mathbb{C} \times D_{r,\ldots,r}$:*

$$\left|\widehat{a}_k^{[m]}(\tau,x)\right| \leqslant C \exp\left(A\,|\tau|^{k}\right).$$

Let us conclude with the following important result which will be very useful for characterizing the k-summability of a formal power series $\widetilde{u}(t,x) \in \mathcal{O}(D_{\rho_1,\ldots,\rho_n})[[t]]$ (see Sect. 9.1) and which can be seen as a reciprocal of Proposition 8.14 on the k-moment-Laplace operators.

Proposition 8.24 *Let $\theta \in \mathbb{R}/2\pi\mathbb{Z}$ be and $u(t,x)$ an analytic function on a domain $\Sigma_{\theta,>\pi s} \times D_{r,\ldots,r}$ for some radius $r > 0$. Assume that $u(t,x)$ is bounded on $\Sigma_{\theta,>\pi s} \times D_{r,\ldots,r}$, that is there exists a positive constant $C > 0$ such that the following estimate holds for all $(t,x) \in \Sigma_{\theta,>\pi s} \times D_{r,\ldots,r}$:*

$$|u(t,x)| \leqslant C.$$

Then, the k-moment-Borel transform $\mathcal{B}_{k;\theta}^{[m;t]}(u)$ of $u(t,x)$ with respect to t in the direction θ is well-defined and defines an analytic function on a domain $\Sigma_{\theta_1,\theta_2}(\infty) \times D_{r,\dots,r}$ for some directions $\theta_1 < \theta < \theta_2$, with an exponential growth of order at most k at infinity with respect to τ, that is, for any $R > 0$, there exist two positive constants $A_R, C_R > 0$ such that the following estimate holds for all $(\tau,x) \in \Sigma_{\theta_1,\theta_2}(\infty) \times D_{r,\dots,r}$ with $|\tau| > R$:

$$\left|\mathcal{B}_{k;\theta}^{[m;t]}(u)(\tau,x)\right| \leqslant C_R \exp\left(A_R |\tau|^k\right).$$

Proof From the Cauchy Theorem, we choose for the path $\gamma_{k;\theta}$ the boundary, oriented negatively, of a proper subsector $\Sigma \Subset \Sigma_{\theta,>\pi s}$, bisected by θ, with radius r_0, and with opening $\pi s + \varepsilon$ for some small enough $\varepsilon \in]0, 2\pi/s - \pi[$.

Let us now fix $\alpha \in]0, \varepsilon s/2[$ and let us prove that Proposition 8.24 holds with the choices $\theta_1 = \theta - \alpha$ and $\theta_2 = \theta + \alpha$. To do that, let us split the path $\gamma_{k;\theta}$ into three pieces: the two radial parts and the circular arc $\gamma_{k;\theta}^c$ of radius r_0, and let us estimate the resulting three integrals.

Let us start with the integral

$$\begin{aligned} I_1(\tau,x) &= -\frac{1}{2i\pi}\int_0^{r_0 e^{i(\theta+(\pi s+\varepsilon)/2)}} u(t,x)E\left(\frac{\tau}{t}\right)\frac{dt}{t} \\ &= -\frac{1}{2i\pi}\int_0^{r_0} u(te^{i(\theta+(\pi s+\varepsilon)/2)},x)E\left(\frac{\tau}{te^{i(\theta+(\pi s+\varepsilon)/2)}}\right)\frac{dt}{t} \end{aligned}$$

along the first radial part $[0, r_0 e^{i(\theta+(\pi s+\varepsilon)/2)}]$. Since $te^{i(\theta+(\pi s+\varepsilon)/2)} \in \Sigma$ and since

$$\arg\left(\frac{\tau}{t}\right) \in \left]-\frac{\pi s}{2} - \left(\alpha - \frac{\varepsilon s}{2}\right), -\frac{\pi s}{2} + \left(\alpha - \frac{\varepsilon s}{2}\right)\right[\subsetneq \left]-2\pi + \frac{\pi s}{2}, -\frac{\pi s}{2}\right[$$

for all $\tau \in \Sigma_{\theta-\alpha,\theta+\alpha}(\infty)$, we deduce from our assumption on $u(t,x)$ and from Condition (2)-(c) of Definition 8.1 that there exists $\beta > 0$ such that, for any $R > 0$, there exists a positive constant $K_R > 0$ such that

$$\left|\frac{1}{t}u(te^{i(\theta+(\pi s+\varepsilon)/2)},x)E\left(\frac{\tau}{te^{i(\theta+(\pi s+\varepsilon)/2)}}\right)\right| \leqslant K_R t^{\beta-1}|\tau|^{-\beta} \leqslant K_R R^{-\beta} t^{\beta-1}$$

for all $(\tau,t,x) \in \Sigma_{\theta-\alpha,\theta+\alpha}(>R) \times [0,r_0] \times D_{r,\dots,r}$, where

$$\Sigma_{\theta-\alpha,\theta+\alpha}(>R) = \{\tau \in \Sigma_{\theta-\alpha,\theta+\alpha}(\infty) \text{ such that } |\tau| > R\}.$$

Therefore, from the Lebesgue's Dominated Convergence Theorem, $I_1(\tau, x)$ defines an analytic function on $\Sigma_{\theta-\alpha,\theta+\alpha}(> R) \times D_{r,\dots,r}$. Moreover,

$$|I_1(\tau, x)| \leqslant \frac{K_R}{2\pi} R^{-\beta} \int_0^{r_0} t^{\beta-1} dt = \frac{K_R r_0^\beta}{2\beta\pi} R^{-\beta}$$

and $I_1(\tau, x)$ is bounded uniformly in $(\tau, x) \in \Sigma_{\theta-\alpha,\theta+\alpha}(> R) \times D_{r,\dots,r}$.

In the same way, the integral

$$I_3(\tau, x) = \frac{1}{2i\pi} \int_0^{r_0 e^{i(\theta-(\pi s+\varepsilon)/2)}} u(t, x) E\left(\frac{\tau}{t}\right) \frac{dt}{t}$$

along the second radial part $[r_0 e^{i(\theta-(\pi s+\varepsilon)/2)}, 0]$ defines also an analytic function on $\Sigma_{\theta-\alpha,\theta+\alpha}(> R) \times D_{r,\dots,r}$ for any $R > 0$, and satisfies on this domain an inequality of the form

$$|I_3(\tau, x)| \leqslant \frac{K'_R r_0^\beta}{2\beta\pi} R^{-\beta}$$

for some convenient positive constant $K'_R > 0$.

Let us now consider the integral

$$\begin{aligned} I_2(\tau, x) &= -\frac{1}{2i\pi} \int_{\gamma_{k;\theta}^c} u(t, x) E\left(\frac{\tau}{t}\right) \frac{dt}{t} \\ &= \frac{1}{2\pi} \int_{\theta-(\pi s+\varepsilon)/2)}^{\theta+(\pi s+\varepsilon)/2} u(r_0 e^{i\beta}, x) E\left(\frac{\tau}{r_0 e^{i\beta}}\right) d\beta \end{aligned}$$

along the circular arc $\gamma_{k;\theta}^c$ of $\gamma_{k;\theta}$. From our assumption on $u(t, x)$ and Condition (2)-(b) of Definition 8.1, we deduce that there exist two positive constants $K', A' > 0$ such that

$$\left| u(r_0 e^{i\beta}, x) E\left(\frac{\tau}{r_0 e^{i\beta}}\right) \right| \leqslant K' \exp\left(\frac{A'}{r_0^k} |\tau|^k\right) \leqslant K' \exp\left(\frac{A'}{r_0^k} R^k\right)$$

for all $(\tau, \beta, x) \in \Sigma_{\theta-\alpha,\theta+\alpha}(R) \times [\theta - (\pi s + \varepsilon)/2), \theta + (\pi s + \varepsilon)/2)] \times D_{r,\dots,r}$, where

$$\Sigma_{\theta-\alpha,\theta+\alpha}(R) = \{\tau \in \Sigma_{\theta-\alpha,\theta+\alpha}(\infty) \text{ such that } |\tau| < R\}.$$

Therefore, applying again the Lebesgue's Dominated Convergence Theorem, $I_2(\tau, x)$ defines an analytic function on $\Sigma_{\theta-\alpha,\theta+\alpha}(R) \times D_{r,\dots,r}$ for any $R > 0$;

hence, on $\Sigma_{\theta-\alpha,\theta+\alpha}(\infty) \times D_{r,\dots,r}$. Moreover, we have

$$|I_2(\tau, x)| \leqslant \frac{K'(\pi s + \varepsilon)}{2\pi} \exp\left(\frac{A'}{r_0^k} |\tau|^k\right)$$

for all $(\tau, x) \in \Sigma_{\theta-\alpha,\theta+\alpha}(\infty) \times D_{r,\dots,r}$.

Finally, by combining the results obtained above for each of the three pieces of $\gamma_{k;\theta}$, we deduce that $\mathcal{B}_{k;\theta}^{[m;t]}(u)(\tau, x)$ defines an analytic function on $\Sigma_{\theta-\alpha,\theta+\alpha}(\infty) \times D_{r,\dots,r}$. Moreover, denoting by ρ the radius of the sector $\Sigma_{\theta,>\pi s}$ and bounding r_0^β by ρ^β in the estimates of the integrals $I_1(\tau, x)$ and $I_3(\tau, x)$, we derive that, for all $R > 0$, there exists a positive constant $C_R > 0$ independent of r_0 such that the following estimate holds for all $(\tau, x) \in \Sigma_{\theta-\alpha,\theta+\alpha}(> R) \times D_{r,\dots,r}$:

$$\left|\mathcal{B}_{k;\theta}^{[m;t]}(u)(\tau, x)\right| \leqslant C_R \exp\left(\frac{A'}{r_0^k} |\tau|^k\right).$$

This completes the proof. ∎

◅ The Case $s \geqslant 2$

Let us now consider a kernel function e of high order $s \geqslant 2$ together with its corresponding moment function m.

Unlike the k-moment-Laplace transform which is valid for any value of s, it is not possible to define here the k-moment-Borel transform directly from the corresponding entire function E given by the relation (8.6) since this one does not have the same properties as for $s < 2$. Indeed, the second condition of Definition 8.1 fails since the set $\Sigma_{\pi,=\pi(2-s)}(\infty)$ does not exist.

To circumvent this difficulty, we proceed as in Proposition 8.7 by considering, for some integer $p \in \mathbb{N}$ satisfying $s/p < 2$, the function $\widetilde{E}_p$ defined by

$$\widetilde{E}_p(t) = \sum_{j \geqslant 0} \frac{t^j}{m(j/p)}.$$

Definition 8.25 The k-moment-Borel transform $\mathcal{B}_{k;\theta}^{[m;t]}(u)$ of a function $u(t, x)$ with respect to t in the direction θ is defined, when the integral exists, by

$$\mathcal{B}_{k;\theta}^{[m;t]}(u)(\tau, x) = -\frac{1}{2i\pi p} \int_{\gamma_{k;\theta}} u(t, x) \widetilde{E}_p\left(\left(\frac{\tau}{t}\right)^{1/p}\right) \frac{dt}{t}, \tag{8.12}$$

where the integration path $\gamma_{k;\theta}$ is as before.

Observe that Definition 8.25 makes sense since the integral in (8.12) does not depend on the choice of p (reason as in the proof of Proposition 8.18).

Observe also that, denoting by $\widetilde{m}_p$ the moment function associated with the pair of kernels $(\widetilde{e}_p, \widetilde{E}_p)$, we have $\mathcal{B}_{k;\theta}^{[m;t]} = \varrho_p \mathcal{B}_{kp;\theta/p}^{[\widetilde{m}_p;t]} \varrho_{1/p}$, where ϱ_p (resp. $\varrho_{1/p}$) stands for the ramification (resp. deramification) operator of order p introduced page 98. In particular, since $\widetilde{m}_p(\lambda) = m(\lambda/p)$ (see Remark 8.8), this proves that $\mathcal{B}_{k;\theta}^{[m;t]}(t^j)(\tau) = \tau^j/m(j)$ for any integer $j \in \mathbb{N}$ (see Proposition 8.19) and, consequently, Definition 8.20 of the formal k-moment-Borel transform $\widehat{\mathcal{B}}_k^{[m;t]}$ still valids for any $s \geqslant 2$. Moreover, the sequence $(m(j))$ being a s-Gevrey-type sequence, all the results stated in Propositions 8.21 and 8.22, and in Corollary 8.23 in the case $s \in]0, 2[$ still also valid for any $s \geqslant 2$.

Observe finally that, due to the properties of the kernel function $\widetilde{E}_p(t)$, one can also prove that the result of Proposition 8.24 remains valid when $s \geqslant 2$ (reason as in the case $s \in]0, 2[$ by splitting the path $\gamma_{k;\theta}$ into three pieces and applying to each of them the corresponding property of $\widetilde{E}_p(t)$).

8.3 Convolution of Kernels

In this section, we consider two kernel functions $e_1(t)$ and $e_2(t)$ of orders $s_1 > 0$ and $s_2 > 0$, and corresponding moment functions $m_1(\lambda)$ and $m_2(\lambda)$ respectively. As before, we set $k_1 = 1/s_1$ and $k_2 = 1/s_2$.

The two following propositions are concerned with the existence of kernels corresponding to the product (resp. quotient) of these two moment functions.

Proposition 8.26 (Product of Moment Functions) *The function e defined by*

$$e(t) = \int_0^{\infty e^{i\theta}} e_1\left(\frac{t}{\tau}\right) e_2(\tau)\frac{d\tau}{\tau} \tag{8.13}$$

for $|\theta| < \pi s_2/2$ and $|\theta - \arg(t)| < \pi s_1/2$, is a kernel function of order $s = s_1 + s_2$ with corresponding moment function $m(\lambda) = m_1(\lambda)m_2(\lambda)$.

Proof Let us first observe that identity (8.13) can be rewritten as

$$e\left(\frac{1}{t}\right) = \mathcal{L}_{k_1;\theta}^{[m_1,\tau]}\left(e_2\left(\frac{1}{\tau}\right)\right)(t).$$

Doing that, it is easy to verify that the function e defined in this way satisfies all the requirements given in Proposition 8.7. Note in particular that the function $\widetilde{E}_p$ can be obtained by applying the k_2-moment-Borel transform $\mathcal{B}_{k_2;\theta}^{[m_2,t]}$ to the entire function $\widetilde{E}_{1,p}$ associated with the kernel function e_1. The corresponding moment function

m is easily obtained by applying the Mellin transformation to the function e and interchanging the order of integration. ∎

Proposition 8.27 (Quotient of Moment Functions) *Assume $s_1 < s_2$. The function e defined by*

$$e(\tau) = \frac{1}{2i\pi p} \int_{\gamma_{k_1;0}^{\infty}} \widetilde{E}_{1,p}\left(\left(\frac{t}{\tau}\right)^{1/p}\right) e_2(t) \frac{dt}{t} \tag{8.14}$$

for $|\arg(\tau)| < \pi s/2$ and $\gamma_{k_1;0}^{\infty}$ a path starting from infinity to the origin along the ray $\arg(t) = -\pi s_2/2 + \varepsilon$ and backing to infinity along the ray $\arg(t) = \pi s_2/2 - \varepsilon$ with $\varepsilon > 0$ sufficiently small, is a kernel function of order $s = s_2 - s_1$ with corresponding moment function $m(\lambda) = m_2(\lambda)/m_1(\lambda)$.

Proof Let us now observe that identity (8.14) can be rewritten as

$$e\left(\frac{1}{\tau}\right) = \mathcal{B}_{k_1;0}^{[m_1,t]}\left(e_2\left(\frac{1}{t}\right)\right)(\tau).$$

The rest of the proof is similar to the one of Proposition 8.26 and is left to the reader. ∎

As an application of these two results, we can construct many more useful kernel functions. Example below provides such kernels.

Example 8.28 Let m be the function defined for $\text{Re}(\lambda) \geqslant 0$ by

$$m(\lambda) = \frac{\Gamma(\alpha_1 + s_1\lambda)\ldots\Gamma(\alpha_\nu + s_\nu\lambda)}{\Gamma(\beta_1 + \sigma_1\lambda)\ldots\Gamma(\beta_\mu + \sigma_\mu\lambda)}$$

with $\alpha_j, \beta_j, s_j, \sigma_j > 0$ and

$$s = \sum_{j=1}^{\nu} s_j - \sum_{j=1}^{\mu} \sigma_j > 0.$$

Then, according to Example 8.3, (2), and Propositions 8.26 and 8.27, m is the moment function associated with a kernel function e of order s which can be represented as multiple integrals involving exponentials. Moreover, the corresponding entire function E can be represented as multiple integrals involving Mittag-Leffler functions.

Observe that, for $s_j = \sigma_k = 1$ and $\nu = \mu + 1$, we have $s = 1$ and the function E is closely related to the generalized confluent hypergeometric function of order $\mu + 1$:

$$E(t) = \sum_{j \geqslant 0} \frac{\Gamma(1+j)\Gamma(\beta_1 + j)\dots\Gamma(\beta_\mu + j)}{\Gamma(\alpha_1 + j)\dots\Gamma(\alpha_{\mu+1} + j)} \frac{t^j}{j!}$$

$$= \frac{\Gamma(\alpha_1)\dots\Gamma(\alpha_{\mu+1})}{\Gamma(\beta_1)\dots\Gamma(\beta_\mu)} {}_{\mu+1}F_{\mu+1}(1, \beta_1, \dots, \beta_\mu; \alpha_1, \dots, \alpha_{\mu+1}; t).$$

Observe also that in the special case where $\nu = \mu = 1$ and $\alpha_1 = \beta_1 = 1$, we find the moment function m_{s_1,s_2} introduced in Example 8.3, (4).

For more details on the convolution of kernels and for other examples, we refer to [6, 8].

Chapter 9
Moment-Borel-Laplace Method and Summability

In this chapter, we are concerned with a generalization of the Borel-Laplace method in terms of the general moment-Borel and moment-Laplace operators. The resulting characterization of the k-summability (see Sect. 9.1) will thus provide an alternative tool to study the k-summability of the formal power series solutions of certain partial differential equations. Some applications are presented in Sect. 9.2.

9.1 Third Characterization of the k-Summability: The Moment-Borel-Laplace Method

Accordingly to the properties of the general k-moment-Borel and k-moment-Laplace operators described in the previous Chap. 8, the classical Borel-Laplace method (see Chap. 7, Proposition 7.18) can be rewritten as follows.

Proposition 9.1 (Moment-Borel-Laplace Method) *Let e be a kernel function of order $s > 0$ and m its corresponding moment function. Let $k = 1/s$.*

Then, a formal power series $\widetilde{u}(t, x) \in \mathcal{O}(D_{\rho_1,\dots,\rho_n})[[t]]$ is k-summable in the direction $\arg(t) = \theta$ if and only if the following two conditions are satisfied:

1. *the formal k-moment-Borel transform $\widehat{u}_k^{[m]}(\tau, x)$ of $\widetilde{u}(t, x)$ with respect to t is convergent at the origin of $\mathbb{C}^{n+1}$;*
2. *the sum $\upsilon_k^{[m]}(\tau, x)$ of $\widehat{u}_k^{[m]}(\tau, x)$ can be analytically continued to a domain $\Sigma_{\theta_1,\theta_2}(\infty) \times D_{r,\dots,r}$ for some directions $\theta_1 < \theta < \theta_2$ and some radius $0 < r \leqslant \min(\rho_1, \dots, \rho_n)$, with a global exponential growth of order at most k at infinity with respect to τ, that is, there exist two positive constants $A, C > 0$ such that the following estimate holds for all $(\tau, x) \in \Sigma_{\theta_1,\theta_2}(\infty) \times D_{r,\dots,r}$:*

$$\left|\upsilon_k^{[m]}(\tau, x)\right| \leqslant C \exp\left(A\,|\tau|^k\right).$$

P. Remy, *Asymptotic Expansions and Summability*, Lecture Notes in Mathematics 2351, https://doi.org/10.1007/978-3-031-59094-8_9

Moreover, the k-sum $u(t,x)$ of $\widetilde{u}(t,x)$ in the direction θ, if any exists, is given by the k-moment-Laplace transform of $v_k^{[m]}(\tau,x)$ with respect to τ in the direction θ:

$$u(t,x) = \mathcal{L}_{k;\theta}^{[m;\tau]}(v_k^{[m]})(t,x) = \int_0^{\infty e^{i\theta}} v_k^{[m]}(\tau,x) e\left(\frac{\tau}{t}\right)\frac{d\tau}{\tau}.$$

Proposition 9.1 can be proved analogously to Proposition 7.18 by using a variant of Nevanlinna's Theorem (Theorem 7.16). Here, we present a more direct approach using the general properties of the k-moment-Borel and k-moment-Laplace operators established in Propositions 8.14 and 8.24.

Proof of Proposition 9.1 Let $\widetilde{u}(t,x) = \sum_{j\geqslant 0} u_j(x)t^j \in \mathcal{O}(D_{\rho_1,\ldots,\rho_n})[[t]]$ be a formal power series and

$$\widehat{u}_k^{[m]}(\tau,x) = \sum_{j\geqslant 0} \frac{u_j(x)}{m(j)}\tau^j \in \mathcal{O}(D_{\rho_1,\ldots,\rho_n})[[\tau]]$$

its formal k-moment-Borel transform with respect to t.

◁ *The* if *part.* By assumption, there exists a radius $R > 0$ and a radius $0 < r' \leqslant r$ such that $\widehat{u}_k^{[m]}(\tau,x)$ converges normally on $D_R \times D_{r',\ldots,r'}$. Let us choose

- a positive real number $\varepsilon \in]0, \min(\pi s/4, (\theta_2-\theta_1)/2)[$;
- a positive real number $\eta \in]\varepsilon, \pi s/2-\varepsilon[$ such that $\theta_1 < \theta-\eta < \theta < \theta+\eta < \theta_2$.

According to the condition (1)-(d) of Definition 8.1 (see also Proposition 8.7), there exist two positive constants $A_\varepsilon, C_\varepsilon > 0$ such that

$$|e(z)| \leqslant C_\varepsilon \exp\left(-A_\varepsilon |z|^k\right)$$

for all $\arg(z) \in]-\pi s/2+\varepsilon, \pi s/2-\varepsilon[$.

Let $d \in \{\theta-\eta, \theta, \theta+\eta\}$. From Proposition 8.14, the k-moment-Laplace transform $\mathcal{L}_{k;d}^{[m;\tau]}(\widehat{u}_k^{[m]})$ of $\widehat{u}_k^{[m]}$ in the direction d defines an analytic function $u_d(t,x)$ on the domain $\mathcal{P}_{s;d}(A,\varepsilon,A_\varepsilon) \times D_{r',\ldots,r'}$, where $\mathcal{P}_{s;d}(A,\varepsilon,A_\varepsilon)$ is the open sector

$$\mathcal{P}_{s;d}(A,\varepsilon,A_\varepsilon) = \left\{t; 0 < |t| < \left(\frac{A_\varepsilon}{A}\right)^s \text{ and } |d-\arg(t)| < \frac{\pi s}{2}-\varepsilon\right\}.$$

Let us now choose $\ell > A$ and let us prove that $u_d(t,x)$ is s-Gevrey asymptotic to $\widetilde{u}(t,x)$ on the subsector

$$\mathcal{P}_{s;d}(\ell,\varepsilon,A_\varepsilon) = \left\{t; 0 < |t| < \left(\frac{A_\varepsilon}{\ell}\right)^s \text{ and } |d-\arg(t)| < \frac{\pi s}{2}-\varepsilon\right\}$$

of $\mathcal{P}_{s;d}(A, \varepsilon, A_\varepsilon)$. To do that, we shall proceed as in the proof of Nevanlinna's Theorem (see Theorem 7.16). Let us fix $b \in D_R$ with $\arg(b) = d$, and let us write $u_d(t, x)$ in the form

$$u_d(t,x) = u_d^b(t,x) + v_d^b(t,x)$$

with

$$\begin{aligned} u_d^b(t,x) &= \int_0^{|b|e^{id}} v_k^{[m]}(\tau, x) e\left(\frac{\tau}{t}\right) \frac{d\tau}{\tau} \\ &= \int_0^{|b|e^{id}} \left(\sum_{j \geqslant 0} \frac{u_j(x)}{m(j)} \tau^{j-1} e\left(\frac{\tau}{t}\right) \right) d\tau \quad \text{and} \\ v_d^b(t,x) &= \int_{|b|e^{id}}^{\infty e^{id}} v_k^{[m]}(\tau, x) e\left(\frac{\tau}{t}\right) \frac{d\tau}{\tau}. \end{aligned}$$

- Let us first prove that $u_d^b(t,x)$ is s-Gevrey asymptotic to $\widetilde{u}(t,x)$ on $\mathcal{P}_{s;d}(\ell, \varepsilon, A_\varepsilon)$. Let $J \geqslant 1$ and $(t,x) \in \mathcal{P}_{s;d}(\ell, \varepsilon, A_\varepsilon) \times D_{r',\dots,r'}$. From Proposition 8.12, we have the relation

$$t^j = \mathcal{L}_{k;d}^{[m;\tau]}\left(\frac{\tau^j}{m(j)}\right)(t) = \int_0^{\infty e^{id}} \frac{\tau^{j-1}}{m(j)} e\left(\frac{\tau}{t}\right) d\tau;$$

hence, the identities

$$\begin{aligned} u_d^b(t,x) - \sum_{j=0}^{J-1} u_j(x) t^j &= \int_0^{|b|e^{id}} \left(\sum_{j \geqslant 0} \frac{u_j(x)}{m(j)} \tau^{j-1} e\left(\frac{\tau}{t}\right) \right) d\tau \\ &\quad - \sum_{j=0}^{J-1} \frac{u_j(x)}{m(j)} \int_0^{\infty e^{id}} \tau^{j-1} e\left(\frac{\tau}{t}\right) d\tau \\ &= \int_0^{|b|e^{id}} \left(\sum_{j \geqslant J} \frac{u_j(x)}{m(j)} \tau^{j-1} e\left(\frac{\tau}{t}\right) \right) d\tau \\ &\quad - \sum_{j=0}^{J-1} \frac{u_j(x)}{m(j)} \int_{|b|e^{id}}^{\infty e^{id}} \tau^{j-1} e\left(\frac{\tau}{t}\right) d\tau. \end{aligned}$$

Since

$$t \in \mathcal{P}_{s;d}(\ell, \varepsilon, A_\varepsilon) \Rightarrow \left|\tau^{j-1} e\left(\frac{\tau}{t}\right)\right| \leqslant C_\varepsilon \, |\tau|^{j-1} \exp\left(-A_\varepsilon \left|\frac{\tau}{t}\right|^k\right) \leqslant C_\varepsilon \, |b|^{j-1}$$

for all $\tau \in [0, |b| \, e^{id}]$ and since $J \geqslant 1$, the series

$$\sum_{j \geqslant J} \frac{u_j(x)}{m(j)} \tau^{j-1} e\left(\frac{\tau}{t}\right)$$

converges normally on $[0, |b| \, e^{id}]$. Therefore, we can permute the sum and the integral, which brings us to the inequality

$$\begin{aligned} \left| u_d^b(t,x) - \sum_{j=0}^{J-1} u_j(x) t^j \right| &\leqslant \sum_{j \geqslant J} \frac{|u_j(x)|}{m(j)} \int_0^{|b|} \tau^{j-1} \left| e\left(\frac{\tau e^{id}}{t}\right)\right| d\tau \\ &\quad + \sum_{j=0}^{J-1} \frac{|u_j(x)|}{m(j)} \int_{|b|}^{+\infty} \tau^{j-1} \left| e\left(\frac{\tau e^{id}}{t}\right)\right| d\tau. \end{aligned}$$

Let us now observe that the inequalities $(\tau/\, |b|)^{j-1} \leqslant (\tau/\, |b|)^{J-1}$ hold both when $\tau \leqslant |b|$ and $j \geqslant J$ and when $\tau \geqslant |b|$ and $j < J$. Hence,

$$\begin{aligned} \left| u_d^b(t,x) - \sum_{j=0}^{J-1} u_j(x) t^j \right| &\leqslant \sum_{j \geqslant J} \frac{|u_j(x)|}{m(j)} |b|^{j-J} \int_0^{|b|} \tau^{J-1} \left| e\left(\frac{\tau e^{id}}{t}\right)\right| d\tau \\ &\quad + \sum_{j=0}^{J-1} \frac{|u_j(x)|}{m(j)} |b|^{j-J} \int_{|b|}^{+\infty} \tau^{J-1} \left| e\left(\frac{\tau e^{id}}{t}\right)\right| d\tau \\ &\leqslant \sum_{j \geqslant 0} \frac{|u_j(x)|}{m(j)} |b|^{j-J} \int_0^{+\infty} \tau^{J-1} \left| e\left(\frac{\tau e^{id}}{t}\right)\right| d\tau \\ &\leqslant C_\varepsilon \sum_{j \geqslant 0} \frac{|u_j(x)|}{m(j)} |b|^{j-J} \\ &\quad \times \int_0^{+\infty} \tau^{J-1} \exp\left(-A_\varepsilon \frac{\tau^k}{|t|^k}\right) d\tau. \end{aligned}$$

Setting then $\eta = A_\varepsilon \tau^k / |t|^k$, we obtain

$$\left| u_d^b(t,x) - \sum_{j=0}^{J-1} u_j(x) t^j \right| \leqslant C_\varepsilon \sum_{j\geqslant 0} \frac{|u_j(x)|}{m(j)} |b|^{j-J} \frac{|t|^J}{A_\varepsilon^{sJ}} \int_0^{+\infty} \eta^{sJ-1} e^{-\eta} d\eta$$

$$\leqslant C_\varepsilon \sum_{j\geqslant 0} \frac{|u_j(x)|}{m(j)} |b|^{j-J} \frac{|t|^J}{A_\varepsilon^{sJ}} \Gamma(sJ).$$

Since $\Gamma(1+sJ) = sJ\Gamma(sJ) \geqslant s\Gamma(sJ)$ and since the choice of b implies

$$\sum_{j\geqslant 0} \frac{|u_j(x)|\,|b|^j}{m(j)} \leqslant C'$$

for some convenient positive constant $C' > 0$ independent of J, t and x, we finally get

$$\left| u_d^b(t,x) - \sum_{j=0}^{J-1} u_j(x) t^j \right| \leqslant C'_d K'^J_d \Gamma(1+sJ) |t|^J,$$

with $C'_d = C_\varepsilon C'/s$ and $K'_d = 1/(|b|\, A_\varepsilon^s)$ independent of t and x.

- On the other hand, we can prove that $v_d^b(t,x)$ is k-exponentially flat on $\mathcal{P}_{s;d}(\ell, \varepsilon, A_\varepsilon)$. Indeed, parameterizing the half-line $[|b|\, e^{id}, \infty e^{id}[$ by $\tau = (|b| + \eta)e^{id}$ with $\eta \geqslant 0$, we get

$$v_d^b(t,x) = \int_0^{+\infty} v_k^{[m]}((|b|+\eta)e^{id}, x) e\left(\frac{(|b|+\eta)e^{id}}{t}\right) \frac{d\eta}{|b|+\eta};$$

hence, using the relations

$$A - \frac{A_\varepsilon}{|t|^k} < A - \ell < 0 \quad \text{and} \quad (|b|+\eta)^k \geqslant |b|^k + \eta^k,$$

the following inequalities hold for all $(t,x) \in \mathcal{P}_{s;d}(\ell, \varepsilon, A_\varepsilon) \times D_{r',\dots,r'}$:

$$\left| v_d^b(t,x) \right| \leqslant \frac{CC_\varepsilon}{|b|} \int_0^{+\infty} \exp\left(\left(A - \frac{A_\varepsilon}{|t|^k}\right)(|b|+\eta)^k\right) d\eta$$

$$\leqslant \frac{CC_\varepsilon}{|b|} \exp\left(\left(A - \frac{A_\varepsilon}{|t|^k}\right)|b|^k\right) \underbrace{\int_0^{+\infty} \exp\left((A-\ell)\eta^k\right) d\eta}_{<+\infty}$$

$$\leqslant \widetilde{C} \exp\left(-\frac{A_\varepsilon\, |b|^k}{|t|^k}\right)$$

with some positive constant $\widetilde{C} > 0$ independent of t and x. Using then the proof of Proposition 4.19, we deduce there exist two positive constants $C''_d, K''_d > 0$ such that

$$\left|v_d^b(t,x)\right| \leqslant C''_d K''^J_d J\Gamma(1+sJ)\,|t|^J$$

for all $(t,x) \in \mathcal{P}_{s;d}(\ell,\varepsilon,A_\varepsilon) \times D_{r',\dots,r'}$ and all $J \geqslant 1$.

- In conclusion, taking $C_d = \max(C'_d, C''_d)$ and $K_d = \max(K'_d, K''_d)$, we have proved that the following estimate holds for all $(t,x) \in \mathcal{P}_{s;d}(\ell,\varepsilon,A_\varepsilon) \times D_{r',\dots,r'}$ and all $J \geqslant 1$:

$$\left|u_d(t,x) - \sum_{j=0}^{J-1} u_j(x)t^j\right| \leqslant C_d K_d^J \Gamma(1+sJ)|t|^J.$$

Now, according to the choice of η, the intersection

$$\mathcal{P}_{s;\theta-\eta}(\ell,\varepsilon,A_\varepsilon) \cap \mathcal{P}_{s;\theta}(\ell,\varepsilon,A_\varepsilon) \cap \mathcal{P}_{s;\theta+\eta}(\ell,\varepsilon,A_\varepsilon)$$

is nonempty and, from Proposition 8.14, the functions $u_{\theta-\eta}(t,x)$, $u_\theta(t,x)$ and $u_{\theta+\eta}(t,x)$ glue together into an analytic function $u(t,x)$ on the domain $\Sigma \times D_{r',\dots,r'}$ with

$$\Sigma = \mathcal{P}_{s;\theta-\eta}(\ell,\varepsilon,A_\varepsilon) \cup \mathcal{P}_{s;\theta}(\ell,\varepsilon,A_\varepsilon) \cup \mathcal{P}_{s;\theta+\eta}(\ell,\varepsilon,A_\varepsilon).$$

Moreover, there exit two positive constants $B, K > 0$ such that

$$\left|u(t,x) - \sum_{j=0}^{J-1} u_j(x)t^j\right| \leqslant BK^J\Gamma(1+sJ)|t|^J \tag{9.1}$$

for all $(t,x) \in \Sigma \times D_{r',\dots,r'}$ and all $J \geqslant 1$. The sector Σ containing a sector $\Sigma_{\theta,>\pi s}$ bisected by θ and with opening larger than πs, inequalities (9.1) show in particular that $u(t,x)$ is the k-sum of $\widetilde{u}(t,x)$ in the direction θ in the sense of Definition 5.1. Hence, $\widetilde{u}(t,x)$ is k-summable in the direction θ and its sum in this direction is given, for instance, by the k-moment-Laplace transform $\mathcal{L}_{k;\theta}^{[m;\tau]}(v_k^{[m]})$.

$\lhd$ *The* only if *part.* Let us now assume that the formal power series $\widetilde{u}(t,x)$ is k-summable in the direction θ, that is, there exist a sector $\Sigma_{\theta,>\pi s}$ bisected by θ and with opening larger than πs, a radius $0 < r < \min(\rho_1,\dots,\rho_n)$ and an analytic function $u(t,x) \in \mathcal{O}(\Sigma_{\theta,>\pi s} \times D_{r,\dots,r})$ which is s-Gevrey asymptotic to $\widetilde{u}(t,x)$ at 0 on $\Sigma_{\theta,>\pi s}$ (see Definition 5.1).

Let $\Sigma' \Subset \Sigma \Subset \Sigma_{\theta,>\pi s}$ be two proper subsectors of $\Sigma_{\theta,>\pi s}$ bisected by θ and with opening larger than πs. By assumption (see Definition 4.1), there exist two positive constants $C_\Sigma, K_\Sigma > 0$ such that the following estimate holds for all $J \geqslant 1$

and all $(t, x) \in \Sigma \times D_{r,\ldots,r}$:

$$\left| u(t,x) - \sum_{j=0}^{J-1} u_j(x) t^j \right| \leqslant C_\Sigma K_\Sigma^J \Gamma(1+sJ) \, |t|^J \, . \tag{9.2}$$

In particular (see Proposition 4.3), we get

$$\left| u_j(x) \right| \leqslant C_\Sigma K_\Sigma^j \Gamma(1+sj)$$

for all $j \geqslant 1$ and all $x \in D_{r,\ldots,r}$; hence, since $(m(j))_{j\geqslant 0}$ is a s-Gevrey-type sequence (see remark page 137), the inequalities

$$\left| u_j(x) \right| \leqslant C'_\Sigma K'^j_\Sigma m(j)$$

for some convenient positive constants $C'_\Sigma, K'_\Sigma > 0$ independent of j and x. Consequently, the formal k-moment-Borel transform $\widehat{u}_k^{[m]}(\tau, x)$ of $\widetilde{u}(t, x)$ defines an analytic function $v_k^{[m]}(\tau, x)$ on $D_{1/K'_\Sigma} \times D_{r,\ldots,r}$, which proves the first point.

To prove the second point, let us first assume $k > 1/2$ (hence, $s \in]0, 2[$) and let us consider the k-moment-Borel transform

$$\mathcal{B}_{k;\theta}^{[m;t]}(u)(\tau, x) = -\frac{1}{2i\pi} \int_{\gamma_{k;\theta}} u(t,x) E\left(\frac{\tau}{t}\right) \frac{dt}{t}$$

of $u(t, x)$ with respect to t in the direction θ, where the path $\gamma_{k;\theta}$ is the boundary, oriented negatively, of a proper subsector of Σ', bisected by θ, with radius r_0, and with opening $\pi s + \varepsilon$ for some small enough $\varepsilon \in]0, 2\pi/s - \pi[$. Since inequality (9.2) implies that $u(t, x)$ is bounded on $\Sigma' \times D_{r,\ldots,r}$ (take $J = 1$ and use the fact that $r < \min(\rho_1, \ldots, \rho_n)$ implies $u_0(x)$ bounded on $D_{r,\ldots,r}$), Proposition 8.24 tells us that $\mathcal{B}_{k;\theta}^{[m;t]}(u)$ is well-defined and defines an analytic function $\Upsilon(\tau, x)$ on $\Sigma_{\theta-\alpha,\theta+\alpha}(\infty) \times D_{r,\ldots,r}$ for any $\alpha \in]0, \varepsilon s/2[$ with an exponential growth of order at most k at infinity with respect to τ: for any $R > 0$, there exist two positive constants $A_R, C_R > 0$ such that

$$|\Upsilon(\tau, x)| \leqslant C_R \exp\left(A_R \, |\tau|^k\right)$$

for all $(\tau, x) \in \Sigma_{\theta-\alpha,\theta+\alpha}(> R) \times D_{r,\ldots,r}$, where $\Sigma_{\theta-\alpha,\theta+\alpha}(> R)$ stands for the set of all the $\tau \in \Sigma_{\theta-\alpha,\theta+\alpha}(\infty)$ satisfying $|\tau| > R$.

Let us now observe that

$$\Upsilon(\tau, x) - \sum_{j=0}^{J-1} \frac{u_j(x)}{m(j)} \tau^j = -\frac{1}{2i\pi} \int_{\gamma_{k;\theta}} \left(u(t,x) - \sum_{j=0}^{J-1} u_j(x) t^j \right) E\left(\frac{\tau}{t}\right) \frac{dt}{t}$$

for all $(\tau, x) \in \Sigma_{\theta-\alpha,\theta+\alpha}(\infty) \times D_{r,\dots,r}$ and all $J \geqslant 1$. Adapting then the proof of Proposition 8.24 with the estimates (9.2), we deduce that, for any $R > 0$, there exist

- a positive constant $A > (K'_\Sigma/K_\Sigma)^k$ independent of R and r_0;
- a positive constant $C'_R > 0$ independent of r_0

such that

$$\left| \Upsilon(\tau, x) - \sum_{j=0}^{J-1} \frac{u_j(x)}{m(j)} \tau^j \right| \leqslant C'_R (K_\Sigma r_0)^J \Gamma(1 + sJ) \exp\left(A \frac{|\tau|^k}{r_0^k}\right) \tag{9.3}$$

for all $(\tau, x) \in \Sigma_{\theta-\alpha,\theta+\alpha}(> R) \times D_{r,\dots,r}$ and all $J \geqslant 1$.

Let us take $0 < R < \tau < 1/(A^s K_\Sigma)$, $x \in D_{r,\dots,r}$ and $J \geqslant 1$, and let us consider the right-hand side of (9.3) as a function of r_0. Since the function $y(r_0) = r_0^J e^{A\tau^k/r_0^k}$ for $r_0 > 0$ reaches its minimal value at $r'_0(J) = (A/(Js))^s \tau$ with $y(r'_0(J)) = A^{Js}(Js)^{-Js} e^{Js} \tau^J$, we choose $j_0 = j_0(\tau)$ so large that $r'_0(j_0) < r_0$. Thereby, for $J \geqslant j_0(\tau)$, we have also $r'_0(J) < r_0$ and, from the Cauchy Theorem, we can always assume that $r_0 = r'_0(J)$ in inequality (9.3). Consequently, these become

$$\left| \Upsilon(\tau, x) - \sum_{j=0}^{J-1} \frac{u_j(x)}{m(j)} \tau^j \right| \leqslant C'_R (A^s K_\Sigma \tau)^J \Gamma(1 + sJ)(Js)^{-Js} e^{Js},$$

with the right-hand side tending to 0 as J tends to infinity. Indeed, $0 < \tau < 1/(A^s K_\Sigma)$ and

$$\Gamma(1 + sJ)(Js)^{-Js} e^{Js} \underset{J \to +\infty}{\sim} \sqrt{2\pi s J}$$

from the Stirling's Formula.

From this, and since $1/(A^s K_\Sigma) < 1/K'_\Sigma$, we first deduce that

$$\Upsilon(\tau, x) = \sum_{j \geqslant 0} \frac{u_j(x)}{m(j)} \tau^j = \widehat{u}_k^{[m]}(\tau, x)$$

for all $(\tau, x) \in]R, 1/(A^s K_\Sigma)[\times D_{r,\dots,r}$ and, thereby, that $\Upsilon(\tau, x)$ and $\upsilon_k^{[m]}(\tau, x)$ are analytic continuations of each other. In particular, they both define the same analytic function on $(D_{1/K'_\Sigma} \cup \Sigma_{\theta-\alpha,\theta+\alpha}(\infty)) \times D_{r,\dots,r}$. Taking then two radii $R' \in]R, 1/K'_\Sigma[$ and $r' \in]0, r[$, and observing that $\upsilon_k^{[m]}(\tau, x)$ is bounded on $(\Sigma_{\theta-\alpha,\theta+\alpha}(\infty) \cap D_{R'}) \times D_{r',\dots,r'}$, we finally obtain that $\upsilon_k^{[m]}(\tau, x)$ can be analytically continued to the domain $\Sigma_{\theta-\alpha,\theta+\alpha}(\infty) \times D_{r',\dots,r'}$ with a global exponential growth of order at most k at infinity with respect to τ, which proves the second point in the case $k > 1/2$.

When $k \leqslant 1/2$ (hence, $s \geqslant 2$), we proceed in a similar way by using the ramification $t \longmapsto t^{1/p}$ for some convenient integer $p \geqslant 1$ such that $s/p < 2$ and by replacing the kernel function E by $\widetilde{E}_p$ (see Definition 8.25).

This completes the proof of Proposition 9.1. ■

9.2 Some Applications to Partial Differential Equations

As we said at the beginning of Part III page 131, the moment-Borel-Laplace method offers a new and flexible tool to study the k-summability of a formal power series by considering, not its formal k-Borel transform as we did in Sect. 7.4 (see Proposition 7.18), but its formal k-moment-Borel transform associated with a convenient moment function. In the case of formal solutions of some partial differential equations, this approach may be more efficient than the classical Borel-Laplace method. Indeed, with a suitable choice of the moment function, it may be easier to obtain the information sought on the formal moment-Borel transform than on the formal Borel transform.

As an illustration, let us consider the homogeneous linear heat equation with constant coefficients

$$\begin{cases} \partial_t u - a\partial_x^2 u = 0 \\ u(0,x) = \varphi(x) \end{cases}, (t,x) \in \mathbb{C}^2, \tag{9.4}$$

where $a \in \mathbb{C}^*$ is a nonzero complex constant and $\varphi(x) \in \mathcal{O}(D_{\rho_1})$, together with its unique formal power series solution

$$\widetilde{u}(t,x) = \sum_{j \geqslant 0} \varphi^{(2j)}(x) \frac{(at)^j}{j!} \in \mathcal{O}(D_{\rho_1})[[t]], \tag{9.5}$$

where $\varphi^{(k)}$ stands for the kth derivative of φ.

From the Borel-Laplace method, we proved in Sect. 7.5.1, Proposition 7.21, the following characterization for the 1-summability of $\widetilde{u}(t,x)$.

Proposition 9.2 *The formal solution $\widetilde{u}(t,x)$ of Eq. (9.4) is 1-summable in the direction θ if and only if the initial data $\varphi(x)$ can be analytically continued to sectors neighboring the directions $\frac{1}{2}(\theta + \arg(a)) \mod \pi$ with a global exponential growth of order at most 2 at infinity.*

We propose here to re-prove this characterization by using, no longer the formal Borel transform $\widetilde{\mathcal{B}}_1^{[t]}(\widetilde{u})$, but the formal moment-Borel transform $\widetilde{\mathcal{B}}_1^{[m;t]}(\widetilde{u})$, where

m is the moment function

$$m(\lambda) = \frac{\Gamma(1+2\lambda)}{\Gamma(1+\lambda)}$$

associated with the kernel function $e_{2,1}(t) = \frac{1}{2\sqrt{\pi}} t^{1/2} e^{-t/4}$ of order 1 (see Example 8.3, (4)). This choice is motivated by the following observation: the formal power series

$$\widetilde{\omega}(\tau, x) = \widetilde{\mathcal{B}}_1^{[m;t]}(\widetilde{u})(\tau^2, x) = \sum_{j \geqslant 0} \varphi^{(2j)}(x) \frac{a^j \tau^{2j}}{(2j)!} \in \mathcal{O}(D_{\rho_1})[[\tau]]$$

is the unique formal power series solution of the homogeneous wave equation

$$\begin{cases} \partial_\tau^2 \omega - a \partial_x^2 \omega = 0 \\ \omega(0, x) = \varphi(x) \\ \partial_\tau \omega(\tau, x)_{|\tau=0} = 0 \end{cases}, (\tau, x) \in \mathbb{C}^2. \tag{9.6}$$

In particular (see Application 3.25), $\widetilde{\omega}(\tau, x)$ defines an analytic function at the origin of $\mathbb{C}^2$ and, from the d'Alembert Formula, its sum $\omega(\tau, x)$ is given by

$$\omega(\tau, x) = \frac{1}{2}\left(\varphi(x + a^{1/2}\tau) + \varphi(x - a^{1/2}\tau)\right) \tag{9.7}$$

for all $(\tau, x) \in D_{r_0, r_1}$ for some convenient radii $r_0, r_1 > 0$ such that $r_1 + r_0\sqrt{|a|} < \rho_1$.

Let us now turn to the proof of Proposition 9.2 (compare it with the proof of Proposition 7.21).

Proof of Proposition 9.2 The case $a = 1$ was proved by S. Michalik in [76]. We extend below its proof to any nonzero constant $a \in \mathbb{C}^*$.

As in the proof of Proposition 7.21, let us set

$$\theta_a = \frac{1}{2}(\theta + \arg(a))$$

and, for any directions $\theta_1 < \theta_2$, let us denote by $\mathcal{C}_{\theta_1, \theta_2}$ the double cone

$$\mathcal{C}_{\theta_1, \theta_2} = \Sigma_{\theta_1, \theta_2}(\infty) \cup \Sigma_{\theta_1+\pi, \theta_2+\pi}(\infty).$$

◁ *The* if *part.* By assumption, there exists $\alpha > 0$ such that $\varphi(x)$ is analytic on the domain $D_{\rho_1} \cup \mathcal{C}_{\theta_a - \alpha, \theta_a + \alpha}$ (see Fig. 7.12 page 122) with a global exponential growth of order at most 2 at infinity.

To prove that $\widetilde{u}(t,x)$ is 1-summable in the direction θ, it is sufficient, according to the characterization of the summability given in Proposition 9.1, to prove that the function $\omega(\tau,x)$ can be analytically continued to a domain $\mathcal{C}_{\theta-\beta,\theta+\beta}\times D_r$ for some small enough $\beta>0$ and some radius $0<r\leqslant\sqrt{\rho_1}$ with a global exponential growth of order at most 2 at infinity with respect to τ, which is obvious thanks to identity (9.7) and our assumption on φ (choose $\beta,r>0$ small enough so that $x\pm a^{1/2}\tau\in D_{\rho_1}\cup\mathcal{C}_{\theta_a-\alpha,\theta_a+\alpha}$ for all $(\tau,x)\in\mathcal{C}_{\theta-\beta,\theta+\beta}\times D_r$).

$\lhd$ *The* only if *part.* Let us now suppose that $\widetilde{u}(t,x)$ is 1-summable in the direction θ. From Proposition 9.1 and the observation above, we deduce that there exist $\alpha>0$ and two radii $R,r>0$ such that the function $\omega(\tau,x)$ is analytic in the domain $(D_R\cup\mathcal{C}_{\theta-\alpha,\theta+\alpha})\times D_r$ with a global exponential growth of order at most 2 at infinity with respect to τ. Under this assumption, we must prove that the initial data $\varphi(x)$ can be analytically continued to sectors neighboring the directions θ_a and $\theta_a+\pi$ with a global exponential growth of order at most 2 at infinity.

Since $\omega(\tau,x)$ is a solution of Eq. (9.6), we can see it as a solution of the Cauchy problem

$$\begin{cases}\partial_\tau^2\omega-a\partial_x^2\omega=0\\ \omega(\tau,0)=\psi_0(\tau)\\ \partial_x\omega(\tau,x)_{|x=0}=\psi_1(\tau)\end{cases},\ (\tau,x)\in\mathbb{C}^2.$$

with two convenient functions $\psi_0(\tau)$ and $\psi_1(\tau)$ which can always be assumed, thanks to our assumption on $\omega(\tau,x)$, analytic on $D_R\cup\mathcal{C}_{\theta-\alpha,\theta+\alpha}$ with a global exponential growth of order at most 2 at infinity (replacing if necessary R and α by $R'<R$ and $\alpha'<\alpha$).

Let us now choose two radii $r_0'\in]0,R[$ and $r_1'\in]0,\min(r,\rho_1)[$ small enough so that $r_0'+r_1'\,|a|^{-1/2}<R$. From the d'Alembert Formula, we get

$$\omega(\tau,x)=\frac{1}{2}\Big(\psi_0(\tau+a^{-1/2}x)+\psi_0(\tau-a^{-1/2}x)\Big)+\frac{a^{1/2}}{2}\int_{\tau-a^{-1/2}x}^{\tau+a^{-1/2}x}\psi_1(\eta)d\eta$$

for all $(\tau,x)\in D_{r_0',r_1'}$; hence, the identity

$$\varphi(x)=\omega(0,x)=\frac{1}{2}\Big(\psi_0(a^{-1/2}x)+\psi_0(-a^{-1/2}x)\Big)+\frac{a^{1/2}}{2}\int_{-a^{-1/2}x}^{a^{-1/2}x}\psi_1(\eta)d\eta$$

for all $x\in D_{r_1'}$. Applying then our assumptions on ψ_0 and ψ_1, we conclude that $\varphi(x)$ can be analytically continued to the domain $\mathcal{C}_{\theta_a-\alpha,\theta_a+\alpha}$ with a global exponential growth of order at most 2 at infinity, which completes the proof. ∎

Remark 9.3 Contrary to the approach developed in Sect. 7.5.1 following the classical Borel-Laplace method which seems hardly applicable to higher spatial-dimensions, the above proof can be easily generalized to the case of the

n-dimensional (homogeneous or not) linear heat equation

$$\begin{cases} \partial_t u - \Delta_x u = \widetilde{f}(t,x) \\ u(0,x) = \varphi(x) \in \mathcal{O}(D_{\rho_1,\ldots,\rho_n}) \end{cases}, (t,x) \in \mathbb{C}^{n+1},$$

providing thus a characterization of the 1-summability of the unique formal power series solution

$$\widetilde{u}(t,x) = \sum_{j \geqslant 0} \Delta_x^{(2j)} \varphi(x) \frac{t^j}{j!} \in \mathcal{O}(D_{\rho_1,\ldots,\rho_n})[[t]]$$

in terms of analytic continuation and appropriate growth condition of some function connected with the inhomogeneity $f(t,x)$ and the initial data $\varphi(x)$. For more details, we refer to [11, 76, 77].

The example of the heat equation (one-dimensional and n-dimensional) that we have just detailed clearly shows the great interest of the moment-summability method, since it allows us to obtain summability results much more efficiently than with the classical Borel-Laplace method. Of course, this approach can also be used, with the same efficiency, in much more general contexts of k-summability studies. In this case, the moment function m which is usually considered is the function

$$m(\lambda) = \frac{\Gamma(1+(s+1)\lambda)}{\Gamma(1+\lambda)} \quad , s = \frac{1}{k}$$

associated with the kernel function

$$\begin{aligned} e_{s+1,1}(t) &= \frac{1}{s+1} t^{1/(s+1)} C_{s+1}(t^{1/(s+1)}) \\ &= \frac{1}{2i\pi} t^{1/(s+1)} \int_\gamma \exp\left(\tau - (t\tau)^{1/(s+1)}\right) d\left(\tau^{1/(s+1)}\right) \end{aligned}$$

of order s (see Example 8.3, (4), and Example 8.9, (2)) so that the formal k-moment-Borel transform $\widetilde{\mathcal{B}}_k^{[m;t]}(\widetilde{u})$ of any formal power series

$$\widetilde{u}(t,x) = \sum_{j \geqslant 0} u_{j,*}(x) \frac{t^j}{j!} \in \mathcal{O}(D_{\rho_1,\ldots,\rho_n})[[t]]$$

is

$$\widetilde{\mathcal{B}}_k^{[m;t]}(\widetilde{u})(\tau,x) = \sum_{j \geqslant 0} u_{j,*}(x) \frac{t^j}{\Gamma(1+(s+1)j)} \in \mathcal{O}(D_{\rho_1,\ldots,\rho_n})[[\tau]].$$

For more details on such situations, we refer for instance to the recent work [82] of S. Michalik in which he studies the k-summability of formal power series solutions of some linear Cauchy-Goursat problems with constant coefficients.

Chapter 10
Linear Moment Partial Differential Equations

Introduced in the 2010 article [13] by W. Balser and M. Yoshino in order to present a new tool to study the classical partial differential equations, the *moment partial differential equations* define actually a new and wide class of functional equations which covers a large number of classical equations: of course the classical partial differential equations, but also the fractional partial differential equations, the q-difference equations, etc. Naturally well-suited to applying the moment-summability method, these equations extend the range of applications of the summability theory, and also offer a wide field of investigation for developing new calculation techniques, as well as new theoretical tools.

The purpose of this chapter is to briefly present some theoretical results that have been obtained in recent years in the linear case. In particular, we shall see how the tools developed in the framework of classical partial differential equations can be extended to the case of moment partial differential equations.

We start with the key notion of *moment derivation* which generalizes the standard derivation acting on the formal power series.

10.1 Moment Derivation

Recall (see remark page 137) that, for any moment function m of order s, the sequence $(m(j))_{j\geqslant 0}$ is a s-Gevrey-type sequence, that is there exist four positive constants $c, C, a, A > 0$ such that the following estimate holds for all $j \geqslant 0$:

$$ca^j\Gamma(1+sj) \leqslant m(j) \leqslant CA^j\Gamma(1+sj). \tag{10.1}$$

Moreover, we have the following.

Lemma 10.1 *Let m a moment function of order $s > 0$. Then, there exist four positive constants $c', C', a'; A' > 0$ such that the following estimate holds for all*

P. Remy, *Asymptotic Expansions and Summability*, Lecture Notes in Mathematics 2351, https://doi.org/10.1007/978-3-031-59094-8_10

$j \geqslant 0$:

$$c'a'^j(j+1)^s \leqslant \frac{m(j+1)}{m(j)} \leqslant C'A'^j(j+1)^s.$$

Proof From the Stirling's Formula, we have

$$\frac{\Gamma(1+s(j+1))}{\Gamma(1+sj)} \underset{j\to+\infty}{\sim} s^s(j+1)^s \tag{10.2}$$

and Lemma 10.1 follows from the estimates (10.1). ■

10.1.1 Some Definitions and Basic Properties

Let us now define the moment derivative and the moment anti-derivative of a formal power series.

Definition 10.2 (Moment Derivation and Moment Anti-derivation) Let m_0 be a moment function of order $s_0 > 0$ and $\widetilde{u}(t,x) \in \mathcal{O}(D_{\rho_1,\ldots,\rho_n})[[t]]$ a formal power series written in the form

$$\widetilde{u}(t,x) = \sum_{j\geqslant 0} u_{j,*}(x)\frac{t^j}{m_0(j)}.$$

Then,

1. The *moment derivative* $\partial_{m_0;t}\widetilde{u}$ *of* $\widetilde{u}(t,x)$ *with respect to* t is the formal power series in $\mathcal{O}(D_{\rho_1,\ldots,\rho_n})[[t]]$ defined by

$$\partial_{m_0;t}\widetilde{u}(t,x) = \sum_{j\geqslant 0} u_{j+1,*}(x)\frac{t^j}{m_0(j)}.$$

2. The *moment anti-derivative* $\partial_{m_0;t}^{-1}\widetilde{u}$ *of* $\widetilde{u}(t,x)$ *with respect to* t is the formal power series in $\mathcal{O}(D_{\rho_1,\ldots,\rho_n})[[t]]$ defined by

$$\partial_{m_0;t}^{-1}\widetilde{u}(t,x) = \sum_{j\geqslant 1} u_{j-1,*}(x)\frac{t^j}{m_0(j)}.$$

Definition 10.2 can be naturally extended to analytic functions at the origin of $\mathbb{C}^{n+1}$ by means of their representation in the form of an infinite series. Moreover, according to Lemma 10.1, their moment derivative and moment anti-derivative are still analytic at the origin.

More generally, we have the following.

Proposition 10.3 *Let m_0 be a moment function of order $s_0 > 0$ and $\widetilde{u}(t,x) \in \mathcal{O}(D_{\rho_1,\dots,\rho_n})[[t]]$ a s-Gevrey formal series with $s \geqslant 0$. Then, the two formal power series $\partial_{m_0;t}\widetilde{u}(t,x)$ and $\partial_{m_0;t}^{-1}\widetilde{(}t,x)$ are still s-Gevrey.*

Proof Let us write $\widetilde{u}(t,x)$ in the form

$$\widetilde{u}(t,x) = \sum_{j\geqslant 0} u_{j,*}(x)\frac{t^j}{m_0(j)}.$$

By assumption (see Definition 3.1), there exist a radius $0 < r \leqslant \min(\rho_1,\dots,\rho_n)$ and two positive constants $C, K > 0$ such that

$$\left|u_{j,*}(x)\right| \leqslant CK^j m_0(j)\Gamma(1+sj) \tag{10.3}$$

for all $x \in D_{r,\dots,r}$ and all $j \geqslant 0$.

From Definition 10.2, we have

$$\partial_{m_0;t}\widetilde{u}(t,x) = \sum_{j\geqslant 0} v_j(x)t^j \quad \text{and} \quad \partial_{m_0;t}^{-1}\widetilde{u}(t,x) = \sum_{j\geqslant 1} w_j(x)t^j$$

with $v_j(x) = u_{j+1,*}(x)/m_0(j)$ and $w_j(x) = u_{j-1,*}(x)/m_0(j)$; hence,

$$\left|v_j(x)\right| \leqslant CK^{j+1}\frac{m_0(j+1)}{m_0(j)}\frac{\Gamma(1+s(j+1))}{\Gamma(1+sj)}\Gamma(1+sj)$$

$$\left|w_j(x)\right| \leqslant CK^{j-1}\frac{m_0(j-1)}{m_0(j)}\frac{\Gamma(1+s(j-1))}{\Gamma(1+sj)}\Gamma(1+sj)$$

for all $x \in D_{r,\dots,r}$ and all j accordingly inequalities (10.3).

Let us now apply Lemma 10.1 and relation (10.2): there exist six positive constants $c', c'', C', C'', a', A' > 0$ such that

$$c'a'^j(j+1)^s \leqslant \frac{m_0(j+1)}{m_0(j)} \leqslant C'A'^j(j+1)^s \quad \text{and}$$

$$c''(j+1)^s \leqslant \frac{\Gamma(1+s(j+1))}{\Gamma(1+sj)} \leqslant C''(j+1)^s$$

for all $j \geqslant 0$. Then,

$$\begin{aligned}|v_j(x)| &\leqslant CC'C''K(KA')^j(j+1)^{2s}\Gamma(1+sj)\\ &\leqslant CC'C''Ke^{2s}(KA'e^{2s})^j\Gamma(1+sj)\end{aligned}$$

$$|w_j(x)| \leqslant \frac{Ca'}{c'c''K}\left(\frac{K}{a'}\right)^j\Gamma(1+sj)$$

for all $x \in D_{r,\dots,r}$ and all j, and Proposition 10.3 follows. ■

The example below presents two classical moment derivation operators $\partial_{m_0;t}$.

Example 10.4

1. For $m_0(\lambda) = \Gamma(1+\lambda)$, the operator $\partial_{m_0;t}$ coincides with the standard derivation operator ∂_t.
2. For $m_0(\lambda) = \Gamma(1+s\lambda)$ with $s \neq 1$, the operator $\partial_{m_0;t}$ is intimately related with, although not equal to, the Caputo s-fractional derivation operator ∂_t^s: for any formal power series $\widetilde{u}(t,x) \in \mathcal{O}(D_{\rho_1,\dots,\rho_n})[[t]]$ as in Definition 10.2, we have

$$\left(\partial_{m_0;t}\widetilde{u}\right)(z^s,x) = \sum_{j\geqslant 0} u_{j+1,*}(x)\frac{z^{sj}}{\Gamma(1+sj)} = \partial_z^s\widetilde{u}(z^s,x).$$

Remark 10.5 The operators $\partial_{m_0;t}$ and $\partial_{m_0;t}^{-1}$ can also be defined for sequences $(m_0(j))_{j\geqslant 0}$ given by moment functions m_0 associated with much more general kernel functions than those presented in Chap. 8 (see Sect. 11.3). The sequence

$$m_0(j) = [j]_q! = \begin{cases} 1 & \text{for } j = 0\\ [j]_q[j-1]_q\dots[1]_q & \text{for } j \geqslant 1\end{cases}$$

with a fixed $q \in]0,1[$ and

$$[j]_q = 1 + q + \dots + q^{j-1} = \frac{q^j - 1}{q - 1}$$

is such a sequence: it is a 0-Gevrey-type sequence in the sense of the inequality (10.1) (we have $1 \leqslant m_0(j) \leqslant (1-q)^{-j}$ for all $j \geqslant 0$) and, therefore, does not fall within the scope of the moment functions we have chosen to deal with in this book; but its associated moment derivation operator $\partial_{m_0;t}$ is well defined and coincides with the q-difference operator $D_{q;t}$. Indeed, since

$$D_{q;t}u(t) = \frac{u(qt) - u(t)}{qt - t},$$

we have

$$D_{q;t}(t^j) = \begin{cases} 0 & \text{for } j = 0 \\ [j]_q t^{j-1} & \text{for } j \geqslant 1 \end{cases};$$

hence,

$$D_{q;t}\left(\sum_{j\geqslant 0} u_{j,*}(x)\frac{t^j}{[j]_q!}\right) = \sum_{j\geqslant 1} u_{j,*}(x)\frac{t^{j-1}}{[j-1]_q!} = \partial_{m_0;t}\left(\sum_{j\geqslant 0} u_{j,*}(x)\frac{t^j}{[j]_q!}\right).$$

We will come back to this particular case in Sect. 10.3.2, Remark 10.24.

In the same way, we define the moment derivation $\partial_{m_j;x_j}$ and the moment anti-derivation $\partial_{m_j;x_j}^{-1}$ with respect to x_j for any moment function m_j of order $s_j > 0$ and any $j \in \{1, \ldots, n\}$. Thereby, for any formal power series $\widetilde{u}(t,x) \in \mathcal{O}(D_{\rho_1,\ldots,\rho_n})[[t]]$ written in the form

$$\widetilde{u}(t,x) = \sum_{j_0,j_1,\ldots,j_n\geqslant 0} u_{j_0,j_1,\ldots,j_n} \frac{t^{j_0}}{m_0(j_0)}\frac{x_1^{j_1}}{m_1(j_1)}\cdots\frac{x_n^{j_n}}{m_n(j_n)},$$

the following identities hold for any $i_0, i_1, \ldots, i_n \geqslant 0$:

$$\partial_{m_0;t}^{i_0}\partial_{m_1;x_1}^{i_1}\ldots\partial_{m_n;x_n}^{i_n}\widetilde{u}(t,x) = \sum_{j_0,\ldots,j_n\geqslant 0} u_{j_0+i_0,j_1+i_1,\ldots,j_n+i_n} \times \frac{t^{j_0}}{m_0(j_0)}\frac{x_1^{j_1}}{m_1(j_1)}\cdots\frac{x_n^{j_n}}{m_n(j_n)},$$

$$\partial_{m_0;t}^{-i_0}\partial_{m_1;x_1}^{-i_1}\ldots\partial_{m_n;x_n}^{-i_n}\widetilde{u}(t,x) = \sum_{j_0\geqslant i_0,\ldots,j_n\geqslant i_n} u_{j_0-i_0,j_1-i_1,\ldots,j_n-i_n} \times \frac{t^{j_0}}{m_0(j_0)}\frac{x_1^{j_1}}{m_1(j_1)}\cdots\frac{x_n^{j_n}}{m_n(j_n)}.$$

Observe that the operators $\partial_{m_0;t}$ and $\partial_{m_0;t}^{-1}$ commute with any operator $\partial_{m_j;x_j}$ and $\partial_{m_j;x_j}^{-1}$, and that the operators $\partial_{m_j;x_j}$ and $\partial_{m_j;x_j}^{-1}$ commute with any operator $\partial_{m_\ell;x_\ell}$ and $\partial_{m_\ell;x_\ell}^{-1}$ as soon as $j \neq \ell$.

Observe also that, as in the case of the standard derivation, the operator $\partial_{m_0;t}^{-1}$ is the right-inverse of the operator $\partial_{m_0;t}$ so that $\partial_{m_0;t}^{i_0}\partial_{m_0;t}^{-i_0'} = \partial_{m_0;t}^{i_0-i_0'}$ for any $i_0, i_0' \in \mathbb{N}$. Of course, the same occurs for the operators $\partial_{m_j;x_j}^{-1}$ and $\partial_{m_j;x_j}$.

Once again, the two previous definitions can be naturally extended to analytic functions at the origin of $\mathbb{C}^{n+1}$ by means of their representation in the form of an infinite series. More precisely, we have the following.

Proposition 10.6 *Let $a(t,x)$ be an analytic function at the origin of $\mathbb{C}^{n+1}$, and let $m_0, m_1, \ldots, m_n$ be $n+1$ moment functions of respective order $s_0, s_1, \ldots, s_n > 0$. Then, for any $i_0, i_1, \ldots, i_n \geqslant 0$, the two formal power series $\partial^{i_0}_{m_0;t}\partial^{i_1}_{m_1;x_1}\ldots\partial^{i_n}_{m_n;x_n}a(t,x)$ and $\partial^{-i_0}_{m_0;t}\partial^{-i_1}_{m_1;x_1}\ldots\partial^{-i_n}_{m_n;x_n}a(t,x)$ define also analytic functions at the origin of $\mathbb{C}^{n+1}$.*

Proof Let us write $a(t,x)$ in the form

$$a(t,x) = \sum_{j_0,j_1,\ldots,j_n \geqslant 0} a_{j_0,j_1,\ldots,j_n} \frac{t^{j_0}}{m_0(j_0)} \frac{x_1^{j_1}}{m_1(j_1)} \cdots \frac{x_n^{j_n}}{m_n(j_n)}.$$

By assumption, there exist $n+2$ positive constants $C, r_0, r_1, \ldots, r_n > 0$ such that

$$\left|\frac{a_{j_0,j_1,\ldots,j_n}}{m_0(j_0)m_1(j_1)\ldots m_n(j_n)}\right| \leqslant C r_0^{j_0} r_1^{j_1} \ldots r_n^{j_n} \tag{10.4}$$

for all $j_0, j_1, \ldots, j_n \geqslant 0$.

From previous definitions, we have

$$\partial^{i_0}_{m_0;t}\partial^{i_1}_{m_1;x_1}\ldots\partial^{i_n}_{m_n;x_n}a(t,x) = \sum_{j_0,\ldots,j_n \geqslant 0} v_{j_0,j_1,\ldots,j_n} t^{j_0} x_1^{j_1} \ldots x_n^{j_n} \text{ and}$$

$$\partial^{-i_0}_{m_0;t}\partial^{-i_1}_{m_1;x_1}\ldots\partial^{-i_n}_{m_n;x_n}a(t,x) = \sum_{j_0 \geqslant i_0,\ldots,j_n \geqslant i_n} w_{j_0,j_1,\ldots,j_n} t^{j_0} x_1^{j_1} \ldots x_n^{j_n}$$

with

$$v_{j_0,j_1,\ldots,j_n} = \frac{a_{j_0+i_0,j_1+i_1,\ldots,j_n+i_n}}{m_0(j_0)m_1(j_1)\ldots m_n(j_n)} \text{ and}$$

$$w_{j_0,j_1,\ldots,j_n} = \frac{a_{j_0-i_0,j_1-i_1,\ldots,j_n-i_n}}{m_0(j_0)m_1(j_1)\ldots m_n(j_n)};$$

hence,

$$\left|v_{j_0,j_1,\ldots,j_n}\right| \leqslant C r_0^{j_0+i_0} r_1^{j_1+i_1} \ldots r_n^{j_n+i_n} \prod_{\ell=0}^{n} \frac{m_\ell(j_\ell+i_\ell)}{m_\ell(j_\ell)} \tag{10.5}$$

$$\left|w_{j_0,j_1,\ldots,j_n}\right| \leqslant C r_0^{j_0-i_0} r_1^{j_1-i_1} \ldots r_n^{j_n-i_n} \prod_{\ell=0}^{n} \frac{m_\ell(j_\ell-i_\ell)}{m_\ell(j_\ell)} \tag{10.6}$$

for all $j_0, j_1, \ldots, j_n$ accordingly inequalities (10.4).

Let us now apply Lemma 10.1: for any $\ell \in \{0, \dots, n\}$, there exist four positive constants $c'_\ell, C'_\ell, a'_\ell, A'_\ell > 0$ such that

$$c'_\ell a'^{j}_\ell (j+1)^{s_\ell} \leqslant \frac{m_\ell(j+1)}{m_\ell(j)} \leqslant C'_\ell A'^{j}_\ell (j+1)^{s_\ell}$$

for all $j \geqslant 0$. Then,

$$\frac{m_\ell(j_\ell+i_\ell)}{m_\ell(j_\ell)} = \begin{cases} 1 & \text{if } i_\ell = 0 \\ \displaystyle\prod_{k=j_\ell}^{j_\ell+i_\ell-1} \frac{m_\ell(k+1)}{m_\ell(k)} & \text{if } i_\ell \geqslant 1 \end{cases} \leqslant C_{\ell,i_\ell} A^{j_\ell}_{\ell,i_\ell}$$

$$\frac{m_\ell(j_\ell-i_\ell)}{m_\ell(j_\ell)} = \begin{cases} 1 & \text{if } i_\ell = 0 \\ \displaystyle\prod_{k=j_\ell-i_\ell+1}^{j_\ell} \frac{m_\ell(k-1)}{m_\ell(k)} & \text{if } i_\ell \geqslant 1 \end{cases} \leqslant c_{\ell,i_\ell} a^{j_\ell}_{\ell,i_\ell}$$

with

$$C_{\ell,i_\ell} = \left(C'_\ell A'^{(i_\ell-1)/2}_\ell e^{s_\ell(i_\ell+1)/2}\right)^{i_\ell}, \quad A_{\ell,i_\ell} = \left(e^{s_\ell} A'_\ell\right)^{i_\ell}$$

$$c_{\ell,i_\ell} = \left(\frac{a'^{(i_\ell+1)/2}_\ell}{c'_\ell}\right)^{i_\ell}, \quad a_{\ell,i_\ell} = \frac{1}{a'^{i_\ell}_\ell}.$$

Combining this with the previous estimates (10.5) and (10.6), we finally get the inequalities

$$\left|v_{j_0,j_1,\dots,j_n}\right| \leqslant \widetilde{C} \widetilde{R}_0^{j_0} \widetilde{R}_1^{j_1} \dots \widetilde{R}_0^{j_n} \quad \text{and} \quad \left|w_{j_0,j_1,\dots,j_n}\right| \leqslant \widetilde{c} \widetilde{r}_0^{j_0} \widetilde{r}_1^{j_1} \dots \widetilde{r}_0^{j_n}$$

with

$$\widetilde{C} = C r_0^{i_0} r_1^{i_1} \dots r_n^{i_n} \prod_{\ell=0}^{n} C_{\ell,i_\ell}, \quad \widetilde{R}_\ell = A_{\ell,i_\ell} r_\ell$$

$$\widetilde{c} = C r_0^{-i_0} r_1^{-i_1} \dots r_n^{-i_n} \prod_{\ell=0}^{n} c_{\ell,i_\ell}, \quad \widetilde{r}_\ell = a_{\ell,i_\ell} r_\ell$$

for all $j_0, j_1, \dots, j_n$, which ends the proof. ■

Remark 10.7 Assume that $a(t,x)$ is analytic on a polydisc $D_{\rho_0,\rho_1,\dots,\rho_n}$. Then, Proposition 10.6 tells us that the function $\partial^{i_0}_{m_0;t} \partial^{i_1}_{m_1;x_1} \dots \partial^{i_n}_{m_n;x_n} a(t,x)$ is analytic on a polydisc $D_{\rho'_0,\rho'_1,\dots,\rho'_n}$ with $0 < \rho'_j \leqslant \rho_j$ for all $j = 0, \dots, n$. Observe here that, for some choices of the moment functions m_j, it is possible to have $\rho'_j < \rho_j$, that is the function $\partial_{m_0;t}^{i_0}\partial_{m_1;x_1}^{i_1}\ldots\partial_{m_n;x_n}^{i_n}a(t,x)$ is analytic on a polydisc smaller than the initial polydisc of analyticity of $a(t,x)$. Indeed, consider for instance a function a independent of x:

$$a(t)=\sum_{j\geqslant 0}a_j\frac{t^j}{m_0(j)},\quad a_j\in\mathbb{C}.$$

We have

$$\partial_{m_0;t}a(t)=\sum_{j\geqslant 0}a_{j+1}\frac{t^j}{m_0(j)}$$

and, from the Cauchy-Hadamard Theorem, the radii ρ_0 and ρ_0' are given by the relations

$$\frac{1}{\rho_0}=\overline{\lim_{j\to+\infty}}\left|\frac{a_j}{m_0(j)}\right|^{\frac{1}{j}}\quad\text{and}$$

$$\frac{1}{\rho_0'}=\overline{\lim_{j\to+\infty}}\left|\frac{a_{j+1}}{m_0(j)}\right|^{\frac{1}{j}}=\overline{\lim_{j\to+\infty}}\left(\left(\left|\frac{a_{j+1}}{m_0(j+1)}\right|^{\frac{1}{j+1}}\right)^{1+\frac{1}{j}}\left(\frac{m_0(j+1)}{m_0(j)}\right)^{\frac{1}{j}}\right)$$

Let us now suppose that the sequence $((m_0(j+1)/m_0(j))^{1/j})_{j\geqslant 1}$ converges to a positive real number $\ell\geqslant 1$. Then,

$$\frac{1}{\rho_0'}=\frac{\ell}{\rho_0}\quad\text{and}\quad\begin{cases}\rho_0'=\rho_0 & \text{if }\ell=1\\ \rho_0'<\rho_0 & \text{if }\ell>1\end{cases}.$$

We shall return to the case $\ell=1$ in the next Sect. 10.3.1, Proposition 10.21.

Of course, the same occurs for the function $\partial_{m_0;t}^{-i_0}\partial_{m_1;x_1}^{-i_1}\ldots\partial_{m_n;x_n}^{-i_n}a(t,x)$.

10.1.2 Integral Representation

Proposition below gives an integral representation of moment derivatives of analytic functions at the origin of $\mathbb{C}^{n+1}$.

Proposition 10.8 *Let $a(t,x)$ be an analytic function at the origin of $\mathbb{C}^{n+1}$, and let $m_0,m_1,\ldots,m_n$ be $n+1$ moment functions of respective order $s_0,s_1,\ldots,s_n>0$ and with respective pair of kernel functions (e_0,E_0), (e_1,E_1), …, (e_n,E_n).*

Let $i_0,i_1,\ldots,i_1\geqslant 0$ and $\rho_0',\rho_1',\ldots,\rho_n'>0$ be such that $\partial_{m_0;t}^{i_0}\partial_{m_1;x_1}^{i_1}\ldots\partial_{m_n;x_n}^{i_n}$ $a(t,x)$ and $a(t,x)$ are analytic on $D_{\rho_0',\ldots,\rho_n'}$.

Let $\varepsilon = (\varepsilon_0, \varepsilon_1, \ldots, \varepsilon_n)$ *be such that* $0 < \varepsilon_\ell < \rho'_\ell$ *for all* $\ell \in \{0, \ldots, n\}$.

Let $I_0, I_1, \ldots, I_n$ *be* $n+1$ *intervals such that* $I_\ell \subsetneq]-\pi s_\ell/2, \pi s_\ell/2[$ *for all* $\ell \in \{0, \ldots, n\}$.

Then, there exist $n+1$ *radii* $0 < r_\ell < \varepsilon_\ell$, $\ell = 0, \ldots, n$, *such that the following identity holds for all* $(t, x) \in D_{r_0,\ldots,r_n}$*:*

$$\partial^{i_0}_{m_0;t}\partial^{i_1}_{m_1;x_1}\ldots\partial^{i_n}_{m_n;x_n}a(t,x) = \frac{1}{(2i\pi)^{n+1}}\int_{(\tau,\xi)\in\gamma(\varepsilon)} a(\tau,\xi)I^{[t,x,\tau,\xi]}_{i_0,i_1,\ldots,i_n}d\tau d\xi, \tag{10.7}$$

where $\gamma(\varepsilon)$ *is the polycircle*

$$\gamma(\varepsilon) = \{(\tau,\xi) \in \mathbb{C}^{n+1};\ |\tau| = \varepsilon_0 \text{ and } |\xi_\ell| = \varepsilon_\ell \text{ for all } \ell \in \{1,\ldots,n\}\}$$

and where

$$I^{[t,x,\tau,\xi]}_{i_0,i_1,\ldots,i_n} = \left(\int_0^{+\infty e^{i\theta_{0,t,\tau}}}\left(\frac{v_0}{\tau}\right)^{i_0}\frac{e_0(v_0)}{v_0\tau}E_0\left(\frac{v_0 t}{\tau}\right)dv_0\right)$$
$$\times\prod_{\ell=1}^{n}\left(\int_0^{+\infty e^{i\theta_{\ell,x_\ell,\xi_\ell}}}\left(\frac{v_\ell}{\xi_\ell}\right)^{i_\ell}\frac{e_\ell(v_\ell)}{v_\ell\xi_\ell}E_\ell\left(\frac{v_\ell x_\ell}{\xi_\ell}\right)dv_\ell\right), \tag{10.8}$$

the directions $\theta_{0,t,\tau}$ *(resp.* $\theta_{\ell,x_\ell,\xi_\ell}$ *for* $\ell = 1,\ldots,n$*) being chosen arbitrarily in the interval* I_0 *(resp.* I_ℓ*).*

Proof Let $(t,x) \in D_{\varepsilon_0,\ldots,\varepsilon_n}$. According to the analyticity of a on $D_{\rho'_0,\ldots,\rho'_n}$, we can write $a(t,x)$ as

$$a(t,x) = \sum_{j_0,\ldots,j_n\geqslant 0}\alpha_{j_0,j_1,\ldots,j_n}\frac{t^{j_0}}{m_0(j_0)}\frac{x_1^{j_1}}{m_1(j_1)}\cdots\frac{x_1^{j_n}}{m_n(j_n)}$$

with

$$\alpha_{j_0,j_1,\ldots,j_n} = m_0(j_0)m_1(j_1)\ldots m_n(j_n)\frac{\partial_t^{j_0}\partial_{x_1}^{j_1}\ldots\partial_{x_n}^{j_n}a(0,0)}{j_0!j_1!\ldots j_n!}.$$

Applying then the Cauchy Integral Formula, we get

$$\alpha_{j_0,j_1,\ldots,j_n} = \frac{m_0(j_0)m_1(j_1)\ldots m_n(j_n)}{(2i\pi)^{n+1}}\int_{(\tau,\xi)\in\gamma(\varepsilon)}\frac{a(\tau,\xi)}{\tau^{j_0+1}\xi_1^{j_1+1}\ldots\xi_n^{j_n+1}}d\tau d\xi;$$

hence, using the definition of the moment function (see Definition 8.1, identity (8.5)) and the fact that e_ℓ is k_ℓ-exponentially flat at infinity for all $\ell \in \{0,\ldots,n\}$, the following identity:

$$\alpha_{j_0,j_1,\ldots,j_n} = \frac{1}{(2i\pi)^{n+1}} \int_{(\tau,\xi)\in\gamma(\varepsilon)} a(\tau,\xi) J_{j_0,\ldots,j_n}^{[t,x,\tau,\xi]} d\tau d\xi$$

with

$$J_{j_0,\ldots,j_n}^{[t,x,\tau,\xi]} = \left(\int_0^{+\infty e^{i\theta_{0,t,\tau}}} \left(\frac{v_0}{\tau}\right)^{j_0} \frac{e_0(v_0)}{v_0\tau} dv_0\right) \times \prod_{\ell=1}^{n} \left(\int_0^{+\infty e^{i\theta_{\ell,x_\ell,\xi_\ell}}} \left(\frac{v_\ell}{\xi_\ell}\right)^{j_\ell} \frac{e_\ell(v_\ell)}{v_\ell \xi_\ell} dv_\ell\right)$$

and arbitrarily directions $\theta_{0,t,\tau} \in I_0$ and $\theta_{\ell,x_\ell,\xi_\ell} \in I_\ell$ for all $\ell = 1,\ldots,n$.

Let us now apply the definition of the moment derivation: we get

$$\begin{aligned} \partial_{m_0;t}^{i_0} \partial_{m_1;x_1}^{i_1} \ldots \partial_{m_n;x_n}^{i_n} a(t,x) &= \sum_{j_0,\ldots,j_n \geqslant 0} \alpha_{j_0+i_0,j_1+i_1,\ldots,j_n+i_n} \frac{t^{j_0}}{m_0(j_0)} \frac{x_1^{j_1}}{m_1(j_1)} \cdots \frac{x_1^{j_n}}{m_n(j_n)} \\ &= \frac{1}{(2i\pi)^{n+1}} \sum_{j_0,\ldots,j_n \geqslant 0} \int_{(\tau,\xi)\in\gamma(\varepsilon)} a(\tau,\xi) \\ &\quad \times K_{j_0,\ldots,j_n}^{[t,x,\tau,\xi,i_0,\ldots,i_n]} d\tau d\xi \end{aligned}$$

with

$$K_{j_0,\ldots,j_n}^{[t,x,\tau,\xi,i_0,\ldots,i_n]} = \left(\int_0^{+\infty e^{i\theta_{0,t,\tau}}} \left(\frac{v_0}{\tau}\right)^{i_0} \frac{e_0(v_0)}{v_0\tau} \frac{(v_0 t/\tau)^{j_0}}{m_0(j_0)} dv_0\right) \times \prod_{\ell=1}^{n} \left(\int_0^{+\infty e^{i\theta_{\ell,x_\ell,\xi_\ell}}} \left(\frac{v_\ell}{\xi_\ell}\right)^{i_\ell} \frac{e_\ell(v_\ell)}{v_\ell\xi_\ell} \frac{(v_\ell x_\ell/\xi_\ell)^{j_\ell}}{m_\ell(j_\ell)} dv_\ell\right),$$

and it is clear that relation (10.7) follows, at least from the formal point of view, from the definition of the entire functions E_ℓ (see Definition 8.1).

To prove that the formal interchange of sums and integrals can be also made with analytic meaning (hence, to prove the existence of the radii $r_0,\ldots,r_n$), we proceed as follows. Thanks to the Beppo-Levi Theorem, let us first observe that

- for all $|t| < \varepsilon_0$ and $|\tau| = \varepsilon_0$:

$$\sum_{j_0 \geqslant 0} \int_0^{+\infty} \left| \left(\frac{v_0 e^{i\theta_{0,t,\tau}}}{\tau}\right)^{i_0} \frac{e_0(v_0 e^{i\theta_{0,t,\tau}})}{v_0 e^{i\theta_{0,t,\tau}}\tau} \frac{(v_0 e^{i\theta_{0,t,\tau}} t/\tau)^{j_0}}{m_0(j_0)} \right| dv_0$$

$$= \frac{1}{\varepsilon_0^{i_0+1}} \sum_{j_0 \geqslant 0} \int_0^{+\infty} v_0^{i_0-1} \left| e_0(v_0 e^{i\theta_{0,t,\tau}}) \right| \frac{(v_0 |t| / \varepsilon_0)^{j_0}}{m_0(j_0)} dv_0$$

$$= \frac{1}{\varepsilon_0^{i_0+1}} \int_0^{+\infty} v_0^{i_0-1} \left| e_0(v_0 e^{i\theta_{0,t,\tau}}) \right| E_0 \left(\frac{v_0 |t|}{\varepsilon_0} \right) dv_0$$

- for all $\ell \in \{1, \dots, n\}$, $|x_\ell| < \varepsilon_\ell$ and $|\xi_\ell| = \varepsilon_\ell$:

$$\sum_{j_\ell \geqslant 0} \int_0^{+\infty} \left| \left(\frac{v_\ell e^{i\theta_{\ell,x_\ell,\xi_\ell}}}{\xi_\ell} \right)^{i_\ell} \frac{e_\ell(v_\ell e^{i\theta_{\ell,x_\ell,\xi_\ell}})}{v_\ell e^{i\theta_{\ell,x_\ell,\xi_\ell}} \xi_\ell} \frac{(v_\ell e^{i\theta_{\ell,x_\ell,\xi_\ell}} x_\ell / \xi_\ell)^{j_\ell}}{m_\ell(j_\ell)} \right| dv_\ell$$

$$= \frac{1}{\varepsilon_\ell^{i_\ell+1}} \sum_{j_\ell \geqslant 0} \int_0^{+\infty} v_\ell^{i_\ell-1} \left| e_\ell(v_\ell e^{i\theta_{\ell,x_\ell,\xi_\ell}}) \right| \frac{(v_\ell |x_\ell| / \varepsilon_\ell)^{j_\ell}}{m_\ell(j_\ell)} dv_\ell$$

$$= \frac{1}{\varepsilon_\ell^{i_\ell+1}} \int_0^{+\infty} v_\ell^{i_\ell-1} \left| e_\ell(v_\ell e^{i\theta_{\ell,x_\ell,\xi_\ell}}) \right| E_\ell \left(\frac{v_\ell |x_\ell|}{\varepsilon_\ell} \right) dv_\ell$$

Now, let us choose, for any $\ell \in \{0, \dots, n\}$, an infinite proper subsector Σ_ℓ of $\Sigma_{0,=\pi s_\ell}$ containing all the directions of I_ℓ. Then, accordingly Definition 8.1 and Remark 8.8, there exist a radius $r'_\ell > 0$ and six positive constants $\alpha_\ell, K_\ell, C_\ell, A_\ell, C'_\ell, A'_\ell > 0$ such that

$$\left| e_\ell(z e^{i\theta_\ell}) \right| \leqslant K_\ell z^{\alpha_\ell} \text{ for all } z \in]0, r'_\ell] \text{ and all directions } \theta_\ell \in I_\ell$$

$$\left| e_\ell(z e^{i\theta_\ell}) \right| \leqslant C_\ell \exp\left(-A_\ell z^{k_\ell} \right) \text{ for all } z \in]0, +\infty[\text{ and all directions } \theta_\ell \in I_\ell$$

$$|E_\ell(z)| \leqslant C'_\ell \exp\left(A'_\ell |z|^{k_\ell} \right) \text{ for all } z \in \mathbb{C}$$

Therefore, choosing for any $\ell \in \{0, \dots, n\}$ a radius r_ℓ such that

$$0 < r_\ell < \min\left(\varepsilon_\ell, \frac{\varepsilon_\ell}{2} \left(\frac{A_\ell}{A'_\ell} \right)^{s_\ell} \right),$$

we get the following estimates:

- *Case* $\ell = 0$. For all $|t| < r_0$ and $|\tau| = \varepsilon_0$:

$$\frac{1}{\varepsilon_0^{i_0+1}} \int_0^{+\infty} v_0^{i_0-1} \left| e_0(v_0 e^{i\theta_{0,t,\tau}}) \right| E_0 \left(\frac{v_0 |t|}{\varepsilon_0} \right) dv_0$$

$$\leqslant \frac{1}{\varepsilon_0^{i_0+1}} \left(K_0 C'_0 \int_0^{r'_0} v_0^{i_0+\alpha_0-1} \exp\left(A'_0 \left(\frac{v_0 |t|}{\varepsilon_0} \right)^{k_0} \right) dv_\ell \right.$$

$$+C_0 C_0' \int_{r_0'}^{+\infty} v_0^{i_0-1} \exp\left(A_0'\left(\frac{v_0\,|t|}{\varepsilon_0}\right)^{k_0} - A_0 v_0^{k_0}\right) dv_0\Bigg)$$

$$\leqslant \frac{1}{\varepsilon_0^{i_0+1}}\left(K_0 C_0' \int_0^{r_0'} v_0^{i_0+\alpha_0-1} \exp\left(\frac{A_0 v_0^{k_0}}{2}\right) dv_0\right.$$

$$\left. +C_0 C_0' \int_{r_0'}^{+\infty} v_0^{i_0-1} \exp\left(-\frac{A_0 v_0^{k_0}}{2}\right) dv_0\right) := B_{0,i_0} < +\infty.$$

- *Case* $\ell \in \{1, \ldots, n\}$. For all $|x_\ell| < r_\ell$ and $|\xi_\ell| = \varepsilon_\ell$:

$$\frac{1}{\varepsilon_\ell^{i_\ell+1}} \int_0^{+\infty} v_\ell^{i_\ell-1} \left|e_\ell(v_\ell e^{i\theta_{\ell,x_\ell,\xi_\ell}})\right| E_\ell\left(\frac{v_\ell\,|x_\ell|}{\varepsilon_\ell}\right) dv_\ell$$

$$\leqslant \frac{1}{\varepsilon_\ell^{i_\ell+1}}\left(K_\ell C_\ell' \int_0^{r_\ell'} v_\ell^{i_\ell+\alpha_\ell-1} \exp\left(A_\ell'\left(\frac{v_\ell\,|x_\ell|}{\varepsilon}\right)^{k_\ell}\right) dv_\ell\right.$$

$$\left. +C_\ell C_\ell' \int_{r_\ell'}^{+\infty} v_\ell^{i_\ell-1} \exp\left(A_\ell'\left(\frac{v_\ell\,|x_\ell|}{\varepsilon}\right)^{k_\ell} - A_\ell v_\ell^{k_\ell}\right) dv_\ell\right)$$

$$\leqslant \frac{1}{\varepsilon_\ell^{i_\ell+1}}\left(K_\ell C_\ell' \int_0^{r_\ell'} v_\ell^{i_\ell+\alpha_\ell-1} \exp\left(\frac{A_\ell v_\ell^{k_\ell}}{2}\right) dv_\ell\right.$$

$$\left. +C_\ell C_\ell' \int_{r_\ell'}^{+\infty} v_\ell^{i_\ell-1} \exp\left(-\frac{A_\ell v_\ell^{k_\ell}}{2}\right) dv_\ell\right) := B_{\ell,i_\ell} < +\infty.$$

Consequently,

$$\sum_{j_0,\ldots,j_n \geqslant 0} \int_{(\tau,\xi)\in\gamma(\varepsilon)} \left|a(\tau,\xi)\; K_{j_0,\ldots,j_n}^{[t,x,\tau,\xi,i_0,\ldots,i_n]}\right| d\tau d\xi$$

$$\leqslant \max_{(\tau,\xi)\in\gamma(\varepsilon)} |a(\tau,\xi)| \times (2\pi)^n \prod_{\ell=0}^{n} \varepsilon_\ell B_{\ell,i_\ell} < +\infty$$

for all $(t,x) \in D_{r_0,\ldots,r_n}$ and any arbitrary directions $\theta_{0,t,\tau} \in I_0$ and $\theta_{\ell,x_\ell,\xi_\ell} \in I_\ell$, $\ell = 1,\ldots,n$, and the interchange between sums and integrals is licit. This ends the proof of Proposition 10.8. ■

Remark 10.9 Formula (10.7) can be seen as a generalization of the classical Cauchy Integral Formula. Indeed, let us choose $m_0(\lambda) = \ldots = m_n(\lambda) = \Gamma(1+\lambda)$ so that the moment derivation $\partial_{m_0;t}$ (resp. $\partial_{m_j;x_j}$ for $j = 1,\ldots,n$) coincides with the standard derivation ∂_t (resp. ∂_{x_j}). Then, the corresponding kernel functions e_j and

E_j are all given by $e_j(t) = te^{-t}$ and $E_j(t) = e^t$ (hence, are all of order $s_\ell = 1$), and, consequently, identity (10.8) becomes

$$I_{i_0,i_1,\ldots,i_n}^{[t,x,\tau,\xi]} = \left(\frac{1}{\tau^{i_0+1}}\int_0^{+\infty e^{i\theta_{0t,\tau}}} v_0^{i_0} e^{-(1-t/\tau)v_0} dv_0\right)$$
$$\times \prod_{\ell=1}^{n}\left(\frac{1}{\xi_\ell^{i_\ell+1}}\int_0^{+\infty e^{i\theta_{\ell,x_\ell,\xi_\ell}}} v_\ell^{i_\ell} e^{-(1-x_\ell/\xi_\ell)v_\ell} dv_\ell\right).$$

Since $|t| < |\tau|$ and $|x_\ell| < |\xi_\ell|$ for all $\ell = 1, \ldots, n$, the exponential $e^{-(1-t/\tau)v_0}$ (resp. $e^{-(1-x_\ell/\xi_\ell)v_\ell}$ for $\ell = 1, \ldots, n$) tends to 0 when v_0 (resp. v_ℓ) goes to infinity along the ray $\theta_{0,t,\tau}$ (resp. $\theta_{\ell,x_\ell,\xi_\ell}$) in $]-\pi/2, \pi/2[$. Thereby,

$$\frac{1}{\tau^{i_0+1}}\int_0^{+\infty e^{i\theta_{0,t,\tau}}} v_0^{i_0} e^{-(1-t/\tau)v_0} dv_0$$
$$= \frac{i_0!}{(\tau-t)^{i_0+1}} \quad \text{(integrate by parts } i_0 \text{ times)}$$
$$\frac{1}{\xi^{i_\ell+1}}\int_0^{+\infty e^{i\theta_{\ell,x_\ell,\xi_\ell}}} v_\ell^{i_\ell} e^{-(1-x_\ell/\xi_\ell)v_\ell} dv_\ell$$
$$= \frac{i_\ell!}{(\xi_\ell - x_\ell)^{i_\ell+1}} \quad \text{(integrate by parts } i_\ell \text{ times)}$$

and Formula (10.7) becomes

$$\partial_t^{i_0}\partial_{x_1}^{i_1}\ldots\partial_{x_n}^{i_n} a(t,x)$$
$$= \frac{i_0!\ldots i_n!}{(2i\pi)^{n+1}}\int_{(\tau,\xi)\in\gamma(\varepsilon)} \frac{a(\tau,\xi)}{(\tau-t)^{i_0+1}(\xi_1-x_1)^{i_1+1}\ldots(\xi_n-x_n)^{i_n+1}} d\tau d\xi$$

which is the classical Cauchy Integral Formula.

As a direct application of Proposition 10.8, we have the following.

Proposition 10.10 *Let $m_1, \ldots, m_n$ be n moment functions of respective order $s_1, \ldots, s_n > 0$ and $\widetilde{u}(t,x) \in \mathcal{O}(D_{\rho_1,\ldots,\rho_n})[[t]]$ a s-Gevrey formal series with $s \geqslant 0$. Then, the formal power series $\partial_{m_1;x_1}^{i_1}\ldots\partial_{m_n;x_n}^{i_n}\widetilde{u}(t,x)$ is still s-Gevrey for any $i_1, \ldots, i_n \geqslant 0$.*

Proof Let us write $\widetilde{u}(t,x)$ in the form

$$\widetilde{u}(t,x) = \sum_{j\geqslant 0} u_j(x)t^j$$

with $u_j(x) \in \mathcal{O}(D_{\rho_1,\dots,\rho_n})$ for all $j \geqslant 0$. By assumption (see Definition 3.1), there exist a radius $0 < r < \min(\rho_1, \dots, \rho_n)$ and two positive constants $C, K > 0$ such that

$$|u_j(x)| \leqslant CK^j\Gamma(1+sj) \tag{10.9}$$

for all $j \geqslant 0$ and all $x \in D_{r,\dots,r}$, and we must prove similar estimates on the analytic coefficients $v_j(x) = \partial_{m_1;x_1}^{i_1} \dots \partial_{m_n;x_n}^{i_n} u_j(x)$ of the formal series $\partial_{m_1;x_1}^{i_1} \dots \partial_{m_n;x_n}^{i_n} \widetilde{u}(t, x)$.

From Proposition 10.6, all the functions $v_j(x)$ are analytic on a common polydisc $D_{\rho_1',\dots,\rho_n'}$ with $0 < \rho_j' \leqslant \rho_j$ for all $j \in \{1, \dots, n\}$. Let us choose a positive real number $0 < \varepsilon < \min(r, \rho_1', \dots, \rho_n')$ and let us denote by $\gamma(\varepsilon)$ the polycircle

$$\gamma(\varepsilon) = \{\xi \in \mathbb{C}^n; |\xi_\ell| = \varepsilon \text{ for all } \ell \in \{1, \dots, n\}\}.$$

Then, from Proposition 10.8 (we choose here for the intervals I_ℓ intervals containing the direction $\theta = 0$ and we always choose $\theta_{0,t,\tau} = \theta_{\ell,x_\ell,\xi_\ell} = 0$), there exists a radius $r' \in]0, \varepsilon[$ and n positive constants $B_{1,i_1}, \dots, B_{n,i_n} > 0$ such that

$$v_j(x) = \frac{1}{(2i\pi)^n} \int_{\xi \in \gamma(\varepsilon)} u_j(\xi) \prod_{\ell=1}^{n} \left(\int_0^{+\infty} \left(\frac{v_\ell}{\xi_\ell}\right)^{i_\ell} \frac{e_\ell(v_\ell)}{v_\ell \xi_\ell} E_\ell\left(\frac{v_\ell x_\ell}{\xi_\ell}\right) dv_\ell \right) d\xi$$

for all $x \in D_{r',\dots,r'}$ and all $j \geqslant 0$, and

$$\left| \int_0^{+\infty} \left(\frac{v_\ell}{\xi_\ell}\right)^{i_\ell} \frac{e_\ell(v_\ell)}{v_\ell \xi_\ell} E_\ell\left(\frac{v_\ell x_\ell}{\xi_\ell}\right) dv_\ell \right| \leqslant B_{\ell,i_\ell}$$

for all $\ell \in \{1, \dots, n\}$, all $x_\ell \in D_{r'}$ and all $|\xi_\ell| = \varepsilon$, where (e_ℓ, E_ℓ) stands for the pair of kernel functions corresponding to the moment function m_ℓ.

Consequently,

$$|v_j(x)| \leqslant \max_{\xi \in \gamma(\varepsilon)} |u_j(\xi)| \times \varepsilon^n \prod_{\ell=1}^{n} B_{\ell,i_\ell}$$

for all $j \geqslant 0$ and all $x \in D_{r',\dots,r'}$; hence, applying estimates (10.9) since $\varepsilon < r$, the inequalities

$$|v_j(x)| \leqslant \left(\prod_{\ell=1}^{n} B_{\ell,i_\ell} \right) \varepsilon^n CK^j\Gamma(1+sj)$$

for all $j \geqslant 0$ and all $x \in D_{r',\dots,r'}$, which ends the proof. ∎

We end this section with a result generalizing Proposition 10.8 to the case of functions with an exponential growth of order at most k at infinity. This will be useful in the next section to show that the k-summability of a formal power series is stable under the moment derivations.

Proposition 10.11 *Let $a(t, x)$ be an analytic function at the origin of $\mathbb{C}^{n+1}$. Assume that $a(t, x)$ can be analytically continued to a domain $\Sigma_{\theta_1,\theta_2}(\infty) \times D_{r,\dots,r}$ for some directions $\theta_1 < \theta_2$ and some radius $r > 0$, with a global exponential growth of order at most k at infinity with respect to t, that is, there exist two positive constants $A, C > 0$ such that the following estimate holds for all $(t, x) \in \Sigma_{\theta_1,\theta_2}(\infty) \times D_{r,\dots,r}$:*

$$|a(t, x)| \leqslant C \exp\left(A\,|t|^k\right).$$

Let $m_0, m_1, \dots, m_n$ be $n + 1$ moment functions of respective order $s_0, s_1, \dots, s_n > 0$.

Let $\theta_1' < \theta_2'$ be two directions in $]\theta_1, \theta_2[$ and $i_0, i_1, \dots, i_n \geqslant 0$. Then,

1. *$\partial_{m_0;t}^{i_0} \partial_{m_1;x_1}^{i_1} \dots \partial_{m_n;x_n}^{i_n} a(t, x)$ is analytic at the origin of $\mathbb{C}^{n+1}$;*
2. *there exists a radius $r' > 0$ such that $\partial_{m_0;t}^{i_0} \partial_{m_1;x_1}^{i_1} \dots \partial_{m_n;x_n}^{i_n} a(t, x)$ can be analytically continued to the domain $\Sigma_{\theta_1',\theta_2'}(\infty) \times D_{r',\dots,r'}$ with a global exponential growth of order at most k at infinity with respect to t.*

Proof The first point stems from Proposition 10.6.

To prove the second point, let us first assume $s_0 < 2$. Let us fix $n + 1$ radii $\rho_0', \dots, \rho_n' > 0$ such that $a(t, x)$ and $\partial_{m_0;t}^{i_0} \partial_{m_1;x_1}^{i_1} \dots \partial_{m_n;x_n}^{i_n} a(t, x)$ are analytic on $D_{\rho_0',\dots,\rho_n'}$. Let us also choose

- a positive real number $0 < \varepsilon < \min(r, \rho_0', \dots, \rho_n')$;
- two directions θ_1'', θ_2'' such that $\theta_1 < \theta_1'' < \theta_1' < \theta_2' < \theta_2'' < \theta_2$ and $\theta_2'' - \theta_2' = \theta_1' - \theta_1'' := \omega \in]0, \pi s_0[$

and let us denote by

- $\gamma(\varepsilon)$ the polycircle

$$\gamma(\varepsilon) = \{(\tau, \xi) \in \mathbb{C}^{n+1};\ |\tau| = \varepsilon \text{ and } |\xi_\ell| = \varepsilon \text{ for all } \ell \in \{1, \dots, n\}\}$$

- $I_0 = \left]-\frac{\pi s_0}{2} + \frac{\omega}{2}, \frac{\pi s_0}{2} - \frac{\omega}{2}\right[\subsetneq \left]-\frac{\pi s_0}{2}, \frac{\pi s_0}{2}\right[$
- $I_\ell \subsetneq \left]-\frac{\pi s_\ell}{2}, \frac{\pi s_\ell}{2}\right[$, $\ell = 1, \dots n$, an interval containing the direction $\theta = 0$
- $I_0' = \left]\frac{\pi s_0}{2} + \frac{\omega}{2}, 2\pi - \frac{\pi s_0}{2} - \frac{\omega}{2}\right[\subsetneq \left]\frac{\pi s_0}{2}, 2\pi - \frac{\pi s_0}{2}\right[$ (which exists due to the hypothesis $s_0 < 2$).

Then, from Proposition 10.8, there exists a radius $r' \in]0, \varepsilon[$ such that the following identity holds for all $(t, x) \in D_{r',\dots,r'}$:

$$\partial_{m_0;t}^{i_0} \partial_{m_1;x_1}^{i_1} \dots \partial_{m_n;x_n}^{i_n} a(t, x) = \frac{1}{(2i\pi)^{n+1}} \int_{(\tau,\xi)\in\gamma(\varepsilon)} a(\tau, \xi) I_{i_0,i_1,\dots,i_n}^{[t,x,\tau,\xi]} d\tau d\xi$$

with

$$I_{i_0,i_1,\dots,i_n}^{[t,x,\tau,\xi]} = \left(\int_0^{+\infty e^{i\theta_{0,t,\tau}}} \left(\frac{v_0}{\tau}\right)^{i_0} \frac{e_0(v_0)}{v_0 \tau} E_0\left(\frac{v_0 t}{\tau}\right) dv_0\right)$$
$$\times \prod_{\ell=1}^{n} \left(\int_0^{+\infty} \left(\frac{v_\ell}{\xi_\ell}\right)^{i_\ell} \frac{e_\ell(v_\ell)}{v_\ell \xi_\ell} E_\ell\left(\frac{v_\ell x_\ell}{\xi_\ell}\right) dv_\ell\right),$$

where (e_ℓ, E_ℓ) stands as before for the pair of kernel functions corresponding to the moment function m_ℓ, and where $\theta_{0,t,\tau}$ is an arbitrary direction in I_0. Recall (see the proof of Proposition 10.8) that the construction of the radius r' implies in particular that there exists, for any $\ell \in \{1, \dots, n\}$, a positive constant $B_{\ell,i_\ell} > 0$ such that

$$\left|\int_0^{+\infty} \left(\frac{v_\ell}{\xi_\ell}\right)^{i_\ell} \frac{e_\ell(v_\ell)}{v_\ell \xi_\ell} E_\ell\left(\frac{v_\ell x_\ell}{\xi_\ell}\right) dv_\ell\right| \leqslant B_{\ell,i_\ell}$$

for all $x_\ell \in D_{r'}$ and all $|\xi_\ell| = \varepsilon$.

Let us now fix $(t,x) \in (\Sigma_{\theta_1',\theta_2'}(\infty)\backslash D_{r'}) \times D_{r',\dots,r'}$. Accordingly Definition 8.1, there exist height positive constants $\alpha_0, \beta_0, K_0, K_0', C_0, C_0' > 0$, $A_0 \in]0,1]$ and $A_0' \geqslant 1$, and two radii $R_0 > 0$ and $r_0 \leqslant \max(\varepsilon R_0/r', (2A_0'/A_0)^{s_0} R_0)$ such that

$$\left|e_0(ze^{i\theta_0})\right| \leqslant K_0 z^{\alpha_0} \text{ for all } z \in]0, r_0] \text{ and all directions } \theta_0 \in I_0$$

$$\left|e_0(ze^{i\theta_0})\right| \leqslant C_0 \exp\left(-A_0 z^{k_0}\right) \text{ for all } z \in]0, +\infty[\text{ and all directions } \theta_0 \in I_0$$

$$|E_0(z)| \leqslant C_0' \exp\left(A_0' |z|^{k_0}\right) \text{ for all } z \in \mathbb{C}$$

$$\left|E_0(ze^{id_0})\right| \leqslant K_0' z^{-\beta_0} \text{ for all } z \in [R_0, +\infty[\text{ and all directions } d_0 \in I_0'.$$

Let us set

$$M_t = \left(\frac{2A_0'}{A_0}\right)^{s_0} |t|$$

so that $M_t > |t|$, and let us deform the path $\gamma(\varepsilon)$ into the path $\gamma_{M_t}(\varepsilon) \times \gamma'(\varepsilon)$, where $\gamma_{M_t}(\varepsilon)$ is the positively oriented path $\gamma_{M_t}(\varepsilon) = \gamma_{M_t}^{[1]}(\varepsilon) + \gamma_{M_t}^{[2]}(\varepsilon) + \gamma_{M_t}^{[3]}(\varepsilon) + \gamma_{M_t}^{[4]}(\varepsilon)$ with (see Fig. 10.1)

- $\gamma_{M_t}^{[1]}(\varepsilon)$ the segment from $\varepsilon e^{i\theta_1''}$ to $M_t e^{i\theta_1''}$,
- $\gamma_{M_t}^{[2]}(\varepsilon)$ the arc connecting the points $M_t e^{i\theta_1''}$ and $M_t e^{i\theta_2''}$,
- $\gamma_{M_t}^{[3]}(\varepsilon)$ the segment from $M_t e^{i\theta_2''}$ to $\varepsilon e^{i\theta_2''}$,
- $\gamma_{M_t}^{[4]}(\varepsilon)$ the arc connecting the points $\varepsilon e^{i\theta_2''}$ and $\varepsilon e^{i\theta_1''}$,

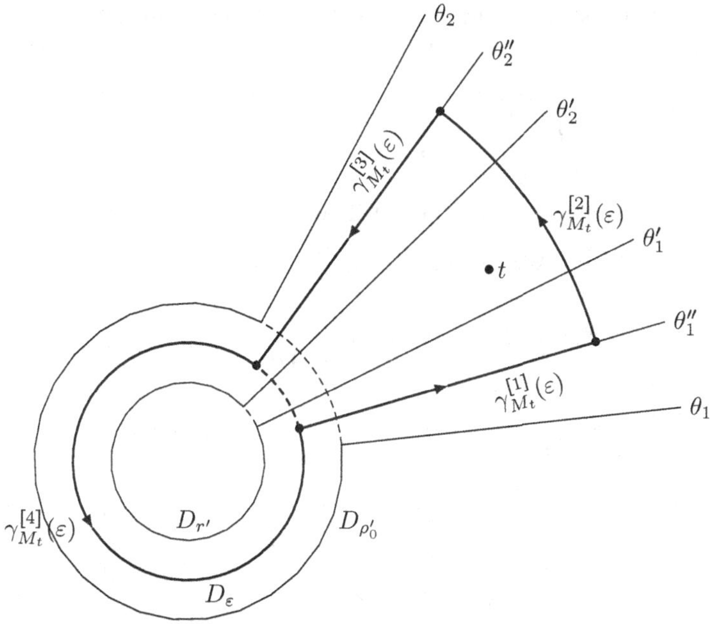

Fig. 10.1 The path $\gamma_{M_t}(\varepsilon)$ (thick black line)

and where $\gamma'(\varepsilon)$ is the polycircle

$$\gamma'(\varepsilon) = \{\xi \in \mathbb{C}^n; |\xi_\ell| = \varepsilon \text{ for all } \ell \in \{1, \ldots, n\}\}.$$

Observing then that, for all $\arg(t) \in]\theta_1', \theta_2'[$ and all $\arg(\tau) \in [\theta_2'', 2\pi + \theta_1'']$ (or $\arg(\tau) = \theta_1''$, or $\arg(\tau) = \theta_2''$), there exist a direction $\theta_{t,\tau} \in I_0$ and a direction $d_{t,\tau} \in I_0'$ such that

$$\theta_{t,\tau} = -\arg(t) + \arg(\tau) + d_{t,\tau} \mod 2\pi, \tag{10.10}$$

we deduce from our assumption on $a(t, x)$ and from the Cauchy Theorem that

$$\partial_{m_0;t}^{i_0} \partial_{m_1;x_1}^{i_1} \ldots \partial_{m_n;x_n}^{i_n} a(t, x) = \frac{1}{(2i\pi)^{n+1}} \int_{\substack{\tau \in \gamma_{M_t}(\varepsilon) \\ \xi \in \gamma'(\varepsilon)}} a(\tau, \xi) I_{i_0,i_1,\ldots,i_n}^{[t,x,\tau,\xi]} d\tau d\xi,$$

where the directions $\theta_{0,t,\tau}$ which occur in $I^{[t,x,\tau,\xi]}_{i_0,i_1,\dots,i_n}$ satisfy the condition (10.10). Notice that, according to the remark just above, it remains to bound the integral

$$\int_0^{+\infty e^{i\theta_{0,t,\tau}}} \left(\frac{v_0}{\tau}\right)^{i_0} \frac{e_0(v_0)}{v_0\tau} E_0\left(\frac{v_0 t}{\tau}\right) dv_0$$

when τ runs through each part of $\gamma_{M_t}(\varepsilon)$. To do that, we proceed as follows.

First Case $\tau \in \gamma^{[1]}_{M_t}(\varepsilon)$ Writing τ as $\tau = \rho e^{i\theta''_1}$ with $\rho \in [\varepsilon, M_t]$, we get

$$\begin{aligned}
&\left|\int_0^{+\infty e^{i\theta_{0,t,\tau}}} \left(\frac{v_0}{\tau}\right)^{i_0} \frac{e_0(v_0)}{v_0\tau} E_0\left(\frac{v_0 t}{\tau}\right) dv_0\right| \\
&\leqslant \frac{1}{\rho^{i_0+1}} \int_0^{+\infty} v_0^{i_0-1} \left|e_0(v_0 e^{i\theta_{0,t,\tau}})\right| \left|E_0\left(\frac{v_0 t e^{i\theta_{0,t,\tau}}}{\rho e^{i\theta''_1}}\right)\right| dv_0 \\
&\leqslant \frac{1}{\varepsilon^{i_0+1}} \int_0^{+\infty} v_0^{i_0-1} \left|e_0(v_0 e^{i\theta_{0,t,\tau}})\right| \left|E_0\left(\frac{v_0 t e^{i\theta_{0,t,\tau}}}{\rho e^{i\theta''_1}}\right)\right| dv_0 \\
&\leqslant \frac{1}{\varepsilon^{i_0+1}} \left(\int_0^{\frac{\rho R_0}{|t|}} v_0^{i_0-1} \left|e_0(v_0 e^{i\theta_{0,t,\tau}})\right| \left|E_0\left(\frac{v_0 t e^{i\theta_{0,t,\tau}}}{\rho e^{i\theta''_1}}\right)\right| dv_0\right. \\
&\quad \left. + \int_{\frac{\rho R_0}{|t|}}^{+\infty} v_0^{i_0-1} \left|e_0(v_0 e^{i\theta_{0,t,\tau}})\right| \left|E_0\left(\frac{v_0 t e^{i\theta_{0,t,\tau}}}{\rho e^{i\theta''_1}}\right)\right| dv_0\right).
\end{aligned}$$

To bound the first integral, we observe that

- $\dfrac{\rho R_0}{|t|} \leqslant \dfrac{M_t R_0}{|t|} = \left(\dfrac{2A'_0}{A_0}\right)^{s_0} R_0 \leqslant r_0$;
- $\left|\dfrac{v_0 t e^{i\theta_{0,t,\tau}}}{\rho e^{i\theta''_1}}\right| = \dfrac{v_0\,|t|}{\rho} \leqslant R_0$ for all $v_0 \in \left]0, \dfrac{\rho R_0}{|t|}\right]$;

hence, for all $i_0 \geqslant 0$,

$$\begin{aligned}
&\int_0^{\frac{\rho R_0}{|t|}} v_0^{i_0-1} \left|e_0(v_0 e^{i\theta_{0,t,\tau}})\right| \left|E_0\left(\frac{v_0 t e^{i\theta_{0,t,\tau}}}{\rho e^{i\theta''_1}}\right)\right| dv_0 \\
&\leqslant K_0 \max_{|z|\leqslant R_0} |E_0(z)| \int_0^{\frac{\rho R_0}{|t|}} v_0^{i_0+\alpha_0-1} dv_0 \\
&\leqslant K_0 \max_{|z|\leqslant R_0} |E_0(z)| \int_0^{r_0} v_0^{i_0+\alpha_0-1} dv_0 < +\infty.
\end{aligned}$$

As for the second integral, we observe now that for all $v_0 \in \left[\frac{\rho R_0}{|t|}, +\infty\right[$:

- $\left|\frac{v_0 t e^{i\theta_{0,t,\tau}}}{\rho e^{i\theta_1''}}\right| = \frac{v_0 |t|}{\rho} \geqslant R_0$;
- $\arg\left(\frac{v_0 t e^{i\theta_{0,t,\tau}}}{\rho e^{i\theta_1''}}\right) = \arg(t) - \theta_1'' + \theta_{0,t,\tau} = d_{0,t,\tau} \mod 2\pi$.

Then, for all $i_0 \geqslant 0$,

$$
\begin{aligned}
&\int_{\frac{\rho R_0}{|t|}}^{+\infty} v_0^{i_0-1} \left|e_0(v_0 e^{i\theta_{0,t,\tau}})\right| \left|E_0\left(\frac{v_0 t e^{i\theta_{0,t,\tau}}}{\rho e^{i\theta_1''}}\right)\right| dv_0 \\
&\leqslant K_0' \max_{v_0 \geqslant \rho R_0/|t|} \left(\frac{v_0 |t|}{\rho}\right)^{-\beta_0} \int_{\frac{\rho R_0}{|t|}}^{+\infty} v_0^{i_0-1} \left|e_0(v_0 e^{i\theta_{0,t,\tau}})\right| dv_0 \\
&\leqslant K_0' R_0^{-\beta_0} \left(K_0 \int_{\frac{\rho R_0}{|t|}}^{r_0} v_0^{i_0+\alpha_0-1} dv_0\right. \\
&\qquad \left. + C_0 \int_{r_0}^{+\infty} v_0^{i_0-1} \exp\left(-A_0 v_0^{k_0}\right) dv_0\right) \\
&\leqslant K_0' R_0^{-\beta_0} \left(K_0 \int_0^{r_0} v_0^{i_0+\alpha_0-1} dv_0\right. \\
&\qquad \left. + C_0 \int_{r_0}^{+\infty} v_0^{i_0-1} \exp\left(-A_0 v_0^{k_0}\right) dv_0\right) < +\infty.
\end{aligned}
$$

Consequently, there exists a positive constant $C^{[1]} > 0$ independent of t and x such that

$$
\begin{aligned}
\left|\int_{\substack{\tau \in \gamma_{M_t}^{[1]}(\varepsilon) \\ \xi \in \gamma'(\varepsilon)}} a(\tau, \xi) I_{i_0, i_1, \ldots, i_n}^{[t,x,\tau,\xi]} d\tau d\xi\right| &\leqslant C^{[1]} \max_{\substack{\tau \in \gamma_{M_t}^{[1]}(\varepsilon) \\ \xi \in \gamma'(\varepsilon)}} |a(\tau, x)| \\
&\leqslant C^{[1]} \max_{\substack{\rho \in [\varepsilon, M_t] \\ \xi \in \gamma'(\varepsilon)}} \left|a(\rho e^{i\theta_1''}, x)\right| \\
&\leqslant C^{[1]} C \exp\left(A M_t^k\right);
\end{aligned}
$$

hence, two positive constants $\widetilde{C}^{[1]}, \widetilde{A}^{[1]} > 0$ independent of t and x such that

$$
\left|\frac{1}{(2i\pi)^{n+1}} \int_{\substack{\tau \in \gamma_{M_t}^{[1]}(\varepsilon) \\ \xi \in \gamma'(\varepsilon)}} a(\tau, \xi) I_{i_0, i_1, \ldots, i_n}^{[t,x,\tau,\xi]} d\tau d\xi\right| \leqslant \widetilde{C}^{[1]} \exp\left(\widetilde{A}^{[1]} |t|^k\right). \tag{10.11}
$$

Second Case $\tau \in \gamma_{M_t}^{[2]}(\varepsilon)$ Writing τ as $\tau = M_t e^{i\theta}$ with $\theta \in [\theta_1'', \theta_2'']$, we get

$$\left|\int_0^{+\infty e^{i\theta_{0,t,\tau}}} \left(\frac{v_0}{\tau}\right)^{i_0} \frac{e_0(v_0)}{v_0\tau} E_0\left(\frac{v_0 t}{\tau}\right) dv_0\right|$$

$$\leqslant \frac{1}{M_t^{i_0+1}} \int_0^{+\infty} v_0^{i_0-1} \left|e_0(v_0 e^{i\theta_{0,t,\tau}})\right| \left|E_0\left(\frac{v_0 t e^{i\theta_{0,t,\tau}}}{\tau}\right)\right| dv_0;$$

hence, since $|t| \geqslant r'$ implies $M_t \geqslant \left(\frac{2A_0'}{A_0}\right)^{s_0} r'$,

$$\left|\int_0^{+\infty e^{i\theta_{0,t,\tau}}} \left(\frac{v_0}{\tau}\right)^{i_0} \frac{e_0(v_0)}{v_0\tau} E_0\left(\frac{v_0 t}{\tau}\right) dv_0\right|$$

$$\leqslant \left(\left(\frac{A_0}{2A_0'}\right)^{s_0} \frac{1}{r'}\right)^{i_0+1} \int_0^{+\infty} v_0^{i_0-1} \left|e_0(v_0 e^{i\theta_{0,t,\tau}})\right| \left|E_0\left(\frac{v_0 t e^{i\theta_{0,t,\tau}}}{\tau}\right)\right| dv_0.$$

Reasoning then as in the proof of Proposition 10.8, we obtain

$$\left|\int_0^{+\infty e^{i\theta_{0,t,\tau}}} \left(\frac{v_0}{\tau}\right)^{i_0} \frac{e_0(v_0)}{v_0\tau} E_0\left(\frac{v_0 t}{\tau}\right) dv_0\right|$$

$$\leqslant \left(\left(\frac{A_0}{2A_0'}\right)^{s_0} \frac{1}{r'}\right)^{i_0+1} \left(K_0 C_0' \int_0^{r_0} v_0^{i_0+\alpha_0-1} \exp\left(\frac{A_0 v_0^{k_0}}{2}\right) dv_0 \right.$$

$$\left. + \, C_0 C_0' \int_{r_0}^{+\infty} v_0^{i_0-1} \exp\left(-\frac{A_0 v_0^{k_0}}{2}\right) dv_0\right).$$

Consequently, there exists a positive constant $C^{[2]} > 0$ independent of t and x such that

$$\left|\int_{\substack{\tau\in\gamma_{M_t}^{[2]}(\varepsilon)\\ \xi\in\gamma'(\varepsilon)}} a(\tau,\xi) I_{i_0,i_1,\ldots,i_n}^{[t,x,\tau,\xi]} \, d\tau d\xi\right| \leqslant C^{[2]} \max_{\substack{\tau\in\gamma_{M_t}^{[2]}(\varepsilon)\\ \xi\in\gamma'(\varepsilon)}} |a(\tau,x)|$$

$$\leqslant C^{[2]} \max_{\substack{\theta\in[\theta_1'',\theta_2'']\\ \xi\in\gamma'(\varepsilon)}} \left|a(M_t e^{i\theta}, x)\right|$$

$$\leqslant C^{[2]} C \exp\left(A M_t^k\right);$$

hence, two positive constants $\widetilde{C}^{[2]}, \widetilde{A}^{[2]} > 0$ independent of t and x such that

$$\left| \frac{1}{(2i\pi)^{n+1}} \int_{\substack{\tau \in \gamma_{M_t}^{[2]}(\varepsilon) \\ \xi \in \gamma'(\varepsilon)}} a(\tau, \xi) I_{i_0,i_1,\ldots,i_n}^{[t,x,\tau,\xi]} d\tau d\xi \right| \leqslant \widetilde{C}^{[2]} \exp\left(\widetilde{A}^{[2]} |t|^k \right). \tag{10.12}$$

Third Case $\tau \in \gamma_{M_t}^{[3]}(\varepsilon)$ Writing τ as $\tau = \rho e^{i\theta_2''}$ with $\rho \in [\varepsilon, M_t]$, we prove analogously to the first case that there exist two positive constants $\widetilde{C}^{[3]}, \widetilde{A}^{[3]} > 0$ independent of t and x such that

$$\left| \frac{1}{(2i\pi)^{n+1}} \int_{\substack{\tau \in \gamma_{M_t}^{[3]}(\varepsilon) \\ \xi \in \gamma'(\varepsilon)}} a(\tau, \xi) I_{i_0,i_1,\ldots,i_n}^{[t,x,\tau,\xi]} d\tau d\xi \right| \leqslant \widetilde{C}^{[3]} \exp\left(\widetilde{A}^{[3]} |t|^k \right). \tag{10.13}$$

Fourth Case $\tau \in \gamma_{M_t}^{[4]}(\varepsilon)$ Let us now write τ as $\tau = \varepsilon e^{i\theta}$ with $\theta \in [\theta_2'', 2\pi + \theta_1'']$. In the same way as in the first case, we have

$$\begin{aligned}
&\left| \int_0^{+\infty e^{i\theta_{0,t,\tau}}} \left(\frac{v_0}{\tau} \right)^{i_0} \frac{e_0(v_0)}{v_0 \tau} E_0\left(\frac{v_0 t}{\tau} \right) dv_0 \right| \\
&\quad \leqslant \frac{1}{\varepsilon^{i_0+1}} \int_0^{+\infty} v_0^{i_0-1} \left| e_0(v_0 e^{i\theta_{0,t,\tau}}) \right| \left| E_0\left(\frac{v_0 t e^{i\theta_{0,t,\tau}}}{\tau} \right) \right| dv_0 \\
&\quad \leqslant \frac{1}{\varepsilon^{i_0+1}} \left(\int_0^{\frac{\varepsilon R_0}{|t|}} v_0^{i_0-1} \left| e_0(v_0 e^{i\theta_{0,t,\tau}}) \right| \left| E_0\left(\frac{v_0 t e^{i\theta_{0,t,\tau}}}{\tau} \right) \right| dv_0 \right. \\
&\qquad \left. + \int_{\frac{\varepsilon R_0}{|t|}}^{+\infty} v_0^{i_0-1} \left| e_0(v_0 e^{i\theta_{0,t,\tau}}) \right| \left| E_0\left(\frac{v_0 t e^{i\theta_{0,t,\tau}}}{\tau} \right) \right| dv_0 \right).
\end{aligned}$$

To bound the first integral, we observe that

- $|t| \geqslant r'$ implies $\dfrac{\varepsilon R_0}{|t|} \leqslant \dfrac{\varepsilon R_0}{r'} \leqslant r_0$;
- $\left| \dfrac{v_0 t e^{i\theta_{0,t,\tau}}}{\tau} \right| = \dfrac{v_0 |t|}{\varepsilon} \leqslant R_0$ for all $v_0 \in \left] 0, \dfrac{\varepsilon R_0}{|t|} \right]$;

hence, for all $i_0 \geqslant 0$,

$$\begin{aligned}
&\int_0^{\frac{\varepsilon R_0}{|t|}} v_0^{i_0-1} \left|e_0(v_0 e^{i\theta_{0,t,\tau}})\right| \left|E_0\left(\frac{v_0 t e^{i\theta_{0,t,\tau}}}{\tau}\right)\right| dv_0 \\
&\quad \leqslant K_0 \max_{|z|\leqslant R_0} |E_0(z)| \int_0^{\frac{\varepsilon R_0}{|t|}} v_0^{i_0+\alpha_0-1} dv_0 \\
&\quad \leqslant K_0 \max_{|z|\leqslant R_0} |E_0(z)| \int_0^{r_0} v_0^{i_0+\alpha_0-1} dv_0 < +\infty.
\end{aligned}$$

As for the second integral, we observe now that for all $v_0 \in \left[\frac{\varepsilon R_0}{|t|}, +\infty\right[$:

- $\left|\frac{v_0 t e^{i\theta_{0,t,\tau}}}{\tau}\right| = \frac{v_0 |t|}{\varepsilon} \geqslant R_0$;
- $\arg\left(\frac{v_0 t e^{i\theta_{0,t,\tau}}}{\tau}\right) = \arg(t) - \arg(\tau) + \theta_{0,t,\tau} = d_{0,t,\tau} \mod 2\pi$.

Then, for all $i_0 \geqslant 0$,

$$\begin{aligned}
&\int_{\frac{\varepsilon R_0}{|t|}}^{+\infty} v_0^{i_0-1} \left|e_0(v_0 e^{i\theta_{0,t,\tau}})\right| \left|E_0\left(\frac{v_0 t e^{i\theta_{0,t,\tau}}}{\tau}\right)\right| dv_0 \\
&\quad \leqslant K_0' \max_{v_0 \geqslant \varepsilon R_0/|t|} \left(\frac{v_0 |t|}{\varepsilon}\right)^{-\beta_0} \int_{\frac{\varepsilon R_0}{|t|}}^{+\infty} v_0^{i_0-1} \left|e_0(v_0 e^{i\theta_{0,t,\tau}})\right| dv_0 \\
&\quad \leqslant K_0' R_0^{-\beta_0} \left(K_0 \int_{\frac{\varepsilon R_0}{|t|}}^{r_0} v_0^{i_0+\alpha_0-1} dv_0\right. \\
&\qquad \left. + C_0 \int_{r_0}^{+\infty} v_0^{i_0-1} \exp\left(-A_0 v_0^{k_0}\right) dv_0\right) \\
&\quad \leqslant K_0' R_0^{-\beta_0} \left(K_0 \int_0^{r_0} v_0^{i_0+\alpha_0-1} dv_0\right. \\
&\qquad \left. + C_0 \int_{r_0}^{+\infty} v_0^{i_0-1} \exp\left(-A_0 v_0^{k_0}\right) dv_0\right) < +\infty.
\end{aligned}$$

Consequently, there exists a positive constant $C^{[4]} > 0$ independent of t and x such that

$$\left| \int_{\substack{\tau\in\gamma^{[4]}_{M_t}(\varepsilon)\\ \xi\in\gamma'(\varepsilon)}} a(\tau,\xi) I^{[t,x,\tau,\xi]}_{i_0,i_1,\ldots,i_n} d\tau d\xi \right| \leqslant C^{[4]} \max_{\substack{\tau\in\gamma^{[4]}_{M_t}(\varepsilon)\\ \xi\in\gamma'(\varepsilon)}} |a(\tau,x)|$$

$$\leqslant C^{[4]} \max_{\substack{\theta\in[\theta''_2, 2\pi+\theta''_1]\\ \xi\in\gamma'(\varepsilon)}} \left| a(\varepsilon e^{i\theta}, x) \right| < +\infty;$$

hence, two positive constants $\widetilde{C}^{[4]}, \widetilde{A}^{[4]} > 0$ independent of t and x such that

$$\left| \frac{1}{(2i\pi)^{n+1}} \int_{\substack{\tau\in\gamma^{[4]}_{M_t}(\varepsilon)\\ \xi\in\gamma'(\varepsilon)}} a(\tau,\xi) I^{[t,x,\tau,\xi]}_{i_0,i_1,\ldots,i_n} d\tau d\xi \right| \leqslant \widetilde{C}^{[4]} \exp\left(\widetilde{A}^{[4]} |t|^k \right). \tag{10.14}$$

Inequalities (10.11), (10.12), (10.13), and (10.14) prove then the second point of Proposition 10.11 when $s_0 < 2$.

Let us now assume $s_0 \geqslant 2$. Accordingly to Proposition 8.7 (see also Remark 8.8), there exist a positive integer p and a pair of kernel function $(\widetilde{e}_0, \widetilde{E}_0)$ of order $s_0/p < 2$ with $\widetilde{e}_0$ as in (8.8). Denoting then by $\widetilde{m}_0$ the corresponding moment function, we have $\widetilde{m}_0(\lambda) = m_0(\lambda/p)$; hence, $m_0(j) = \widetilde{m}_0(pj)$.

Let us set $z = t^{1/p}$ and $\widetilde{a}(z,x) = a(z^p, x) = a(t,x)$. Then, $\widetilde{a}(z,x)$ is an analytic function at the origin of $\mathbb{C}^{n+1}$ and, writing $a(t,x)$ as

$$a(t,x) = \sum_{j_0,j_1,\ldots,j_n\geqslant 0} a_{j_0,j_1,\ldots,j_n} \frac{t^{j_0}}{m_0(j_0)} \frac{x_1^{j_1}}{m_1(j_1)} \cdots \frac{x_n^{j_n}}{m_n(j_n)},$$

we get

$$\begin{aligned}
\widetilde{a}(z,x) &= \sum_{j_0,j_1,\ldots,j_n\geqslant 0} a_{j_0,j_1,\ldots,j_n} \frac{z^{pj_0}}{m_0(j_0)} \frac{x_1^{j_1}}{m_1(j_1)} \cdots \frac{x_n^{j_n}}{m_n(j_n)} \\
&= \sum_{j_0,j_1,\ldots,j_n\geqslant 0} a_{j_0,j_1,\ldots,j_n} \frac{z^{pj_0}}{\widetilde{m}_0(pj_0)} \frac{x_1^{j_1}}{m_1(j_1)} \cdots \frac{x_n^{j_n}}{m_n(j_n)} \\
&= \sum_{j'_0,j_1,\ldots,j_n\geqslant 0} \widetilde{a}_{j'_0,j_1,\ldots,j_n} \frac{z^{j'_0}}{\widetilde{m}_0(j'_0)} \frac{x_1^{j_1}}{m_1(j_1)} \cdots \frac{x_n^{j_n}}{m_n(j_n)}
\end{aligned}$$

with, for all $j_1, \ldots, j_n \geqslant 0$,

$$\widetilde{a}_{j'_0, j_1, \ldots, j_n} = \begin{cases} a_{j_0, j_1, \ldots, j_n} & \text{if } j'_0 = pj_0 \\ 0 & \text{otherwise} \end{cases}.$$

Consequently,

$$\begin{aligned}
&\partial^{pi_0}_{\widetilde{m}_0;z} \partial^{i_1}_{m_1;x_1} \ldots \partial^{i_n}_{m_n;x_n} \widetilde{a}(z, x) \\
&= \sum_{j'_0, \ldots, j_n \geqslant 0} \widetilde{a}_{j'_0 + pi_0, j_1 + i_1, \ldots, j_n + i_n} \frac{z^{j'_0}}{\widetilde{m}_0(j'_0)} \frac{x_1^{j_1}}{m_1(j_1)} \cdots \frac{x_n^{j_n}}{m_n(j_n)} \\
&= \sum_{j_0, \ldots, j_n \geqslant 0} a_{j_0 + i_0, j_1 + i_1, \ldots, j_n + i_n} \frac{z^{pj_0}}{\widetilde{m}_0(pj_0)} \frac{x_1^{j_1}}{m_1(j_1)} \cdots \frac{x_n^{j_n}}{m_n(j_n)} \\
&= \sum_{j_0, \ldots, j_n \geqslant 0} a_{j_0 + i_0, j_1 + i_1, \ldots, j_n + i_n} \frac{t^{j_0}}{m_0(j_0)} \frac{x_1^{j_1}}{m_1(j_1)} \cdots \frac{x_n^{j_n}}{m_n(j_n)} \\
&= \partial^{i_0}_{m_0;t} \partial^{i_1}_{m_1;x_1} \ldots \partial^{i_n}_{m_n;x_n} a(t, x).
\end{aligned}$$

On the other hand, according to our assumption on $a(t, x)$, the function $\widetilde{a}(z, x)$ can be analytically continued to the domain $\Sigma_{\theta_1/p, \theta_2/p}(\infty) \times D_{r,\ldots,r}$ with a global exponential growth of order at most kp at infinity with respect to z. Then, from the calculations made in the case $s_0 < 2$, we deduce that there exists a radius $r' > 0$ such that $\partial^{pi_0}_{\widetilde{m}_0;z} \partial^{i_1}_{m_1;x_1} \ldots \partial^{i_n}_{m_n;x_n} \widetilde{a}(z, x)$ can be analytically continued to the domain $\Sigma_{\theta'_1/p, \theta'_2/p}(\infty) \times D_{r',\ldots,r'}$ with a global exponential growth of order at most kp at infinity with respect to z. Hence, thanks to the identities just above, $\partial^{i_0}_{m_0;t} \partial^{i_1}_{m_1;x_1} \ldots \partial^{i_n}_{m_n;x_n} a(t, x)$ can be analytically continued to the domain $\Sigma_{\theta'_1, \theta'_2}(\infty) \times D_{r',\ldots,r'}$ with a global exponential growth of order at most k at infinity with respect to t. This completes the proof of Proposition 10.11. ■

10.1.3 Action of the Formal Moment-Borel and Laplace Operators

We end this section devoted to the moment derivation by describing the action of the formal moment-Borel and formal moment-Laplace operators on the moment derivation and moment anti-derivation operators.

Proposition 10.12 *Let $m, m_0, m_1, \ldots, m_n$ be $n+2$ moment functions of respective order $s, s_0, s_1, \ldots, s_n > 0$. Let $k = 1/s$.*

1. Action of the formal moment-Borel transform $\widetilde{\mathcal{B}}_k^{[m;t]}$. *For any formal power series $\widetilde{u}(t,x) \in \mathcal{O}(D_{\rho_1,\ldots,\rho_n})[[t]]$:*

   (a) $\widetilde{\mathcal{B}}_k^{[m;t]}\left(\partial_{m_0;t}\widetilde{u}\right)(\tau,x) = \partial_{mm_0;\tau}\left(\widetilde{\mathcal{B}}_k^{[m;t]}(\widetilde{u})\right)(\tau,x)$

   (b) $\widetilde{\mathcal{B}}_k^{[m;t]}\left(\partial_{m_0;t}^{-1}\widetilde{u}\right)(\tau,x) = \partial_{mm_0;\tau}^{-1}\left(\widetilde{\mathcal{B}}_k^{[m;t]}(\widetilde{u})\right)(\tau,x)$

   (c) $\widetilde{\mathcal{B}}_k^{[m;t]}$ *commutes with* $\partial_{m_j;x_j}$ *and* $\partial_{m_j;x_j}^{-1}$ *for all* $j \in \{1,\ldots,n\}$*:*

      (i) $\widetilde{\mathcal{B}}_k^{[m;t]}\left(\partial_{m_j;x_j}\widetilde{u}\right)(\tau,x) = \partial_{m_j;x_j}\left(\widetilde{\mathcal{B}}_k^{[m;t]}(\widetilde{u})\right)(\tau,x)$

      (ii) $\widetilde{\mathcal{B}}_k^{[m;t]}\left(\partial_{m_j;x_j}^{-1}\widetilde{u}\right)(\tau,x) = \partial_{m_j;x_j}^{-1}\left(\widetilde{\mathcal{B}}_k^{[m;t]}(\widetilde{u})\right)(\tau,x)$

2. Action of the formal moment-Laplace transform $\widetilde{\mathcal{L}}_k^{[m;\tau]}$. *Assume $s_0 < s$. For any formal power series $\widetilde{u}(\tau,x) \in \mathcal{O}(D_{\rho_1,\ldots,\rho_n})[[\tau]]$:*

   (a) $\widetilde{\mathcal{L}}_k^{[m;\tau]}\left(\partial_{m_0;\tau}\widetilde{u}\right)(t,x) = \partial_{m/m_0;t}\left(\widetilde{\mathcal{L}}_k^{[m;\tau]}(\widetilde{u})\right)(t,x)$

   (b) $\widetilde{\mathcal{L}}_k^{[m;\tau]}\left(\partial_{m_0;\tau}^{-1}\widetilde{u}\right)(t,x) = \partial_{m/m_0;t}^{-1}\left(\widetilde{\mathcal{L}}_k^{[m;\tau]}(\widetilde{u})\right)(t,x)$

   (c) $\widetilde{\mathcal{L}}_k^{[m;\tau]}$ *commutes with* $\partial_{m_j;x_j}$ *and* $\partial_{m_j;x_j}^{-1}$ *for all* $j \in \{1,\ldots,n\}$*:*

      (i) $\widetilde{\mathcal{L}}_k^{[m;\tau]}\left(\partial_{m_j;x_j}\widetilde{u}\right)(t,x) = \partial_{m_j;x_j}\left(\widetilde{\mathcal{L}}_k^{[m;\tau]}(\widetilde{u})\right)(t,x)$

      (ii) $\widetilde{\mathcal{L}}_k^{[m;\tau]}\left(\partial_{m_j;x_j}^{-1}\widetilde{u}\right)(t,x) = \partial_{m_j;x_j}^{-1}\left(\widetilde{\mathcal{L}}_k^{[m;\tau]}(\widetilde{u})\right)(t,x)$

Recall (see Propositions 8.26 and 8.27) that mm_0 (resp. m/m_0) is a moment function of order $s+s_0$ (resp. $s-s_0$).

Proof Let us write the formal power series $\widetilde{u}(t,x) \in \mathcal{O}(D_{\rho_1,\ldots,\rho_n})[[t]]$ in the form

$$\widetilde{u}(t,x) = \sum_{j\geqslant 0} u_{j,*}(x)\frac{t^j}{m_0(j)}.$$

Then, identity (1)-(a) is straightforward from the two following calculations:

$$\widetilde{\mathcal{B}}_k^{[m;t]}\left(\partial_{m_0;t}\widetilde{u}\right)(\tau,x) = \widetilde{\mathcal{B}}_k^{[m;t]}\left(\sum_{j\geqslant 0}\frac{u_{j+1,*}(x)t^j}{m_0(j)}\right)(\tau,x) = \sum_{j\geqslant 0}\frac{u_{j+1,*}(x)\tau^j}{m(j)m_0(j)}$$

$$\partial_{mm_0;\tau}\left(\widetilde{\mathcal{B}}_k^{[m;t]}(\widetilde{u})\right)(\tau,x) = \partial_{mm_0;\tau}\left(\sum_{j\geqslant 0}\frac{u_{j,*}(x)\tau^j}{m(j)m_0(j)}\right) = \sum_{j\geqslant 0}\frac{u_{j+1,*}(x)\tau^j}{m(j)m_0(j)}.$$

The other identities are proved in the same way and are left to the reader. ∎

Combining Propositions 10.11 and 10.12, we can now prove that the set of all the k-summable formal power series in a given direction θ is closed under the moment derivations.

Proposition 10.13 *Let* $\widetilde{u}(t,x) \in \mathcal{O}(D_{\rho_1,\dots,\rho_n})\{t\}_{k;\theta}$ *be a k-summable formal power series in a given direction* θ.

Let $m_0, m_1, \dots, m_n$ *be* $n+1$ *moment functions of respective order* $s_0, s_1, \dots, s_n > 0$.

Let $i_0, i_1, \dots, i_n \geqslant 0$. *Then,* $\partial_{m_0;t}^{i_0} \partial_{m_1;x_1}^{i_1} \dots \partial_{m_n;x_n}^{i_n} \widetilde{u}(t,x) \in \mathcal{O}(D_{\rho_1',\dots,\rho_n'})\{t\}_{k;\theta}$ *for some convenient radii* $0 < \rho_j' \leqslant \rho_j$ *for all* $j = 1, \dots, n$.

Proof Let us set $\widetilde{v}(t,x) = \partial_{m_0;t}^{i_0} \partial_{m_1;x_1}^{i_1} \dots \partial_{m_n;x_n}^{i_n} \widetilde{u}(t,x)$. From Proposition 10.6, there exist n radii $\rho_1', \dots, \rho_n' > 0$ such that $\rho_j' \leqslant \rho_j$ for all $j = 1, \dots, n$ and $\widetilde{v}(t,x) \in \mathcal{O}(D_{\rho_1',\dots,\rho_n'})[[t]]$.

Let us now choose a moment function m of order $s = 1/k$. From Proposition 10.12, we have

$$\widetilde{\mathcal{B}}_k^{[m;t]}(\widetilde{v})(\tau,x) = \partial_{mm_0;\tau}^{i_0} \partial_{m_1;x_1}^{i_1} \dots \partial_{m_n;x_n}^{i_n} \left(\widetilde{\mathcal{B}}_k^{[m;t]}(\widetilde{u})\right)(\tau,x),$$

where, thanks to Proposition 9.1, the formal power series $\widetilde{\mathcal{B}}_k^{[m;t]}(\widetilde{u})(\tau,x)$ defines an analytic function at the origin of $\mathbb{C}^{n+1}$ which can be analytically continued to a domain $\Sigma_{\theta_1,\theta_2}(\infty) \times D_{r,\dots,r}$ for some directions $\theta_1 < \theta < \theta_2$ and some radius $0 < r \leqslant \min(\rho_1', \dots, \rho_n')$, with a global exponential growth of order at most k at infinity with respect to τ. Applying then Proposition 10.11, we deduce that $\widetilde{\mathcal{B}}_k^{[m;t]}(\widetilde{v})(\tau,x)$ also defines an analytic function at the origin of $\mathbb{C}^{n+1}$ which can be analytically continued to a domain $\Sigma_{\theta_1',\theta_2'}(\infty) \times D_{r',\dots,r'}$ for some directions $\theta_1' < \theta < \theta_2'$ and some radius $0 < r' \leqslant \min(\rho_1', \dots, \rho_n')$, with a global exponential growth of order at most k at infinity with respect to τ. Hence, $\widetilde{v}(t,x) \in \mathcal{O}(D_{\rho_1',\dots,\rho_n'})\{t\}_{k;\theta}$ from Proposition 9.1, which achieves the proof. ∎

As a consequence of Proposition 10.13 and the definition of the k-sum of a formal power series, we can extend the definition of the moment derivations as follows.

Definition 10.14 Let $\widetilde{u}(t,x) \in \mathcal{O}(D_{\rho_1,\dots,\rho_n})\{t\}_{k;\theta}$ be a k-summable formal power series in a given direction θ. Let $u(t,x)$ be the k-sum of $\widetilde{u}(t,x)$ in the direction θ.

Let $m_0, m_1, \dots, m_n$ be $n+1$ moment functions of respective order $s_0, s_1, \dots, s_n > 0$.

Let $i_0, i_1, \dots, i_n \geqslant 0$. Then, the moment derivative $\partial_{m_0;t}^{i_0} \partial_{m_1;x_1}^{i_1} \dots \partial_{m_n;x_n}^{i_n} u(t,x)$ of the function $u(t,x)$ is given by

$$\partial_{m_0;t}^{i_0} \partial_{m_1;x_1}^{i_1} \dots \partial_{m_n;x_n}^{i_n} u(t,x) = \mathcal{S}_{k;\theta}(\partial_{m_0;t}^{i_0} \partial_{m_1;x_1}^{i_1} \dots \partial_{m_n;x_n}^{i_n} \widetilde{u})(t,x),$$

where $\mathcal{S}_{k;\theta}$ stands for the operator of k-summation (see Corollary 5.5).

Doing so, observe in particular that this definition implies that $\mathcal{S}_{k;\theta}$ commutes with any moment derivation.

10.2 The General Problem

Let us now give ourselves $n + 1$ moment functions $m_0, m_1, \dots, m_n$ of respective order $s_0, s_1, \dots, s_n > 0$, and let us consider the general inhomogeneous linear moment partial differential equation in 1-dimensional time variable $t \in \mathbb{C}$ and n-dimensional spatial variable $x = (x_1, \dots, x_n) \in \mathbb{C}^n$ of the form

$$\begin{cases} \partial_{m_0;t}^{\kappa} u - \sum_{i \in \mathcal{K}} \sum_{q \in Q_i} a_{i,q}(t,x) \partial_{m_0;t}^{i} \partial_{m;x}^{q} u = \widetilde{f}(t,x) \\ \partial_{m_0;t}^{j} u(t,x)_{|t=0} = \varphi_j(x),\ j = 0, \dots, \kappa - 1 \end{cases} \tag{10.15}$$

where

- $\kappa \geqslant 1$ is a positive integer;
- $\mathcal{K}$ is a nonempty subset of $\{0, \dots, \kappa - 1\}$;
- Q_i is a nonempty finite subset of $\mathbb{N}^n$ for all $i \in \mathcal{K}$;
- the coefficients $a_{i,q}(t,x)$ are analytic and not identically zero on a polydisc $D_{\rho_0,\rho_1,\dots,\rho_n}$ centered at the origin of $\mathbb{C}^{n+1}$ for all $i \in \mathcal{K}$ and all $q \in Q_i$;
- $\partial_{m;x}^{q}$ denotes the moment derivation $\partial_{m_1;x_1}^{q_1} \dots \partial_{m_n;x_n}^{q_n}$ while $q = (q_1, \dots, q_n) \in \mathbb{N}^n$;
- the inhomogeneity $\widetilde{f}(t,x)$ belongs to $\mathcal{O}(D_{\rho_1,\dots,\rho_n})[[t]]$;
- the initial conditions $\varphi_j(x)$ are analytic on $D_{\rho_1,\dots,\rho_n}$ for all $j = 0, \dots, \kappa - 1$.

Let us also assume that $\partial_{m;x}^{q} a \in \mathcal{O}(D_{\rho_1,\dots,\rho_n})$ for all $q \in \mathbb{N}^n$ and all $a \in \mathcal{O}(D_{\rho_1,\dots,\rho_n})$.

Observe that in the special case where $m_0(\lambda) = \dots = m_n(\lambda) = \Gamma(1 + \lambda)$, Eq. (10.15) is reduced to a classical inhomogeneous linear partial differential equation which we have already studied in Sects. 3.2, 6.2, and 7.5.

Observe also that, as for the latter, Eq. (10.15) is formally well-posed.

Lemma 10.15 *Equation (10.15) admits a unique formal solution* $\widetilde{u}(t,x) \in \mathcal{O}(D_{\rho_1,\dots,\rho_n})[[t]]$.

Proof Let us write the coefficients $a_{i,q}(t,x)$ and the inhomogeneity $\widetilde{f}(t,x)$ in the form

$$a_{i,q}(t,x) = \sum_{j \geqslant 0} a_{i,q;j,*}(x) \frac{t^j}{m_0(j)} \quad \text{and} \quad \widetilde{f}(t,x) = \sum_{j \geqslant 0} f_{j,*}(x) \frac{t^j}{m_0(j)}$$

with $a_{i,q;j,*}(x), f_{j,*}(x) \in \mathcal{O}(D_{\rho_1,\dots,\rho_n})$ for all $j \geqslant 0$. Looking for $\widetilde{u}(t,x)$ in the same type:

$$\widetilde{u}(t,x) = \sum_{j \geqslant 0} u_{j,*}(x) \frac{t^j}{m_0(j)} \quad \text{with } u_{j,*}(x) \in \mathcal{O}(D_{\rho_1,\dots,\rho_n}) \text{ for all } j \geqslant 0,$$

one easily checks that the coefficients $u_{j,*}(x)$ are uniquely determined for all $j \geqslant 0$ by the recurrence relations

$$u_{j+\kappa,*}(x) = f_{j,*}(x) + \sum_{i\in\mathcal{K}}\sum_{q\in Q_i}\sum_{\ell=0}^{j} \frac{m_0(j)}{m_0(\ell)m_0(j-\ell)} a_{i,q;\ell,*}(x)\partial_{m;x}^q u_{j-\ell+i,*}(x) \tag{10.16}$$

together with the initial conditions $u_{j,*}(x) = \varphi_j(x)$ for all $j = 0, \ldots, \kappa - 1$, which ends the proof. ■

As in the case of classical partial differential equations (see Sect. 3.2), the Gevrey regularity of the formal solution $\widetilde{u}(t,x) \in \mathcal{O}(D_{\rho_1,\ldots,\rho_n})[[t]]$ of Eq. (10.15) depends both on the Gevrey regularity of the inhomogeneity $f(t,x)$ and on the structure of Eq. (10.15) provided by its associated linear moment operator

$$L = \partial_{m_0;t}^{\kappa} - \sum_{i\in\mathcal{K}}\sum_{q\in Q_i} a_{i,q}(t,x)\partial_{m_0;t}^{i}\partial_{m;x}^{q}. \tag{10.17}$$

As before, one can describe this structure by means of a Newton polygon, called *moment Newton polygon*. This notion, introduced by S. Michalik in [81] for constant coefficients and generalized later by S. Michalik and M. Suwińska in [83, 122] for variable coefficients, extends naturally the notion of Newton polygon of classical linear partial differential operators (see Definition 3.14) in order to take into account the various orders $s_0, \ldots, s_n > 0$ which are not necessarily equal to 1 anymore.

Definition 10.16 (Moment Newton Polygon) We call *moment Newton polygon at $t = 0$ of the linear moment operator (10.17)* the domain $\mathcal{N}_{m_0,\ldots,m_n;t}(L)$ defined as the convex hull of

$$C(s_0\kappa, -\kappa) \cup \bigcup_{i\in\mathcal{K}}\bigcup_{q\in Q_i} C\left(\sum_{j=1}^{n} s_j q_j + s_0 i,\, v_{i,q} - i\right),$$

where

- $C(a,b) = \{(x,y) \in \mathbb{R}^2; x \leqslant a \text{ and } y \geqslant b\}$ for all $a, b \in \mathbb{R}$;
- $v_{i,q} \geqslant 0$ stands for the valuation at $t = 0$ of $a_{i,q}(t,x)$: $a_{i,q}(t,x) = t^{v_{i,q}} a'_{i,q}(t,x)$ with $a'_{i,q}(0,x) \not\equiv 0$.

The following proposition, which generalizes the result of Proposition 3.15, describes the general geometric structure of $\mathcal{N}_{m_0,\ldots,m_n;t}(L)$.

Proposition 10.17 *Let*

$$\mathcal{S}_{m_0,\ldots,m_n} = \left\{ (i,q) \text{ such that } i \in \mathcal{K},\ q \in Q_i \text{ and } \sum_{j=1}^{n} s_j q_j > s_0(\kappa - i) \right\}.$$

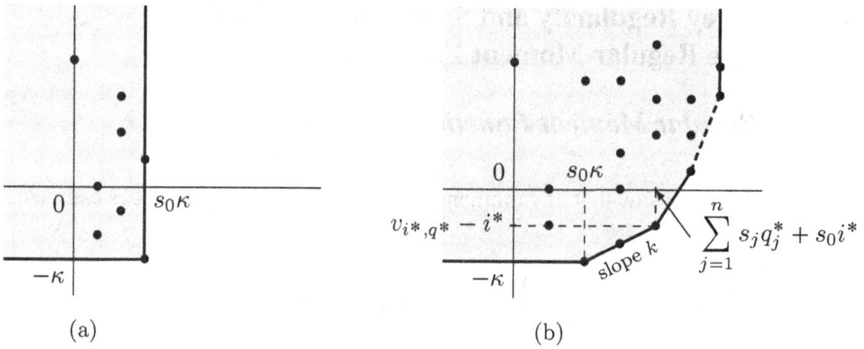

Fig. 10.2 The moment Newton polygon $\mathcal{N}_{m_0,\dots,m_n;t}(L)$. (**a**) Case $\mathcal{S}_{m_0,\dots,m_n} = \emptyset$. (**b**) Case $\mathcal{S}_{m_0,\dots,m_n} \neq \emptyset$

1. *Suppose* $\mathcal{S}_{m_0,\dots,m_n} = \emptyset$. *Then,* $\mathcal{N}_{m_0,\dots,m_n;t}(L) = C(s_0\kappa, -\kappa)$. *In particular,* $\mathcal{N}_{m_0,\dots,m_n;t}(L)$ *has no side with a positive slope (see Fig. 10.2a).*
2. *Suppose* $\mathcal{S}_{m_0,\dots,m_n} \neq \emptyset$. *Then,* $\mathcal{N}_{m_0,\dots,m_n;t}(L)$ *has at least one side with a positive slope. Moreover, its smallest positive slope k is given by*

$$k = \min_{(i,q)\in\mathcal{S}_{m_0,\dots,m_n}} \left(\frac{\kappa - i + v_{i,q}}{\sum_{j=1}^{n} s_j q_j - s_0(\kappa - i)} \right) = \frac{\kappa - i^* + v_{i^*,q^*}}{\sum_{j=1}^{n} s_j q_j^* - s_0(\kappa - i^*)},$$

where the pair $(i^*, q^*) \in \mathcal{S}_{m_0,\dots,m_n}$ *stands for any convenient pair, chosen and fixed once and for all, so that the edge of* $\mathcal{N}_{m_0,\dots,m_n;t}(L)$ *with slope k be the segment with end points* $(s_0\kappa, -\kappa)$ *and* $\left(\sum_{j=1}^{n} s_j q_j^* + s_0 i^*, v_{i^*,q^*} - i^*\right)$ *(see Fig. 10.2b).*

Proof The proof is similar to the one of Proposition 3.15 and is left to the reader. ∎

In the following two sections, we shall now present some results about the Gevrey regularity and the summability of the formal solution of Eq. (10.15). For the functions $m_0, \dots, m_n$, we first choose them from a particularly important class of moment functions: the so-called *regular moment functions*.

10.3 Gevrey Regularity and Summability: The Case of the Regular Moment Functions

10.3.1 Regular Moment Functions

In Lemma 10.1, we saw that any moment function of order $s > 0$ induces estimates of the form

$$c'a'^j(j+1)^s \leqslant \frac{m(j+1)}{m(j)} \leqslant C'A'^j(j+1)^s \quad \text{for all } j \geqslant 0 \tag{10.18}$$

with convenient positive constants $c', C', a', A' > 0$. In this section, we focus on some particular moment functions of order s, called *s-regular moment functions*, for which we impose strong restrictive conditions in the estimates (10.18).

Definition 10.18 (s-Regular Moment Function) A moment function m of order $s > 0$ is called *a s-regular moment function* if there exist two positive constants $c, C > 0$ such that the following estimate holds for all $j \geqslant 0$:

$$c(j+1)^s \leqslant \frac{m(j+1)}{m(j)} \leqslant C(j+1)^s.$$

The following proposition is straightforward from the definition.

Proposition 10.19 *Let m_1 and m_2 be two regular moment functions of respective order s_1 and s_2. Then,*

1. *the moment function m_1m_2 is a (s_1+s_2)-regular moment function;*
2. *if $s_1 > s_2$, the moment function m_1/m_2 is a (s_1-s_2)-regular moment function.*

The following example gives us some classical regular moment functions.

Example 10.20

1. Let $s, \alpha > 0$. From the Stirling's Formula, we have

$$\frac{\Gamma(\alpha+s(j+1))}{\Gamma(\alpha+sj)} \underset{j\to+\infty}{\sim} s^s(j+1)^s$$

   and, consequently, the moment function $m(\lambda) = \Gamma(\alpha+s\lambda)$ (see Examples 8.3, (2) and 8.9, (1)) is a s-regular moment function.
2. As an application of Proposition 10.19, the moment function

$$m_{s_1,s_2}(\lambda) = \frac{\Gamma(1+s_1\lambda)}{\Gamma(1+s_2\lambda)}, \; s_1 > s_2 > 0$$

   introduced in Examples 8.3, (4) and 8.9, (2) is a (s_1-s_2)-regular moment function.

3. More generally (see Example 8.28), the moment function

$$m(\lambda) = \frac{\Gamma(\alpha_1 + s_1\lambda)...\Gamma(\alpha_\nu + s_\nu\lambda)}{\Gamma(\beta_1 + \sigma_1\lambda)...\Gamma(\beta_\mu + \sigma_\mu\lambda)}$$

with $\alpha_j, \beta_j, s_j, \sigma_j > 0$ and

$$s = \sum_{j=1}^{\nu} s_j - \sum_{j=1}^{\mu} \sigma_j > 0$$

is a s-regular moment function.

We end this section by a result specifying the statement of Proposition 10.6 on the moment (anti)-derivatives of analytic functions at the origin of $\mathbb{C}^{n+1}$.

Proposition 10.21 *Let $a(t,x) \in \mathcal{O}(D_{\rho_0,\rho_1,...,\rho_n})$ be an analytic function on $D_{\rho_0,\rho_1,...,\rho_n}$, and let $m_0, m_1, \ldots, m_n$ be $n+1$ regular moment functions of respective orders $s_0, s_1, \ldots, s_n > 0$. Then, for any $i_0, i_1, \ldots, i_n \geqslant 0$, the two formal power series $\partial_{m_0;t}^{i_0}\partial_{m_1;x_1}^{i_1}...\partial_{m_n;x_n}^{i_n} a(t,x)$ and $\partial_{m_0;t}^{-i_0}\partial_{m_1;x_1}^{-i_1}...\partial_{m_n;x_n}^{-i_n} a(t,x)$ define also analytic functions on $D_{\rho_0,\rho_1,...,\rho_n}$.*

Proof Let us write $a(t,x)$ in the form

$$a(t,x) = \sum_{j_0,j_1,...,j_n \geqslant 0} a_{j_0,j_1,...,j_n} \frac{t^{j_0}}{m_0(j_0)} \frac{x_1^{j_1}}{m_1(j_1)} \cdots \frac{x_n^{j_n}}{m_n(j_n)}$$

so that

$$\partial_{m_0;t}^{i_0}\partial_{m_1;x_1}^{i_1}...\partial_{m_n;x_n}^{i_n} a(t,x) = \sum_{j_0,...,j_n \geqslant 0} v_{j_0,j_1,...,j_n} t^{j_0} x_1^{j_1} ... x_n^{j_n} \text{ and}$$

$$\partial_{m_0;t}^{-i_0}\partial_{m_1;x_1}^{-i_1}...\partial_{m_n;x_n}^{-i_n} a(t,x) = \sum_{j_0 \geqslant i_0,...,j_n \geqslant i_n} w_{j_0,j_1,...,j_n} t^{j_0} x_1^{j_1} ... x_n^{j_n}$$

with

$$v_{j_0,j_1,...,j_n} = \frac{a_{j_0+i_0,j_1+i_1,...,j_n+i_n}}{m_0(j_0)m_1(j_1)...m_n(j_n)} \text{ and}$$

$$w_{j_0,j_1,...,j_n} = \frac{a_{j_0-i_0,j_1-i_1,...,j_n-i_n}}{m_0(j_0)m_1(j_1)...m_n(j_n)}.$$

For any $\ell \in \{0,\dots,n\}$, let us choose two radii $r_\ell, r'_\ell > 0$ such that $r_\ell < r'_\ell < \rho_\ell$. By assumption, there exists a positive constant $C > 0$ such that

$$\left|\frac{a_{j_0,j_1,\dots,j_n}}{m_0(j_0)m_1(j_1)\dots m_n(j_n)}\right| \leqslant C\left(\frac{1}{r'_0}\right)^{j_0}\left(\frac{1}{r'_1}\right)^{j_1}\dots\left(\frac{1}{r'_n}\right)^{j_n}$$

for all $j_0, j_1, \dots, j_n \geqslant 0$. Then, for all $|t| \leqslant r_0$ and all $|x_\ell| \leqslant r_\ell$, $\ell = 1,\dots,n$, we get

$$\left|v_{j_0,j_1,\dots,j_n} t^{j_0} x_1^{j_1}\dots x_n^{j_n}\right| \leqslant \frac{C}{r_0'^{i_0} r_1'^{i_1}\dots r_n'^{i_n}}\left(\prod_{\ell=0}^{n}\left(\frac{r_\ell}{r'_\ell}\right)^{j_\ell}\right)\left(\prod_{\ell=0}^{n}\frac{m_\ell(j_\ell+i_\ell)}{m_\ell(j_\ell)}\right)$$

$$\left|w_{j_0,j_1,\dots,j_n} t^{j_0} x_1^{j_1}\dots x_n^{j_n}\right| \leqslant C r_0'^{i_0} r_1'^{i_1}\dots r_n'^{i_n}\left(\prod_{\ell=0}^{n}\left(\frac{r_\ell}{r'_\ell}\right)^{j_\ell}\right)\left(\prod_{\ell=0}^{n}\frac{m_\ell(j_\ell-i_\ell)}{m_\ell(j_\ell)}\right)$$

for all $j_0, j_1, \dots, j_n$. Since m_ℓ is a s_ℓ-regular moment function, there exist two positive constants $c_\ell, C_\ell > 0$ such that

$$c_\ell (j+1)^{s_\ell} \leqslant \frac{m_\ell(j+1)}{m_\ell(j)} \leqslant C_\ell (j+1)^{s_\ell}$$

for all $j \geqslant 0$. Then,

$$\frac{m_\ell(j_\ell+i_\ell)}{m_\ell(j_\ell)} = \begin{cases} 1 & \text{if } i_\ell = 0 \\ \displaystyle\prod_{k=j_\ell}^{j_\ell+i_\ell-1}\frac{m_\ell(k+1)}{m_\ell(k)} & \text{if } i_\ell \geqslant 1 \end{cases} \leqslant C_\ell^{i_\ell}(j_\ell+1)^{s_\ell}\dots(j_\ell+i_\ell)^{s_\ell}$$

$$\frac{m_\ell(j_\ell-i_\ell)}{m_\ell(j_\ell)} = \begin{cases} 1 & \text{if } i_\ell = 0 \\ \displaystyle\prod_{k=j_\ell-i_\ell+1}^{j_\ell}\frac{m_\ell(k-1)}{m_\ell(k)} & \text{if } i_\ell \geqslant 1 \end{cases} \leqslant \left(\frac{1}{c_\ell}\right)^{i_\ell}$$

and the previous estimates become

$$\left|v_{j_0,j_1,\dots,j_n} t^{j_0} x_1^{j_1}\dots x_n^{j_n}\right| \leqslant \frac{C}{r_0'^{i_0} r_1'^{i_1}\dots r_n'^{i_n}}\left(\prod_{\ell=0}^{n} C_\ell^{i_\ell}(j_\ell+1)^{s_\ell}\dots(j_\ell+i_\ell)^{s_\ell}\left(\frac{r_\ell}{r'_\ell}\right)^{j_\ell}\right)$$

$$\left|w_{j_0,j_1,\dots,j_n} t^{j_0} x_1^{j_1}\dots x_n^{j_n}\right| \leqslant C r_0'^{i_0} r_1'^{i_1}\dots r_n'^{i_n}\left(\prod_{\ell=0}^{n}\frac{1}{c_\ell^{i_\ell}}\left(\frac{r_\ell}{r'_\ell}\right)^{j_\ell}\right)$$

for all $j_0, j_1, \dots, j_n$, all $|t| \leqslant r_0$ and all $|x_\ell| \leqslant r_\ell$.

Since $r_\ell < r'_\ell$, these inequalities prove in particular that the formal power series $\partial_{m_0;t}^{i_0}\partial_{m_1;x_1}^{i_1}\ldots\partial_{m_n;x_n}^{i_n}a(t,x)$ and $\partial_{m_0;t}^{-i_0}\partial_{m_1;x_1}^{-i_1}\ldots\partial_{m_n;x_n}^{-i_n}a(t,x)$ are normally convergent on the closed polydisc $\overline{D}_{r_0}\times\overline{D}_{r_1}\times\ldots\times\overline{D}_{r_n}$; hence, on all the compact sets of $D_{\rho_0,\rho_1,\ldots,\rho_n}$. Consequently, they define analytic functions on $D_{\rho_0,\rho_1,\ldots,\rho_n}$, which completes the proof. ∎

Let us now turn to the study of Eq. (10.15) in which the functions $m_0,\ldots,m_n$ are all regular functions. Before stating various general results, let us first consider, as in the case of classical partial differential equations (see Sects. 3.2, 6.2, and 7.5), a simple example to show how the reasoning and tools developed previously can be adapted to the case of moment partial differential equations. We choose here to treat a moment analog of the heat equation, the so-called *s_0-regular moment heat equation*, so as to be able to easily compare our present calculations with those already carried out within the framework of the classical heat equation (see Sects. 3.2.1, 6.2.1, 7.5.1, and 9.2).

10.3.2 Example: The s_0-Regular Moment Heat Equation

Let us consider the inhomogeneous linear moment partial differential equation in two variables

$$\begin{cases}\partial_{m_0;t}u - a(t,x)\partial_x^2 u = \widetilde{f}(t,x)\\ u(0,x)=\varphi(x)\end{cases}, (t,x)\in\mathbb{C}^2, \tag{10.19}$$

where m_0 is a s_0-regular moment function, the coefficient $a(t,x)\in\mathcal{O}(D_{\rho_0}\times D_{\rho_1})$ satisfies $a(0,x)\not\equiv 0$, $\varphi(x)\in\mathcal{O}(D_{\rho_1})$, and where $\widetilde{f}(t,x)\in\mathcal{O}(D_{\rho_1})[[t]]$ (notice that the moment function m_1 is defined here as $m_1(\lambda)=\Gamma(1+\lambda)$ so that $\partial_{m_1;x}=\partial_x$).

Let L_{m_0} denote the linear moment operator associated with Eq. (10.19):

$$L_{m_0}=\partial_{m_0;t}-a(t,x)\partial_x^2.$$

Since, $a(0,x)\not\equiv 0$, the valuation at $t=0$ of a is 0 and, consequently, the moment Newton polygon of L_{m_0} is given by Fig. 10.3. In particular, it admits no positive slope when $s_0\geqslant 2$, and a unique positive slope otherwise, the latter being equal to $1/(2-s_0)$.

As for the classical heat equation (see Sect. 3.2, Proposition 3.9), the Gevrey regularity of the unique formal solution of Eq. (10.19) follows a noteworthy dichotomy with respect to the s-Gevrey regularity of the inhomogeneity $\widetilde{f}(t,x)$, the "*frontier*" value being entirely determined by the structure of the moment operator L_{m_0}.

Proposition 10.22 *Let s_c be the nonnegative real number equal to 0 if the moment Newton polygon of L_{m_0} has no positive slope, and equal to the inverse of its positive*

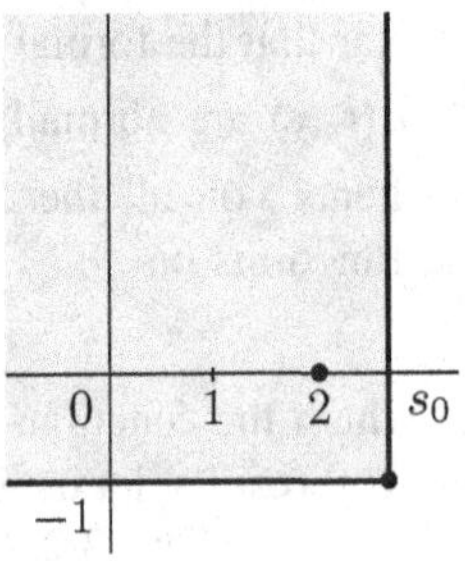

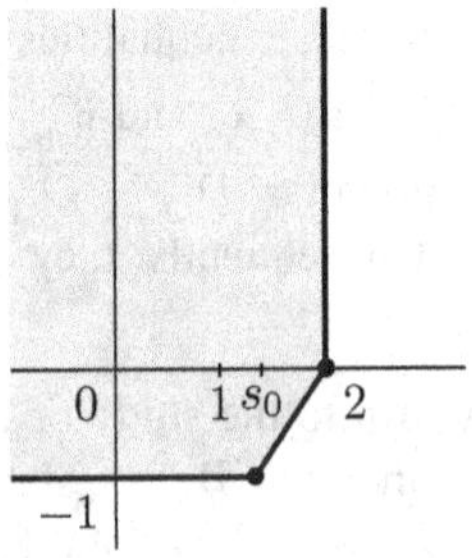

Fig. 10.3 The moment Newton polygon of L_{m_0}. **(a)** Case $s_0 \geqslant 2$. **(b)** Case $s_0 \in]0, 2[$

slope otherwise:

$$s_c = \begin{cases} 0 & \text{if } s_0 \geqslant 2 \\ 2 - s_0 & \text{if } s_0 \in]0, 2[\end{cases} = \max(0, 2 - s_0).$$

Let $\widetilde{u}(t, x) \in \mathcal{O}(D_{\rho_1})[[t]]$ be the formal solution of the s_0-regular moment heat equation (10.19). Then,

1. *$\widetilde{u}(t, x)$ and $\widetilde{f}(t, x)$ are simultaneously s-Gevrey for any $s \geqslant s_c$.*
2. *$\widetilde{u}(t, x)$ is generically s_c-Gevrey while $\widetilde{f}(t, x)$ is s-Gevrey with $s < s_c$.*

Observe that we find the result of Proposition 3.9 when we choose $m_0(\lambda) = \Gamma(1 + \lambda)$. Indeed, we have in this case $s_0 = 1$ and $\partial_{m_0;t} = \partial_t$; hence, Eqs. (10.19) and (3.4) coincide.

Observe also that the formal power series being written in the form

$$\widetilde{u}(t, x) = \sum_{j \geqslant 0} u_{j,*}(x) \frac{t^j}{m_0(j)}$$

accordingly the definition of the moment derivation $\partial_{m_0;t}$, the factor $\Gamma(1+sj)$ which occurs in the definition of the s-Gevrey formal series (see Definition 3.1) must be replaced by $m_0(j)\Gamma(1 + sj)$ (compare with Remark 3.6).

Proposition 10.22 is proved in a similar way as Proposition 3.9.

Proof of the First Point of Proposition 10.22 According to Propositions 10.3 and 10.10, it is clear that

$$\widetilde{u}(t, x) \in \mathcal{O}(D_{\rho_1})[[t]]_s \Rightarrow \widetilde{f}(t, x) \in \mathcal{O}(D_{\rho_1})[[t]]_s.$$

Reciprocally, let us fix $s \geqslant s_c$ and let us assume that the inhomogeneity $\widetilde{f}(t, x)$ of Eq. (10.19) is s-Gevrey. By assumption, its coefficients $f_{j,*}(x) \in \mathcal{O}(D_{\rho_1})$ satisfy the following condition (see Definition 3.1 and the remark just above): there exist three positive constants $0 < r < \rho_1$, $C > 0$ and $K > 0$ such that the inequalities

$$\left|f_{j,*}(x)\right| \leqslant CK^j m_0(j)\Gamma(1+sj) \tag{10.20}$$

hold for all $|x| \leqslant r$ and all $j \geqslant 0$, and we must prove that the coefficients $u_{j,*}(x) \in \mathcal{O}(D_{\rho_1})$ of the formal solution $\widetilde{u}(t,x)$ satisfy similar inequalities. From the general recurrence relations (10.16), we first derive, for all $|x| \leqslant r$ and all $j \geqslant 0$, the identities

$$\begin{aligned}\frac{u_{j+1,*}(x)}{m_0(j+1)\Gamma(1+s(j+1))} &= \frac{f_{j,*}(x)}{m_0(j+1)\Gamma(1+s(j+1))}\\ &+\sum_{k=0}^{j}\frac{m_0(j)}{m_0(k)m_0(j-k)}\frac{a_{j-k,*}(x)\partial_x^2 u_{k,*}(x)}{m_0(j+1)\Gamma(1+s(j+1))}\end{aligned}$$

together with the initial condition $u_{0,*}(x) = \varphi(x)$. Applying then the Nagumo norms of indices $((s+s_0)(j+1), r)$, we deduce successively from Property 1 and Properties 4-5 of Proposition 12.2 the inequalities

$$\begin{aligned}\frac{\left\|u_{j+1,*}\right\|_{(s+s_0)(j+1),r}}{m_0(j+1)\Gamma(1+s(j+1))} &\leqslant \frac{\left\|f_{j,*}\right\|_{(s+s_0)(j+1),r}}{m_0(j+1)\Gamma(1+s(j+1))}\\ &+\sum_{k=0}^{j}\frac{m_0(j)}{m_0(k)m_0(j-k)}\frac{\left\|a_{j-k,*}\partial_x^2 u_{k,*}\right\|_{(s+s_0)(j+1),r}}{m_0(j+1)\Gamma(1+s(j+1))}\\ &\leqslant \frac{\left\|f_{j,*}\right\|_{(s+s_0)(j+1),r}}{m_0(j+1)\Gamma(1+s(j+1))}\\ &+\sum_{k=0}^{j}A_{s,s_0,j,k}\frac{\left\|u_{k,*}\right\|_{(s+s_0)k,r}}{m_0(k)\Gamma(1+sk)}\end{aligned}$$

for all $j \geqslant 0$, where the terms $A_{s,s_0,j,k}$ are nonnegative and defined by

$$\begin{aligned}A_{s,s_0,j,k} = &\frac{e^2\left\|a_{j-k,*}\right\|_{(s+s_0)(j-k)+s+s_0-2,r}}{m_0(j-k)\Gamma(1+s(j-k))}\frac{m_0(j)}{m_0(j+1)}\frac{\Gamma(1+sj)}{\Gamma(1+s(j+1))}\\ &\times\frac{((s+s_0)k+2)((s+s_0)k+1)}{\binom{sj}{sk}}.\end{aligned}$$

Observe that all the norms written in these inequalities, and especially the norms $\left\|a_{j-k,*}\right\|_{(s+s_0)(j-k)+s+s_0-2,r}$, are well-defined. Indeed, the assumption $s \geqslant s_c$ and the condition $s_c+s_0 \geqslant 2$ imply $(s+s_0)(j-k)+s+s_0-2 \geqslant 0$ for all $k \in \{0,\dots,j\}$.

Observe also that, since m_0 is a s_0-regular moment function and $\lambda \longmapsto \Gamma(1+s\lambda)$ is a s-regular moment function, there exists a positive constant $C' > 0$ such that

$$\frac{m_0(j)}{m_0(j+1)} \frac{\Gamma(1+sj)}{\Gamma(1+s(j+1))} \leqslant \frac{C'}{(j+1)^{s+s_0}} \tag{10.21}$$

for all $j \geqslant 0$. Therefore, using also the inequalities $s + s_0 \geqslant s_c + s_0 \geqslant 2$ and $\binom{sj}{sk} \geqslant 1$ (see Proposition 13.5), we get the following estimates for all $j \geqslant 0$ and all $k \in \{0, \dots, j\}$:

$$\begin{aligned} A_{s,s_0,j,k} &\leqslant \frac{C'(e(s+s_0))^2 \left\| a_{j-k,*} \right\|_{(s+s_0)(j-k)+s+s_0-2,r}}{m_0(j-k)\Gamma(1+s(j-k))} \frac{(k+1)^2}{(j+1)^{s+s_0}} \\ &\leqslant \frac{C'(e(s+s_0))^2 \left\| a_{j-k,*} \right\|_{(s+s_0)(j-k)+s+s_0-2,r}}{m_0(j-k)\Gamma(1+s(j-k))} (j+1)^{2-(s+s_0)} \\ &\leqslant \frac{C'(e(s+s_0))^2 \left\| a_{j-k,*} \right\|_{(s+s_0)(j-k)+s+s_0-2,r}}{m_0(j-k)\Gamma(1+s(j-k))} \end{aligned}$$

and, consequently, the inequalities

$$\frac{\left\| u_{j+1,*} \right\|_{(s+s_0)(j+1),r}}{m_0(j+1)\Gamma(1+s(j+1))} \leqslant g_{s,s_0,j} + \sum_{k=0}^{j} \alpha_{s,s_0,j-k} \frac{\left\| u_{k,*} \right\|_{(s+s_0)k,r}}{m_0(k)\Gamma(1+sk)},$$

for all $j \geqslant 0$, where we set

$$g_{s,s_0,j} = \frac{\left\| f_{j,*} \right\|_{(s+s_0)(j+1),r}}{m_0(j+1)\Gamma(1+s(j+1))} \quad \text{and}$$

$$\alpha_{s,s_0,j} = \frac{C'(e(s+s_0))^2 \left\| a_{j,*} \right\|_{(s+s_0)j+s+s_0-2,r}}{m_0(j)\Gamma(1+sj)}.$$

We shall now bound the Nagumo norms $\left\| u_{j,*} \right\|_{(s+s_0)j,r}$ for any $j \geqslant 0$ by using a technique of majorant series. To do that, let us consider the formal power series

$$v(X) = \sum_{j \geqslant 0} v_j X^j \in \mathbb{R}^+[[X]],$$

where the coefficients v_j are recursively determined from the initial condition $v_0 = \left\|u_{0,*}\right\|_{0,r} = \|\varphi\|_{0,r}$ by the relations

$$v_{j+1} = g_{s,s_0,j} + \sum_{k=0}^{j} \alpha_{s,s_0,j-k} v_k$$

for all $j \geqslant 0$. By construction, we clearly have

$$0 \leqslant \frac{\left\|u_{j,*}\right\|_{(s+s_0)j,r}}{m_0(j)\Gamma(1+sj)} \leqslant v_j \quad \text{for all } j \geqslant 0.$$

On the other hand, according to the assumption on the coefficients $f_{j,*}(x)$ (see inequality (10.20)) and the analyticity of the function $a(t,x)$ at the origin $(0,0) \in \mathbb{C}^2$, we derive from the definition of the Nagumo norms, from inequality (10.21), and from the fact that the sequence $(1/\Gamma(1+sj))_{j\geqslant 0}$ is bounded by a convenient positive constant $C''' > 0$ the relations

$$0 \leqslant g_{s,s_0,j} \leqslant \frac{CK^j m_0(j)\Gamma(1+sj)r^{(s+s_0)(j+1)}}{m_0(j+1)\Gamma(1+s(j+1))} \leqslant CC'r^{s+s_0}\left(Kr^{s+s_0}\right)^j$$

and

$$\begin{aligned} 0 \leqslant \alpha_{s,s_0,j} &\leqslant \frac{C'(e(s+s_0))^2 C'' K''^j m_0(j) r^{(s+s_0)j+s+s_0-2}}{m_0(j)\Gamma(1+sj)} \\ &\leqslant C'C''C'''(e(s+s_0))^2 r^{s+s_0-2}\left(K''r^{s+s_0}\right)^j \end{aligned}$$

for all $j \geqslant 0$, with two convenient constants $C'', K'' > 0$ independent of j. Consequently, the series

$$g_{s,s_0}(X) = \sum_{j\geqslant 0} g_{s,s_0,j} X^j \quad \text{and} \quad \alpha_{s,s_0}(X) = \sum_{j\geqslant 0} \alpha_{s,s_0,j} X^j$$

are convergent. Thereby, the formal series $v(X)$ satisfying the relation

$$(1 - X\alpha_{s,s_0}(X))v(X) = \|\varphi\|_{0,r} + Xg_{s,s_0}(X), \tag{10.22}$$

it is also convergent, and there exist two positive constants $C_1, K_1 > 0$ such that $v_j \leqslant C_1 K_1^j$ for all $j \geqslant 0$. Hence,

$$\left\|u_{j,*}\right\|_{(s+s_0)j,r} \leqslant C_1 K_1^j m_0(j)\Gamma(1+sj) \quad \text{for all } j \geqslant 0.$$

We deduce from this similar estimates on the sup-norm of the $u_{j,*}(x)$ by shrinking the closed disc $|x| \leqslant r$. Let $0 < r' < r$. Then, for all $x \in D_{r'}$ and all $j \geqslant 0$, we have

$$\begin{aligned}|u_{j,*}(x)| = \left| u_{j,*}(x) d_r(x)^{(s+s_0)j} \frac{1}{d_r(x)^{(s+s_0)j}} \right| &\leqslant \frac{|u_{j,*}(x) d_r(x)^{(s+s_0)j}|}{(r-r')^{(s+s_0)j}} \\ &\leqslant \frac{\|u_{j,*}\|_{(s+s_0)j,r}}{(r-r')^{(s+s_0)j}}\end{aligned}$$

and, consequently,

$$\sup_{x \in D_{r'}} |u_{j,*}(x)| \leqslant C_1 \left(\frac{K_1}{(r-r')^{s+s_0}} \right)^j m_0(j) \Gamma(1+sj);$$

which ends the proof of the first point of Proposition 10.22. ∎

Proof of the Second Point of Proposition 10.22 Let us now fix $s < s_c$, which implies in particular $s_0 \in]0, 2[$ and $s_c = 2 - s_0$. According to the filtration of the s-Gevrey spaces $\mathcal{O}(D_{\rho_1})[[t]]_s$ (see page 17) and the first point of Proposition 10.22, it is clear that the following implications hold:

$$\widetilde{f}(t,x) \in \mathcal{O}(D_{\rho_1})[[t]]_s \Rightarrow \widetilde{f}(t,x) \in \mathcal{O}(D_{\rho_1})[[t]]_{s_c} \Rightarrow \widetilde{u}(t,x) \in \mathcal{O}(D_{\rho_1})[[t]]_{s_c}. \tag{10.23}$$

To conclude that we can not say better about the Gevrey order of $\widetilde{u}(t,x)$, that is $\widetilde{u}(t,x)$ is *generically s_c-Gevrey*, we need to find an example for which the formal solution $\widetilde{u}(t,x)$ of Eq. (10.19) is s'-Gevrey for no $s' < s_c$. To do that, let us consider the following moment analog of Eq. (3.10) we previously used in the case of the classical heat equation:

$$\begin{cases} \partial_{m_0;t} u - a\partial_x^2 u = \widetilde{f}(t,x) & , a \in \mathbb{C}^* \\ u(0,x) = \varphi(x) = \dfrac{1}{1-x} \in \mathcal{O}(D_1) \end{cases} \tag{10.24}$$

where $\widetilde{f}(t,x) \in \mathcal{O}(D_1)[[t]]_s$ satisfies the condition $\partial_x^\ell f_{j,*}(0) \geqslant 0$ for all $\ell, j \geqslant 0$. The coefficients $u_{j,*}(x) \in \mathcal{O}(D_1)$ of the formal solution $\widetilde{u}(t,x)$ of Eq. (10.24) are now recursively determined by the initial condition $u_{0,*}(x) = \varphi(x)$ and, for all $j \geqslant 0$, by the relations

$$u_{j+1,*}(x) = f_{j,*}(x) + a\partial_x^2 u_{j,*}(x).$$

We derive straightaway from this that

$$u_{j,*}(x) = a^j \partial_x^{2j} \varphi(x) + \sum_{k=0}^{j-1} a^k \partial_x^{2k} f_{j-1-k,*}(x)$$

for all $x \in D_1$ and all $j \geqslant 0$, with the classical convention that the sum is 0 when $j = 0$. Hence, due to our assumption on the inhomogeneity $\widetilde{f}(t, x)$, the inequalities

$$u_{j,*}(0) \geqslant a^j \Gamma(1+2j) \quad \text{for all } j \geqslant 0. \tag{10.25}$$

Let us now suppose that $\widetilde{u}(t, x)$ is s'-Gevrey for some $s' < s_c$. Then, inequalities (10.25) above and the fact that m_0 is a moment function of order s_0 (see inequality (10.1)) imply the relations

$$1 \leqslant C \left(\frac{K}{a}\right)^j \frac{m_0(j)\Gamma(1+s'j)}{\Gamma(1+2j)} \leqslant CC' \left(\frac{KK'}{a}\right)^j \frac{\Gamma(1+s_0 j)\Gamma(1+s'j)}{\Gamma(1+2j)} \tag{10.26}$$

for all $j \geqslant 0$ and some convenient positive constants $C, C', K, K' > 0$ independent of j. Applying then the Stirling's Formula, we get

$$CC' \left(\frac{KK'}{a}\right)^j \frac{\Gamma(1+s_0 j)\Gamma(1+s'j)}{\Gamma(1+2j)} \underset{j \to +\infty}{\sim} C'' \left(\frac{K''}{j^{s_c - s'}}\right)^j \sqrt{j}$$

with

$$C'' = CC' \sqrt{\pi s_0 s'} \quad \text{and} \quad K'' = \frac{KK' s_0^{s_0} s'^{s'} e^{s_c - s'}}{4a}.$$

We conclude that inequalities (10.26) are impossible since the right hand-side goes to 0 when j tends to infinity. Hence, $\widetilde{u}(t, x)$ is s'-Gevrey for no $s' < s_c$ and, consequently, $\widetilde{u}(t, x)$ is exactly s_c-Gevrey thanks to (10.23). This achieves the proof of the second point of Proposition 10.22. ■

Remark 10.23

1. The assumption "m_0 *is a* s_0*-regular moment function*" plays a fundamental role in the proof of the first point of Proposition 10.22. Without it, that is for a general moment function m_0 of order s_0, we get a geometric additional term A^j with $A > 1$ in the estimate of $A_{s,s_0,j,k}$, which brings us to consider a A-difference equation instead of the functional relation (10.22) and our conclusion fails with this method.
2. Our initial choice on the moment function m_1 to have $\partial_{m_1;x} = \partial_x$ allows us to use the classical Nagumo norms we introduced in Sect. 3.2.1. With a more general moment function m_1, these norms can no longer be used because they

are not compatible with the moment derivations. However, this problem can be circumvented by defining some convenient modified Nagumo norms (see for instance the recent work [122] of M. Suwińka).

3. Apart from the previous point, the example of the Eq. (10.19) clearly shows how the calculations initially developed in the framework of partial differential equations are naturally modified when the term $j!$ is changed to $m_0(j)$ in the writing of the formal solution $\widetilde{u}(t,x)$. Comparing the above proof with that of Proposition 3.9 page 24, we observe in particular the following two fundamental changes:

   - the term $\Gamma(1+(s+1)j)$ used to control the Gevrey order of $\widetilde{u}(t,x)$ (see inequality (3.8)) is replaced by the term $m_0(j)\Gamma(1+sj)$ (see inequality (10.20));
   - the binomial coefficient $\binom{j}{k} = \dfrac{j!}{k!(j-k)!}$ is replaced by its moment analogue $\dfrac{m_0(j)}{m_0(k)m_0(j-k)}$.

   Using the properties of m_0 (here, its regularity), we can conclude in the classic way. It should also be noted that in the case where the standard derivation ∂_x is replaced by a moment derivation $\partial_{m_1;x}$, the use of the modified Nagumo norms does not fundamentally alter these observations, since these norms are constructed to have the same properties as the classical Nagumo norms.

As we said in the introduction of this chapter, one of the interests of the moment partial differential equations is that they can handle many classical functional equations using the same general form. Here, for instance, the choice $m_0(\lambda) = \Gamma(1+\lambda)$ gives the classical inhomogeneous heat equation $\partial_t u - a(t,x)\partial_x^2 u = \widetilde{f}(t,x)$ back, whereas the choice $m_0(\lambda) = \Gamma(1+s\lambda)$ with $s \neq 1$ leads to a s-fractional-type analogue. Remark 10.24 below even shows, at least for the study of the Gevrey regularity of a certain type of moment partial differential equations of which Eq. (10.19) is one, that we can also deal with much more general functional equations.

Remark 10.24 The above calculations remain valid when the sequence $(m_0(j))$ is a 0-regular sequence, that is a 0-Gevrey-type sequence satisfying the following property (compare with Definition 10.18): there exist two positive constants $c, C > 0$ such that

$$c \leqslant \frac{m_0(j+1)}{m_0(j)} \leqslant C$$

for all $j \geqslant 0$ (we have then $s_0 = 0$). The 0-Gevrey-type sequence $([j]_q!)$ introduced in Remark 10.5 page 170 being of this type (we clearly have $1 \leqslant [j+1]_q \leqslant (1-q)^{-1}$ for all $j \geqslant 0$), we can then choose $m_0(j) = [j]_q!$ and, consequently, Proposition 10.22 applied, providing thus the Gevrey regularity of the formal solution of the inhomogeneous q-difference-differential equation $D_{q;t}u - a(t,x)\partial_x^2 u = \widetilde{f}(t,x)$.

The study of the summability of the formal solution $\widetilde{u}(t,x)$ of Eq. (10.19) is much more complicated than in the case of the classical heat equation (see Sects. 6.2.1, 7.5.1, and 9.2). To simplify our discussion, let us consider the simpler case of the equation

$$\begin{cases} \partial_{m_0;t}u - a\partial_x^2 u = 0 \\ u(0,x) = \varphi(x) \end{cases}, (t,x) \in \mathbb{C}^2, \tag{10.27}$$

where m_0 is a s_0-regular moment function with $s_0 \in]0,2[$, $a \in \mathbb{C}^*$ is a nonzero complex constant, and where $\varphi(x) \in \mathcal{O}(D_{\rho_1})$. For this equation, the formal solution $\widetilde{u}(t,x)$ is given by

$$\widetilde{u}(t,x) = \sum_{j\geqslant 0} \varphi^{(2j)}(x)\frac{(at)^j}{m_0(j)},$$

where $\varphi^{(k)}$ stands for the kth derivative of φ (compare with the formal solution (9.5) of Eq. (9.4)). Since $s_0 \in]0,2[$, Proposition 10.22 above tells us that $\widetilde{u}(t,x)$ is generically $(2-s_0)$-Gevrey (hence, divergent) and the question of its summability is natural. As in the case of the classical heat equation, many ways can be considered. For instance, one can use an approach similar to the one developed in Sect. 6.2.1 by replacing the successive derivations ∂_t^ℓ by the successive moment derivations $\partial_{m_0;t}^\ell$ (we refer for instance to [123] for such an approach), or use an approach based on the Borel-Laplace method as in Sect. 7.5.1.

Here below, we prefer to adapt the approach of Sect. 9.2 based on the characterization of the summability in terms of the moment-Borel-Laplace method, this method being particularly well suited to moment partial differential equation. This leads us to the following result, whose statement and proof will be compared with those of Proposition 9.2.

Proposition 10.25 *Let $k = 1/(2-s_0)$. Then, the formal solution $\widetilde{u}(t,x) \in \mathcal{O}(D_{\rho_1})[[t]]$ of the s_0-regular moment heat equation (10.27) is k-summable in the direction θ if and only if the initial data $\varphi(x)$ can be analytically continued to sectors neighboring the directions $\frac{1}{2}(\theta + \arg(a)) \mod \pi$ with a global exponential growth of order at most $2k$ at infinity.*

Observe that in the case where $m_0(\lambda) = \Gamma(1+\lambda)$ (hence, $s_0 = 1$ and $\partial_{m_0;t} = \partial_t$), Eq. (10.27) coincides with Eq. (9.4), and we find the result of Proposition 9.2.

Proof Let us consider the moment function m of order $2-s_0$ defined by

$$m(\lambda) = \frac{\Gamma(1+2\lambda)}{m_0(\lambda)}.$$

Then, the formal moment-Borel transform $\widetilde{\mathcal{B}}_k^{[m;t]}(\widetilde{u})$ of $\widetilde{u}$ is defined by

$$\widetilde{\mathcal{B}}_k^{[m;t]}(\widetilde{u})(\tau,x)=\sum_{j\geqslant 0}\varphi^{(2j)}(x)\frac{(a\tau)^j}{m_0(j)m(j)}=\sum_{j\geqslant 0}\varphi^{(2j)}(x)\frac{(a\tau)^j}{(2j)!}$$

and the formal power series

$$\widetilde{\omega}(\tau,x)=\widetilde{\mathcal{B}}_k^{[m;t]}(\widetilde{u})(\tau^2,x)=\sum_{j\geqslant 0}\varphi^{(2j)}(x)\frac{a^j\tau^{2j}}{(2j)!}\in\mathcal{O}(D_{\rho_1})[[\tau]]$$

is the unique formal power series solution of the homogeneous wave equation (9.6). The rest of the proof is similar to the one of Proposition 9.2 and is left to the reader. ■

Let us now present some other general results.

10.3.3 More General Results

Let us first look at the Gevrey regularity of the formal solution $\widetilde{u}(t,x)$ of our general Eq. (10.15). As in the case of the s_0-regular moment heat equation (see Proposition 10.22), we denote by s_c the nonnegative real number equal to 0 if the moment Newton polygon of the linear moment operator (10.17) associated with Eq. (10.15) has no positive slope, and equal to the inverse of its smallest positive slope otherwise (see Proposition 10.17):

$$s_c=\max\left(0,\max_{\substack{i\in\mathcal{K}\\ q\in Q_i}}\left(\frac{\sum_{j=1}^{n}s_jq_j-s_0(\kappa-i)}{\kappa-i+v_{i,q}}\right)\right).$$

Reasoning then as in the proof of Proposition 10.22, we can prove the following.

Theorem 10.26 ([122]) *Let $\widetilde{u}(t,x)$ be the formal solution in $\mathcal{O}(D_{\rho_1,\dots,\rho_n})[[t]]$ of Eq. (10.15). Assume that the moment functions $m_0,m_1,\dots,m_n$ are regular moment functions of respective orders $s_0>0$ and $s_1,\dots,s_n\geqslant 1$.*

Then, $\widetilde{u}(t,x)$ and the inhomogeneity $\widetilde{f}(t,x)$ are simultaneously s_c-Gevrey.

Observe that in the case where $m_0(\lambda)=\dots=m_n(\lambda)=\Gamma(1+\lambda)$, that is the moment derivations $\partial_{m_0;t}$ and $\partial_{m_j;x_j}$ for all $j=1,\dots,n$ coincide respectively with the usual derivations ∂_t and ∂_{x_j}, we find the result already stated in Proposition 3.19 for the classical linear partial differential equations.

Concerning the summability of $\widetilde{u}(t,x)$, there are at present very few results when the moment functions $m_0,\ldots,m_n$ of Eq. (10.15) are regular. These are essentially given in the framework of fractional linear partial differential equations which are classically obtained by choosing for m_j the functions $m_j(\lambda)=\Gamma(1+s_j\lambda)$ with a positive real number s_j (see Example 10.4).

As an example, let us consider the equation

$$\begin{cases} \left(\partial_t^{1/k_0}\right)^{\kappa} u - \sum\limits_{i\in\mathcal{K}}\sum\limits_{q\in Q_i} a_{i,q}\left(\partial_t^{1/k_0}\right)^{i}\left(\partial_x^{1/k_1}\right)^{q} u = 0 \\ \left(\partial_t^{1/k_0}\right)^{j} u(t,x)_{|t=0} = 0,\ j=0,\ldots,\kappa-2 \\ \left(\partial_t^{1/k_0}\right)^{\kappa-1} u(t,x)_{|t=0} = \varphi(x) \end{cases} \tag{10.28}$$

in two variables $(t,x)\in\mathbb{C}^2$, where

- $\kappa\geqslant 1$ is a positive integer;
- $k_0,k_1\geqslant 1$ are two positive integers satisfying $k_0>k_1$;
- $\mathcal{K}$ is a nonempty subset of $\{0,\ldots,\kappa-1\}$ containing 0;
- Q_i is a nonempty finite subset of $\mathbb{N}$ for all $i\in\mathcal{K}$ and satisfies the following condition:
  - $\kappa\in Q_0$ and $q\leqslant\kappa$ for all $q\in Q_0$;
  - $q\leqslant\kappa-i$ for all $i\in\mathcal{K}\backslash\{0\}$, if any exist, and all $q\in Q_i$.
- $a_{i,q}\in\mathbb{C}^*$ is a nonzero complex constant for all $i\in\mathcal{K}$ and all $q\in Q_i$;
- the initial data $\varphi(x)$ is $1/k_1$-analytic on D_{ρ_1}, that is the function $x\longmapsto\varphi(x^{k_1})$ is analytic for all $x^{k_1}\in D_{\rho_1}$ (in other words, the function φ is an analytic function defined on the Riemann surface of x^{1/k_1}).

According to our assumptions on the sets $\mathcal{K}$ and Q_i, the moment Newton polygon of its linear moment operator (written in terms of moment functions!) admits a single positive slope k, the latter being given by the segment with end points $(\kappa/k_0,-\kappa)$ and $(\kappa/k_1,0)$; hence $k=k_0k_1/(k_0-k_1)$. Note here that, in order to be compatible with the fractional derivation ∂_t^{1/k_0}, we now consider formal power series in t^{1/k_0} and not formal power series in t. Consequently, we are interested in their k/k_0-summability and not in their k-summability as before (see Corollary 5.7 for the relationship between summability and reduced series).

As in the case of Eq. (10.15), Eq. (10.28) is formally well-posed: looking for $\widetilde{u}(t,x)$ in the form

$$\widetilde{u}(t,x)=\sum_{j\geqslant 0} u_{j,*}(x)\frac{t^{j/k_0}}{\Gamma(1+j/k_0)} \tag{10.29}$$

with $u_{j,*}(x)$ $1/k_0$-analytic on D_{ρ_1} for all $j \geqslant 0$, one easily checks that its coefficients are uniquely determined for all $j \geqslant 0$ by the recurrence relations

$$u_{j+\kappa,*}(x) = \sum_{i \in \mathcal{K}} \sum_{q \in Q_i} a_{i,q} \left(\partial_x^{1/k_1}\right)^q u_{j+i,*}(x)$$

together with the initial conditions $u_{j,*}(x) = \varphi_j(x)$ for all $j = 0, \ldots, \kappa - 1$. Adapting then the methods presented in the previous sections, one can prove the following.

Theorem 10.27 ([78, 80]) *The formal solution (10.29) of Eq. (10.28) is $k_1/(k_0 - k_1)$-summable in the direction θ if and only if the initial data $\varphi(x)$ can be $1/k_1$-analytically continued to sectors neighboring the directions $k_1(\theta/k_0 + \arg(\lambda))$ with a global exponential growth of order at most $k_0/(k_0 - k_1)$ at infinity for all $\lambda \in \mathbb{C}^*$ solution of the characteristic equation*

$$\lambda^\kappa - \sum_{i \in \mathcal{K}} p_i \lambda^i = 0 \quad \textit{with } p_i = \lim_{\xi \to +\infty} \frac{1}{\xi^{\kappa - i}} \sum_{q \in Q_i} a_{i,q} \xi^q.$$

For more details on the fractional calculus, the reader can consult for instance [3, 36, 52].

10.4 Gevrey Regularity and Summability: The General Case

Let us conclude this chapter on moment linear partial differential equations with some comments on the general case. To simplify our discussion, we consider only the case of two variables and we also assume that Eq. (10.15) has constant coefficients. In other words, it can be rewritten as

$$\begin{cases} P(\partial_{m_0;t}, \partial_{m_1;x})u = \widetilde{f}(t,x) \\ \partial_{m_0;t}^j u(t,x)_{|t=0} = \varphi_j(x),\ j = 0, \ldots, \kappa - 1 \end{cases}, \tag{10.30}$$

where $P(X_0, X_1) \in \mathbb{C}[X_0, X_1]$ is a polynomial in two variables of degree κ with respect to X_0. More precisely, $P(X_0, X_1) = X_0^\kappa + Q(X_0, X_1)$ with $Q(X_0, X_1) \in \mathbb{C}[X_0, X_1]$ of degree $\leqslant \kappa - 1$ with respect to X_0.

Equation (10.30) was first investigated by W. Balser and M. Yoshino in [13]. In this paper, the authors characterized the Gevrey regularity of its formal solution $\widetilde{u}(t,x)$ in terms of the Gevrey regularity of the inhomogeneity $\widetilde{f}(t,x)$ and of the orders of the moment functions m_0 and m_1. Moreover, they introduced some various tools, including a generalization of Toeplitz operators, in order to studied the summability of $\widetilde{u}(t,x)$ when $\widetilde{f}(t,x)$ is analytic at the origin of $\mathbb{C}^2$.

More recently, S. Michalik proposed in [80, 81] another interesting way to investigate the Gevrey regularity and the summability of $\widetilde{u}(t,x)$. Assuming first $\widetilde{f}(t,x) \equiv 0$ and using the factorization of P in the form

$$P(X_0, X_1) = (X_0 - \lambda_1(X_1))^{r_1} \ldots (X_0 - \lambda_\ell(X_1))^{r_\ell},$$

where $\lambda_1(\zeta), \ldots, \lambda_\ell(\zeta)$ are the roots of the characteristic equation $P(\lambda, \zeta) = 0$ with multiplicities $r_1, \ldots, r_\ell \geqslant 1$ $(r_1 + \ldots + r_\ell = \kappa)$, he showed that $\widetilde{u}(t,x)$ can be written as

$$\widetilde{u}(t,x) = \sum_{\alpha=1}^{\ell} \sum_{\beta=1}^{r_\alpha} \widetilde{u}_{\alpha,\beta}(t,x),$$

$\widetilde{u}_{\alpha,\beta}(t,x)$ being the formal power series solution of the equation

$$\begin{cases} (\partial_{m_0;t} - \lambda_\alpha(\partial_{m_1;x}))^{\beta} u_{\alpha,\beta} = 0 \\ \partial_{m_0;t}^{j} u_{\alpha,\beta}(t,x)_{|t=0} = 0,\ j = 0, \ldots, \beta - 2 \\ \partial_{m_0;t}^{\beta-1} u_{\alpha,\beta}(t,x)_{|t=0} = \lambda_\alpha^{\beta-1}(\partial_{m_1;x})\varphi_{\alpha,\beta}(x) \end{cases} \tag{10.31}$$

for some analytic function $\varphi_{\alpha,\beta}(x)$ entirely determined from the initial data $\varphi_j(x)$. Then, he proved that the Gevrey order of $\widetilde{u}_{\alpha,\beta}(t,x)$ depends both on the order q_α of the pole of $\lambda_\alpha(\zeta)$ at infinity, and on the orders s_0 and s_1 of the moment functions m_0 and m_1.

Proposition 10.28 ([80]) *With previous notations:*

1. *Assume $s_0 \geqslant q_\alpha s_1$. Then, $\widetilde{u}_{\alpha,\beta}(t,x)$ defined an analytic function at the origin of $\mathbb{C}^2$ and its sum is an analytic solution of Eq. (10.31).*
2. *Assume $s_0 < q_\alpha s_1$. Then, $\widetilde{u}_{\alpha,\beta}(t,x)$ is a $(q_\alpha s_1 - s_0)$-Gevrey formal series.*

In the last case, he characterized besides the $(q_\alpha s_1 - s_0)^{-1}$-summability of $\widetilde{u}_{\alpha,\beta}(t,x)$ in terms of analytic continuation properties of the initial data of Eq. (10.31), thus joining the results already stated previously.

Finally, returning to the formal solution $\widetilde{u}(t,x)$ of the initial Eq. (10.30), he described the summability, as well as the multi-summability, of $\widetilde{u}(t,x)$ in terms of the initial data $\varphi_j(x)$.

In the general case $\widetilde{f}(t,x) \not\equiv 0$, the above calculations remain valid, but with some adaptations. For instance, Eq. (10.31) becomes

$$\begin{cases} (\partial_{m_0;t} - \lambda_\alpha(\partial_{m_1;x}))^{\beta} u_{\alpha,\beta} = \widetilde{f}_{\alpha,\beta}(t,x) \\ \partial_{m_0;t}^{j} u_{\alpha,\beta}(t,x)_{|t=0} = 0,\ j = 0, \ldots, \beta - 2 \\ \partial_{m_0;t}^{\beta-1} u_{\alpha,\beta}(t,x)_{|t=0} = \lambda_\alpha^{\beta-1}(\partial_{m_1;x})\varphi_{\alpha,\beta}(x) \end{cases}$$

for some formal power series $\widetilde{f}_{\alpha,\beta}(t,x) \in \mathcal{O}(D_{\rho_1})[[t]]$ entirely determined from $\widetilde{f}(t,x)$ and, consequently, the Gevrey order of $\widetilde{u}_{\alpha,\beta}(t,x)$ also depends on the Gevrey regularity of $\widetilde{f}(t,x)$.

Proposition 10.29 ([80]) *Assume that $\widetilde{f}(t,x)$ is a s-Gevrey formal series. Then, with previous notations, $\widetilde{u}_{\alpha,\beta}$ is a Gevrey formal series of order* $\max(q_\alpha s_1 - s_0, s)$.

Of course, adaptations of same type occur in the study of the summability of $\widetilde{u}_{\alpha,\beta}(t,x)$; hence, of the formal solution $\widetilde{u}(t,x)$. We refer to [80, 81] for more details.

Part IV
Appendices

Chapter 11
Some Related Equations

In this chapter, we shall briefly present some other applications and extensions of the theory of the summability to formal power series solution of equations other than partial differential equations or moment partial differential equations. As for the latter, the efforts to explore such applications are far from being complete and shall provide an excellent chance for future research.

11.1 Integro-Differential Equations

Let us first look at the unique formal power series $\widetilde{u}(t,x) \in \mathcal{O}(D_{\rho_1,\dots,\rho_n})[[t]]$ solution of the integro-differential equation

$$u - G(t,x,(\partial_t^{-i}\partial_x^q u)_{(i,q)\in\Lambda}) = \widetilde{f}(t,x) \tag{11.1}$$

in 1-dimensional time variable $t \in \mathbb{C}$ and n-dimensional spatial variable $x \in \mathbb{C}^n$, where

- Λ is a nonempty finite subset of $\mathbb{N}^* \times \mathbb{N}^n$;
- $G(t,x,(Z_{i,q})_{(i,q)\in\Lambda})$ is analytic in a polydisc $D_{\rho_0} \times D_{\rho_1,\dots,\rho_n} \times \prod_{(i,q)\in\Lambda} D_{\rho_{i,q}}$ centered at the origin of $\mathbb{C}^{n+1+\#\Lambda}$, where $\#\Lambda$ stands for the cardinal of Λ;
- the inhomogeneity $\widetilde{f}(t,x) \in \mathcal{O}(D_{\rho_1,\dots,\rho_n})[[t]]$ is a formal power series in t with analytic coefficients in $D_{\rho_1,\dots,\rho_n}$.

Although very similar to Problem 1.1 stated in the framework of partial differential equations and for which we have given various important results in the previous sections, there are very few results dealing directly with the Gevrey regularity and summability of $\widetilde{u}(t,x)$, except in the linear case (see for instance [87, 104, 135]). For the latter, the results on the Gevrey regularity and the summability of $\widetilde{u}(t,x)$ are comparable to those given in Proposition 3.19 and 6.12 (see also Remark 6.14) for

P. Remy, *Asymptotic Expansions and Summability*, Lecture Notes in Mathematics 2351, https://doi.org/10.1007/978-3-031-59094-8_11

linear partial differential equations. For example, for the Gevrey regularity, we have the following.

Proposition 11.1 ([87, 135]) *Let s be the nonnegative rational number equal to the inverse of the smallest positive slope of the Newton polygon of the associated linear operator of Eq. (11.1) if any exists, and equal to 0 otherwise.*

Then, $\widetilde{u}(t,x)$ and $\widetilde{f}(t,x)$ are simultaneously s-Gevrey. Moreover, if $\widetilde{f}(t,x)$ is convergent, then $\widetilde{u}(t,x)$ is generically s-Gevrey.

In the nonlinear case, only S. Malek stated a summability result under some restrictive additional conditions on G [70].

As for partial differential equations, the same problem can be formulated using moment anti-derivations $\partial_{m_0;t}^{-1}$ instead of the classical anti-derivation ∂_t^{-1}. We refer to the very recent work [123] of M. Suwińka for an example of such an equation in the linear case.

11.2 Singular Partial Differential Equations

Let us now consider singular partial differential equations in 1-dimensional time variable $t \in \mathbb{C}$ and n-dimensional spatial variable $x \in \mathbb{C}^n$ of the form

$$(t\partial_t)^\kappa u = G(t, x, ((t\partial_t)^i \partial_x^q u)_{(i,q)\in\Lambda}) \tag{11.2}$$

where

- $\kappa \geqslant 1$ is a positive integer;
- Λ is a nonempty finite subset of $\{0, \dots, \kappa - 1\} \times \mathbb{N}^n$;
- $G(t, x, (Z_{i,q})_{(i,q)\in\Lambda})$ is analytic on a polydisc centered at the origin of $\mathbb{C}^{n+1+\#\Lambda}$, where $\#\Lambda$ stands for the cardinal of Λ;
- $G(0, x, 0) \equiv 0$ near $x = 0$.

In the 1998 article [34], R. Gérard and H. Tahara considered this equation in the case $\kappa = 1$ (also called, *of first order*) and $n = 1$ with additional restrictive assumptions on G. In particular, they gave sufficient conditions for it to admit formal power series solutions and studied, in the case where they exist, their Gevrey regularity, including their convergence. These results were then generalized, still in the case $\kappa = 1$, by M. Miyake and A. Shirai in a series of papers [89–91, 115–120] (note that, for some of these papers, the equation studied is a little more general than Eq. (11.2) since they allow the multidimensional case for the time variable t).

In the general case $\kappa \geqslant 1$, and still under some additional assumption on G, A. Shirai proved in the 2001 article [114] that the Gevrey regularity of the formal solutions of Eq. (11.2), if any exist, is entirely determined by a convenient Newton polygon, which generalizes to singular partial differential equations the one we introduced in Sect. 3.2.2. In the recent work [55], A. Lastra and H. Tahara extended this result to the very general Eq. (11.2), assuming however that it is of totally characteristic type.

If the Gevrey regularity of the formal power series solutions of singular equations of type (11.2) is now relatively well known, it is not the same for their summability. There are indeed very few results on this subject. Let us nevertheless mention the important works in this direction of M. Hibino in the linear case of first order [40–42], of S. Ouchi in the semilinear case [97], and of Z. Luo, H. Chen and C. Zhang in the degenerate nonlinear case [66].

More recently, several authors have been interested in singular partial differential equations of another form than (11.2), where the singular locus is no longer given by $t = 0$, but by the zero set of an analytic map P satisfying $P(0) = 0$. For such equations, the formal solutions are then sought, not in the form of a formal power series in t, but in the form of a formal power series with respect to the germ P. In doing so, the notions of Gevrey asymptotics and summability presented in the previous sections are extended to be adapted to this new type of series [23–25, 92]. For some types of these equations, some results on the Gevrey regularity of formal solutions have been obtained by S. A. Carrillo, C. A. Hurtado and A. Lastra [21, 22], as well as on their summability [20].

11.3 Generalized Moment Partial Differential Equations

The notions of Gevrey asymptotics and summability, including moment summability, of a formal power series $\widetilde{u}(t,x) \in \mathcal{O}(D_{\rho_1,\dots,\rho_n})[[t]]$ which we presented in this work can be extended to formal power series whose coefficients (resp. remainders) are governed by sequences more general than, but sharing their fundamental properties with, the sequences $(\Gamma(1+sj))_{j\geqslant 0}$ for some $s \geqslant 0$ or, equivalently by the Stirling's Formula, the powers of the factorial $j!$.

This extension concerns the consideration of Carleman ultraholomorphic classes in sectors, which consist of holomorphic functions whose the derivatives of order $j \geqslant 0$ are uniformly bounded there by, essentially, the values $j!M_j$, where $\mathbb{M} = (M_j)_{j\geqslant 0}$ is a sequence of positive real numbers. In order to obtain good properties for these classes, the sequence $\mathbb{M}$ is usually subject to some standard conditions. In particular, it is generally considered as *strongly regular*.

Definition 11.2 ([126]) A sequence $\mathbb{M} = (M_j)_{j\geqslant 0}$ of positive real numbers is said to be *strongly regular* if the following four conditions hold:

1. $M_0 = 1$;
2. $\mathbb{M}$ is *logarithmically convex*:

$$M_j^2 \leqslant M_{j-1}M_{j+1} \quad \text{for all } j \geqslant 1;$$

3. $\mathbb{M}$ is of *moderate growth*: there exists a positive constant $A > 0$ such that

$$M_{j+\ell} \leqslant A^{j+\ell} M_j M_\ell \quad \text{for all } j, \ell \geqslant 0;$$

4. $\mathbb{M}$ satisfies the following *strong non-quasianalyticity condition*: there exists a positive constant $B > 0$ such that

$$\sum_{\ell \geqslant j} \frac{M_\ell}{(\ell+1)M_{\ell+1}} \leqslant B \frac{M_j}{M_{j+1}} \quad \text{for all } j \geqslant 0.$$

Example 11.3 One of the best known examples of strongly regular sequences is the sequence $\mathbb{M}_s = (\Gamma(1+sj))_{j\geqslant 0}$ appearing in the Gevrey classes.

Under these standard conditions on $\mathbb{M}$, flat functions in the class can be constructed on sectors of optimal opening and, adapting the work of W. Balser on moment summability (see Part III), suitable kernels and Laplace and Borel-type transforms can be introduced to lead to a tractable concept of $\mathbb{M}$-*summability*. For more details on this topic, we refer for instance to [112, 113] (see also [47–50, 56] and [60] for an application to q-difference equations).

As in Part III, moments functions can be associated to these generalized kernels, leading thus to *generalized moment partial differential equations* and *generalized moment integro-differential equations*. For some studies of these equations, we refer to the recent works [57–59] of A. Lastra and S. Michalik and M. Suwińska.

Chapter 12
Nagumo Norms

In this chapter, we recall the definition of the Nagumo norms [19, 94, 130] and some of their properties.

Definition 12.1 Let $f \in \mathcal{O}(D_{\rho_1,\ldots,\rho_n})$, $p \geqslant 0$ and $0 < r < \min(\rho_1, \ldots, \rho_n)$ be. Then, the *Nagumo norm* $\|f\|_{p,r}$ *with indices* (p, r) *of* f is defined by

$$\|f\|_{p,r} := \sup_{\|x\| \leqslant r} \left| f(x) d_r(x)^p \right|,$$

where $\|x\|$ stands for the maximum of the $|x_\ell|$'s: $\|x\| = \max(|x_1|, \ldots, |x_n|)$, and where $d_r(x)$ denotes the Euclidian distance $d_r(x) = r - \|x\|$.

Following Proposition 12.2 gives us some properties of the Nagumo norms.

Proposition 12.2 *Let* $f, g \in \mathcal{O}(D_{\rho_1,\ldots,\rho_n})$, $p, p' \geqslant 0$ *and* $0 < r < \min(\rho_1, \ldots, \rho_n)$ *be. Then,*

1. $\|\cdot\|_{p,r}$ *is a norm on* $\mathcal{O}(D_{\rho_1,\ldots,\rho_n})$.
2. $|f(x)| \leqslant \|f\|_{p,r} d_r(x)^{-p}$ *for all* $\|x\| < r$.
3. $\|f\|_{0,r} = \sup_{\|x\| \leqslant r} |f(x)|$ *is the usual sup-norm on the polydisc* $D_{r,\ldots,r}$.
4. $\|fg\|_{p+p',r} \leqslant \|f\|_{p,r} \|g\|_{p',r}$.
5. $\left\| \partial_{x_\ell} f \right\|_{p+1,r} \leqslant e(p+1) \|f\|_{p,r}$ *for all* $\ell \in \{1, \ldots, n\}$.
6. $\left\| \partial_{x_\ell}^{-1} f \right\|_{p,r} \leqslant r \|f\|_{p,r}$.

Proof Properties 1–4 are straightforward and are left to the reader.

P. Remy, *Asymptotic Expansions and Summability*, Lecture Notes in Mathematics 2351, https://doi.org/10.1007/978-3-031-59094-8_12

To prove Property 5, we proceed as follows. Let $\ell \in \{1, \ldots, n\}$ be, $x \in \mathbb{C}^n$ such that $\|x\| < r$, and $0 < R < d_r(x)$. Using the Cauchy Integral Formula, we have

$$\partial_{x_\ell} f(x) = \frac{1}{(2i\pi)^n} \int_{\gamma(x)} \frac{f(x')}{(x'_\ell - x_\ell)^2 \prod_{\substack{k=1 \\ k \neq \ell}}^{n} (x'_k - x_k)} dx',$$

where $\gamma(x) := \{x' = (x'_1, \ldots, x'_n) \in \mathbb{C}^n; \left|x'_k - x_k\right| = R \text{ for all } k \in \{1, \ldots, n\}\}$. Since

$$x' \in \gamma(x) \Rightarrow \left\|x'\right\| < r,$$

we can apply Property 2 of Proposition 12.2; hence, the inequalities

$$\begin{aligned}
\left|\partial_{x_\ell} f(x)\right| &\leqslant \frac{1}{R} \max_{x' \in \gamma(x)} \left|f(x')\right| \leqslant \frac{1}{R} \|f\|_{p,r} \max_{x' \in \gamma(x)} d_r(x')^{-p} \\
&= \frac{1}{R} \|f\|_{p,r} (d_r(x) - R)^{-p}.
\end{aligned}$$

Observe that the last equality stems from the relations

$$d_r(x') = r - \left\|x'\right\| = r - \left\|x + x' - x\right\| \geqslant d_r(x) - \left\|x' - x\right\| = d_r(x) - R > 0.$$

When $p = 0$, the choice $R = \dfrac{d_r(x)}{e}$ implies the inequality

$$\left|\partial_{x_\ell} f(x)\right| \leqslant e \|f\|_{0,r} \, d_r(x)^{-1};$$

hence, the inequality

$$\left|\partial_{x_\ell} f(x)\right| d_r(x) \leqslant e \|f\|_{0,r} \,. \tag{12.1}$$

When $p > 0$, the choice $R = \dfrac{d_r(x)}{p+1}$ and the relations

$$\left(1 - \frac{1}{p+1}\right)^{-p} = \left(1 + \frac{1}{p}\right)^{p} < e,$$

brings us to the inequalities

$$\begin{aligned}\left|\partial_{x_\ell} f(x)\right| &\leqslant \|f\|_{p,r}\, d_r(x)^{-p-1}(p+1)\left(1-\frac{1}{p+1}\right)^{-p}\\ &\leqslant e(p+1)\,\|f\|_{p,r}\, d_r(x)^{-p-1}\end{aligned}$$

and then to the inequality

$$\left|\partial_{x_\ell} f(x)\right| d_r(x)^{p+1} \leqslant e(p+1)\,\|f\|_{p,r}\,. \tag{12.2}$$

Property 5 follows since inequalities (12.1) and (12.2) remain valid when $\|x\| = r$.

We are left to prove Property 6. Let $\ell \in \{1, \dots, n\}$ be, $x \in \mathbb{C}^n$ such that $\|x\| < r$, and $\alpha_{x,\ell}$ the nonnegative real number defined by

$$\alpha_{x,\ell} = \begin{cases} 0 & \text{if } n = 1 \\ \max(|x_1|, \dots |x_{\ell-1}|, |x_{\ell+1}|, \dots, |x_n|) & \text{if } n \geqslant 2 \end{cases}.$$

Notice that $\|x\| = \max(\alpha_{x,\ell}, |x_\ell|)$ and, consequently, $\alpha_{x,\ell} = \|x\|$ if $|x_\ell| \leqslant \alpha_{x,\ell}$ and $\alpha_{x,\ell} < \|x\| = |x_\ell|$ otherwise. Furthermore, we have $r - \alpha_{x,\ell} \geqslant d_r(x)$. Applying now Property 2, we get

$$\left|\partial_{x_\ell}^{-1} f(x)\right| = \left|\int_0^{x_\ell} f(x_1, \dots, x_{\ell-1}, t, x_{\ell+1}, \dots, x_n)dt\right| \leqslant \|f\|_{p,r}\, I_{\ell,p} \tag{12.3}$$

for all $p \geqslant 0$, where

$$I_{\ell,p} = \int_0^{|x_\ell|} \frac{du}{(r - \max(\alpha_{x,\ell}, u))^p}$$

with

$$\max(\alpha_{x,\ell}, u) = \begin{cases} \alpha_{x,\ell} = \|x\| \;\text{ for all } u \in [0, |x_\ell|] & \text{if } |x_\ell| \leqslant \alpha_{x,\ell} \\ \begin{cases} \alpha_{x,\ell} & \text{for all } u \in [0, \alpha_{x,\ell}] \\ u & \text{for all } u \in [\alpha_{x,\ell}, |x_\ell|] \end{cases} & \text{if } |x_\ell| > \alpha_{x,\ell} \end{cases}. \tag{12.4}$$

This brings then us to the following discussion.

- *Case* $p = 0$. Due to inequality (12.3) above, we straightaway have

$$\begin{aligned}\left|\partial_{x_\ell}^{-1} f(x)\right| &\leqslant \|f\|_{0,r}\, I_{\ell,0} = \|f\|_{0,r} \int_0^{|x_\ell|} du = |x_\ell|\, \|f\|_{0,r} \leqslant \|x\|\, \|f\|_{0,r}\\ &\leqslant r\, \|f\|_{0,r}\,;\end{aligned}$$

hence, the inequality

$$\left|\partial_{x_\ell}^{-1} f(x)\right| d_r(x)^0 \leqslant r \, \|f\|_{0,r} \, . \tag{12.5}$$

- *Case* $p = 1$. Thanks to relations (12.4) and inequality $\ln(t) \leqslant t$ for all $t > 0$, we first have

$$I_{\ell,1} = \int_0^{|x_\ell|} \frac{du}{r - \|x\|} = \frac{|x_\ell|}{d_r(x)} \leqslant \frac{\|x\|}{d_r(x)} \leqslant \frac{r}{d_r(x)}$$

when $|x_\ell| \leqslant \alpha_{x,\ell}$, and

$$\begin{aligned} I_{\ell,1} = \int_0^{\alpha_{x,\ell}} \frac{du}{r - \alpha_{x,\ell}} + \int_{\alpha_{x,\ell}}^{|x_\ell|} \frac{du}{r - u} &= \frac{\alpha_{x,\ell}}{r - \alpha_{x,\ell}} + \ln\left(\frac{r - \alpha_{x,\ell}}{r - |x_\ell|}\right) \\ &\leqslant \frac{\alpha_{x,\ell}}{r - \|x\|} + \frac{r - \alpha_{x,\ell}}{r - \|x\|} = \frac{r}{d_r(x)} \end{aligned}$$

when $|x_\ell| > \alpha_{x,\ell}$. Applying then inequality (12.3), we finally get

$$\left|\partial_{x_\ell}^{-1} f(x)\right| d_r(x) \leqslant r \, \|f\|_{1,r} \, . \tag{12.6}$$

- *Case* $p \geqslant 2$. Applying first relations (12.4), we get

$$I_{\ell,p} = \int_0^{|x_\ell|} \frac{du}{(r - \|x\|)^p} = \frac{|x_\ell|}{d_r(x)^p} \leqslant \frac{\|x\|}{d_r(x)^p} \leqslant \frac{r}{d_r(x)^p}$$

when $|x_\ell| \leqslant \alpha_{x,\ell}$, and

$$\begin{aligned} I_{\ell,p} &= \int_0^{\alpha_{x,\ell}} \frac{du}{(r - \alpha_{x,\ell})^p} + \int_{\alpha_{x,\ell}}^{|x_\ell|} \frac{du}{(r - u)^p} \\ &= \frac{\alpha_{x,\ell}}{(r - \alpha_{x,\ell})^p} + \frac{1}{(p-1)(r - |x_\ell|)^{p-1}} - \frac{1}{(p-1)(r - \alpha_{x,\ell})^{p-1}} \\ &\leqslant \frac{\|x\|}{d_r(x)^p} + \frac{1}{d_r(x)^{p-1}} = \frac{r}{d_r(x)^p} \end{aligned}$$

when $|x_\ell| > \alpha_{x,\ell}$. Inequality (12.3) implies then

$$\left|\partial_{x_\ell}^{-1} f(x) d_r(x)^p\right| \leqslant r \, \|f\|_{p,r} \, . \tag{12.7}$$

Property 6 follows since inequalities (12.5), (12.6), and (12.7) remain valid when $\|x\| = r$. This achieves the proof of Proposition 12.2. ∎

Remark 12.3 Inequalities 4–6 of Proposition 12.2 are the most important properties. Observe besides that the same index r occurs on their both sides, allowing thus to get estimates for the product fg in terms of f and g, for the derivatives $\partial_{x_\ell} f$ for any $\ell \in \{1, \ldots, n\}$ in terms of f, and for the anti-derivatives $\partial_{x_\ell}^{-1} f$ for any $\ell \in \{1, \ldots, n\}$ in terms of f without having to shrink the polydisc $D_{r,\ldots,r}$.

Chapter 13
Generalized Binomial and Multinomial Coefficients

In combinatorial analysis, the binomial coefficients $\binom{n}{m}$ and the multinomial coefficients $\binom{n}{n_1, \ldots, n_q}$ are defined for any nonnegative integers $0 \leqslant m \leqslant n$ and any tuples $(n, n_1, \ldots, n_q)$ of nonnegative integers satisfying $q \geqslant 2$ and $n_1 + \ldots + n_q = n$ by the relations

$$\binom{n}{m} = \frac{n!}{m!(n-m)!} \quad \text{and} \quad \binom{n}{n_1, \ldots, n_q} = \frac{n!}{n_1! \ldots n_q!}.$$

They respectively denote the number of ways of choosing m objects from a collection of n distinct objects without regard to order, and the number of ways of putting $n = n_1 + \ldots + n_q$ different objects into q different boxes with n_i in the i-th box for all $i = 1, \ldots, q$.

Using the fact that $n! = \Gamma(1+n)$ for any integer $n \geqslant 0$, one can easily extend the definitions of these coefficients to the case where their terms are no longer necessarily integers by setting

$$\binom{a}{b} = \frac{\Gamma(1+a)}{\Gamma(1+b)\Gamma(1+a-b)} \tag{13.1}$$

for any nonnegative real numbers $0 \leqslant b \leqslant a$ and

$$\binom{a}{a_1, \ldots, a_q} = \frac{\Gamma(1+a)}{\Gamma(1+a_1) \ldots \Gamma(1+a_q)} = \frac{\Gamma(1+a)}{\prod_{i=1}^{q} \Gamma(1+a_i)} \tag{13.2}$$

P. Remy, *Asymptotic Expansions and Summability*, Lecture Notes in Mathematics 2351, https://doi.org/10.1007/978-3-031-59094-8_13

for any tuples $(a, a_1, \ldots, a_q)$ of nonnegative real numbers satisfying $q \geqslant 2$ and $a_1 + \ldots + a_q = a$. Observe that all these coefficients are positive. Observe also that one has the following decomposition

$$\binom{a}{a_1, \ldots, a_q} = \prod_{i=2}^{q} \binom{a_1 + \ldots + a_i}{a_1 + \ldots + a_{i-1}}. \tag{13.3}$$

The five propositions below extend to the generalized binomial coefficients (13.1) and the generalized multinomial coefficients (13.2) some well-known results in combinatorial analysis.

Proposition 13.1 (Pascal Formula) *Let $0 \leqslant b \leqslant a$ be two nonnegative real numbers and $1 \leqslant m \leqslant n$ two nonnegative integers. Then,*

$$\binom{a+n+1}{b+m} = \binom{a+n}{b+m} + \binom{a+n}{b+m-1}. \tag{13.4}$$

Proof We compute:

$$\begin{aligned}
\binom{a+n}{b+m} + \binom{a+n}{b+m-1} &= \frac{\Gamma(1+a+n)}{\Gamma(1+b+m)\Gamma(1+a-b+n-m)} \\
&\quad + \frac{\Gamma(1+a+n)}{\Gamma(1+b+m-1)\Gamma(1+a-b+n-m+1)} \\
&= \frac{(a-b+n-m+1)\Gamma(1+a+n)}{\Gamma(1+b+m)\Gamma(1+a-b+n-m+1)} \\
&\quad + \frac{(b+m)\Gamma(1+a+n)}{\Gamma(1+b+m)\Gamma(1+a-b+n-m+1)} \\
&= \frac{(a+n+1)\Gamma(1+a+n)}{\Gamma(1+b+m)\Gamma(1+a-b+n-m+1)} \\
&= \frac{\Gamma(1+a+n+1)}{\Gamma(1+b+m)\Gamma(1+a-b+n-m+1)} \\
&= \binom{a+n+1}{b+m};
\end{aligned}$$

hence, the identity (13.4). ■

Proposition 13.2 (Vandermonde Inequality)

1. *(Binomial case) Let $0 \leqslant b \leqslant a$ be two nonnegative real numbers and $0 \leqslant m \leqslant n$ two nonnegative integers. Then,*

$$\binom{a+n}{b+m} \geqslant \binom{a}{b}\binom{n}{m}. \tag{13.5}$$

2. *(Multinomial case) Let $q \geqslant 2$ be an integer, $(a, a_1, \ldots, a_q)$ a tuple of nonnegative real numbers and $(n, n_1, \ldots, n_q)$ a tuple of nonnegative integers such that $a_1 + \ldots + a_q = a$ and $n_1 + \ldots + n_q = n$. Then,*

$$\binom{a+n}{a_1+n_1, \ldots, a_q+n_q} \geqslant \binom{a}{a_1, \ldots, a_q}\binom{n}{n_1, \ldots, n_q}. \tag{13.6}$$

Proof ⋆ *First point.* The inequality (13.5) is clear for $n = m = 0$. Let us now fix $0 \leqslant b \leqslant a$ and let us prove by induction on $n \geqslant 1$ the property

$$(\mathcal{P}_n) : \forall m \in \{0, \ldots, n\}, \binom{a+n}{b+m} \geqslant \binom{a}{b}\binom{n}{m}.$$

A direct calculation gives us the property $(\mathcal{P}_1)$:

$$\binom{a+1}{b} = \frac{\Gamma(1+a+1)}{\Gamma(1+b)\Gamma(1+a+1-b)} = \frac{a+1}{a+1-b}\binom{a}{b} \geqslant \binom{a}{b} = \binom{a}{b}\binom{1}{0},$$

$$\binom{a+1}{b+1} = \frac{\Gamma(1+a+1)}{\Gamma(1+b+1)\Gamma(1+a-b)} = \frac{a+1}{b+1}\binom{a}{b} \geqslant \binom{a}{b} = \binom{a}{b}\binom{1}{1}.$$

Assuming now the property $(\mathcal{P}_n)$ for a certain $n \geqslant 1$, let us prove the property $(\mathcal{P}_{n+1})$. As for the property $(\mathcal{P}_1)$, the sought inequality stems from a direct calculation when $m = 0$ and $m = n+1$:

$$\begin{aligned}\binom{a+n+1}{b} = \frac{\Gamma(1+a+n+1)}{\Gamma(1+b)\Gamma(1+a+n+1-b)} &= \left(\prod_{k=1}^{n+1} \frac{a+k}{a+k-b}\right)\binom{a}{b} \\ &\geqslant \binom{a}{b} = \binom{a}{b}\binom{n+1}{0},\end{aligned}$$

$$\begin{aligned}\binom{a+n+1}{b+n+1} = \frac{\Gamma(1+a+n+1)}{\Gamma(1+b+n+1)\Gamma(1+a-b)} &= \left(\prod_{k=1}^{n+1} \frac{a+k}{b+k}\right)\binom{a}{b} \\ &\geqslant \binom{a}{b} = \binom{a}{b}\binom{n+1}{n+1}.\end{aligned}$$

When $m \in \{1, \dots, n\}$, it stems from Proposition 13.1 and the property $(\mathcal{P}_n)$ as follows:

$$\begin{aligned}\binom{a+n+1}{b+m} = \binom{a+n}{b+m} + \binom{a+n}{b+m-1} &\geqslant \binom{a}{b}\binom{n}{m} + \binom{a}{b}\binom{n}{m-1} \\ &= \binom{a}{b}\left(\binom{n}{m} + \binom{n}{m-1}\right) \\ &= \binom{a}{b}\binom{n+1}{m}.\end{aligned}$$

This ends the induction and achieves then the proof of the first point of Proposition 13.2.

⋆ *Second point.* Let us apply the relation (13.3) and the inequality (13.5) to each factor of the product. We get

$$\begin{aligned}\binom{a+n}{a_1+n_1, \dots, a_q+n_q} &= \prod_{i=2}^{q} \binom{a_1+\dots+a_i+n_1+\dots+n_i}{a_1+\dots+a_{i-1}+n_1+\dots+n_{i-1}} \\ &\geqslant \prod_{i=2}^{q} \binom{a_1+\dots+a_i}{a_1+\dots+a_{i-1}}\binom{n_1+\dots+n_i}{n_1+\dots+n_{i-1}} \\ &= \left(\prod_{i=2}^{q} \binom{a_1+\dots+a_i}{a_1+\dots+a_{i-1}}\right)\left(\prod_{i=2}^{q} \binom{n_1+\dots+n_i}{n_1+\dots+n_{i-1}}\right).\end{aligned}$$

The inequality (13.6) follows then by applying again the relation (13.3), which ends the proof of the second point of Proposition 13.2. ■

Remark 13.3 When all the terms a, b, n and m are nonnegative integers, inequality (13.5) is also a direct consequence of the Chu-Vandermonde Identity

$$\binom{a+n}{b+m} = \sum_{k+\ell=b+m} \binom{a}{k}\binom{n}{\ell}, \tag{13.7}$$

since all the coefficients in the sum are nonnegative. However, this proof fails in our general case where a and b are no longer integers, since some terms in the sum of (13.7) may now be negative.

Proposition 13.4 (Variations of the Binomial Coefficients)

1. *Let* $b \geqslant 0$. *Then, the function* $\mathrm{Bin}_b : a \in [b, +\infty[\longmapsto \binom{a}{b}$ *is increasing on* $[b, +\infty[$.

2. *Let* $a > 0$. *Then, the function* $\mathrm{bin}_a : b \in [0, a] \longmapsto \binom{a}{b}$ *is increasing on* $\left[0, \frac{a}{2}\right]$ *and decreasing on* $\left[\frac{a}{2}, a\right]$. *In particular,* $\mathrm{bin}_a\,(a/2)$ *is the maximum of* bin_a *on* $[0, a]$.

Proof The derivatives of Bin_b and bin_a are respectively defined for all $a \in [b, +\infty[$ and all $b \in [0, a]$ by the identities

$$\mathrm{Bin}'_b(a) = \binom{a}{b}(\Psi(1+a) - \Psi(1+a-b)) \quad \text{and}$$

$$\mathrm{bin}'_a(b) = \binom{a}{b}(\Psi(1+a-b) - \Psi(1+b)),$$

where $\Psi = \Gamma'/\Gamma$ is the Psi (or Digamma) function. Since the function $\ln \Gamma$ is convex on $]0, +\infty[$, then the function Ψ is increasing on $]0, +\infty[$ and Proposition 13.4 follows from Lagrange Theorem. ∎

Proposition 13.5 (Bounds of the Binomial Coefficients) *Let* $0 \leqslant b \leqslant a$ *and* A *a nonnegative integer greater than* $a/2$. *Then,*

$$1 \leqslant \binom{a}{b} \leqslant 4^A. \tag{13.8}$$

Proof The first inequality of (13.8) is straightforward from the first point of Proposition 13.4:

$$\binom{a}{b} \geqslant \binom{b}{b} = 1.$$

To prove the second inequality of (13.8), we successively apply the first point and the second point of Proposition 13.4, then the classical bound on the binomial coefficients with nonnegative integer terms:

$$\binom{a}{b} \leqslant \binom{2A}{b} \leqslant \binom{2A}{A} \leqslant 2^{2A} = 4^A.$$

This achieves the proof of Proposition 13.5. ∎

Proposition 13.6 (Sum of the Inverses of Binomial and Multinomial Coefficients) *Let* $s > 0$ *be and let us set* $C_s = s'(2 + \Gamma(ss'))$, *where* s' *is the positive integer* $\geqslant 1$ *defined by*

$$s' = \begin{cases} 1 & \text{if } s \geqslant 1 \\ \left\lfloor \frac{1}{s} \right\rfloor + 1 & \text{if } s < 1 \end{cases},$$

where $\lfloor a \rfloor$ *stands for the floor of* $a \in \mathbb{R}$.

1. *(Binomial case) The following inequality holds for all integers $n \geqslant 0$:*

$$\sum_{m=0}^{n} \frac{1}{\binom{sn}{sm}} \leqslant C_s. \tag{13.9}$$

2. *(Multinomial case) The following inequality holds for all integers $q \geqslant 2$ and $n \geqslant 0$:*

$$\sum_{n_1+\ldots+n_q=n} \frac{1}{\binom{sn}{sn_1,\ldots,sn_q}} \leqslant C_s^{q-1}. \tag{13.10}$$

Proof $\star$ *First point.* The inequality (13.9) is straightforward from the first inequality of Proposition 13.5 when $n < 2s'$ since we have the relations

$$\sum_{m=0}^{n} \frac{1}{\binom{sn}{sm}} \leqslant n+1 \leqslant 2s' \leqslant C_s.$$

Let us now assume $n \geqslant 2s'$ and let us write the left hand-side of (13.9) in the form

$$\sum_{m=0}^{n} \frac{1}{\binom{sn}{sm}} = \sum_{m=0}^{s'-1} \frac{1}{\binom{sn}{sm}} + \sum_{m=s'}^{n-s'} \frac{1}{\binom{sn}{sm}} + \sum_{m=n-s'+1}^{n} \frac{1}{\binom{sn}{sm}}.$$

Due to the first inequality of Proposition 13.5, the first and the third sums of the right hand-side are both $\leqslant s'$. Applying next the second point of Proposition 13.4, all the terms $\binom{sn}{sm}$ of the second sum are $\geqslant \binom{sn}{ss'}$. This brings us to the following relations:

$$\begin{aligned}
\sum_{m=0}^{n} \frac{1}{\binom{sn}{sm}} &\leqslant 2s' + \frac{n-2s'+1}{\binom{sn}{ss'}} \\
&= 2s' + \frac{(n-2s'+1)\Gamma(1+ss')\Gamma(1+sn-ss'))}{\Gamma(1+sn)} \\
&= 2s' + s'\Gamma(ss')\frac{n-2s'+1}{n}\frac{\Gamma(1+sn-ss')}{\Gamma(1+sn-1)} \\
&\leqslant 2s' + s'\Gamma(ss')\frac{\Gamma(1+sn-ss')}{\Gamma(1+sn-1)}.
\end{aligned}$$

Inequality (13.9) follows then from the increase of the Gamma function on $[2, +\infty[$. Indeed, the inequality $ss' \geqslant 1$ implies $2 \leqslant 1 + sn - ss' \leqslant 1 + sn - 1$, and thereby $\dfrac{\Gamma(1+sn-ss')}{\Gamma(1+sn-1)} \leqslant 1$. This completes the proof of the first point.

⋆ *Second point.* Applying the relation (13.3) and setting $n'_k = n_1 + ... + n_k$ for all $k = 1, \ldots, q-1$, we first get the identities

$$\begin{aligned}
&\sum_{n_1+...+n_q=n} \frac{1}{\dbinom{sn}{sn_1, \ldots, sn_q}} \\
&= \sum_{n_1+...+n_{q-1}\leqslant n} \frac{1}{\dbinom{sn}{s(n_1+...+n_{q-1})} \cdots \dbinom{s(n_1+n_2)}{sn_1}} \\
&= \sum_{n'_{q-1}=0}^{n} \sum_{n'_{q-2}=0}^{n'_{q-1}} \cdots \sum_{n'_1=0}^{n'_2} \frac{1}{\dbinom{sn}{sn'_{q-1}} \dbinom{sn'_{q-1}}{sn'_{q-2}} \cdots \dbinom{sn'_2}{sn'_1}}.
\end{aligned}$$

Inequality (13.10) stems then from the inequality (13.9) which we apply $q-1$ times, which ends the proof of the second point. ■

Chapter 14
Mittag-Leffler's Function

The Mittag-Leffler function

$$E_\alpha(t) = \sum_{j \geqslant 0} \frac{t^j}{\Gamma(1 + \alpha j)} \quad , \alpha > 0$$

and its generalized form

$$E_{\alpha,\beta}(t) = \sum_{j \geqslant 0} \frac{t^j}{\Gamma(\beta + \alpha j)} \quad , \alpha, \beta > 0,$$

respectively introduced by the Swedish mathematicians M. G. Mittag-Leffler in 1903 [84, 85] and A. Wiman in 1905 [132, 133] in connection with methods of summation of some divergent series, arise naturally in the solution of fractional order integral equations or fractional order differential equations, and especially in the investigations of the fractional generalization of the kinetic equations, random walks, Levy flights, superdiffusive transport and in the study of complex systems. Therefore, they have been the subject of numerous investigations and many properties of these functions have been established.

In this chapter, we recall only the results we need in Chap. 8. For more details on these functions, we refer to the original papers of M. G. Mittag-Leffler [84–86] and A. Wiman [132, 133], and to the review article [39] of H. J. Haubold, A. M. Mathai and R. K. Saxena which compiles all the properties and references about E_α and $E_{\alpha,\beta}$.

Let us start with the following proposition which proves that the Mittag-Leffler functions are entire functions of finite order.

Proposition 14.1 *Let* $\alpha, \beta > 0$*. Then, the function* $E_{\alpha,\beta}$ *is entire on* $\mathbb{C}$ *with a global exponential growth of order at most* $1/\alpha$ *at infinity.*

P. Remy, *Asymptotic Expansions and Summability*, Lecture Notes in Mathematics 2351, https://doi.org/10.1007/978-3-031-59094-8_14

Proof Let us set $u_{\alpha,\beta,j} = 1/\Gamma(\beta+\alpha j)$. From the Stirling Formula

$$\Gamma(\beta+\alpha j) \underset{j\to+\infty}{\sim} \sqrt{2\pi}\,e^{-\alpha j}(\alpha j)^{\alpha j+\beta-1/2}$$

we easily get the limits

$$\lim_{j\to+\infty} u_{\alpha,\beta,j}^{1/j} = 0 \quad \text{and} \quad \lim_{j\to+\infty} \frac{j\ln j}{\ln(1/u_{\alpha,\beta,j})} = \frac{1}{\alpha}.$$

Proposition 14.1 follows then from the Cauchy-Hadamard Theorem and from identity (7.2) page 103. ∎

The following result, which is a direct consequence of the Hankel Formula for the inverse of the Gamma function, provides an important representation of $E_{\alpha,\beta}$ in terms of integral.

Proposition 14.2 *Let $\alpha, \beta > 0$ and $t \in \mathbb{C}$. Then,*

$$E_{\alpha,\beta}(t) = \frac{1}{2i\pi}\int_\gamma e^\tau \frac{\tau^{\alpha-\beta}}{\tau^\alpha - t}d\tau,$$

where γ is a Hankel-type path starting from infinity along the ray $\arg(\tau) = -\pi$, circling the origin in the counterclockwise at a distance greater than $|t|^{1/\alpha}$ and backing at infinity along the ray $\arg(\tau) = \pi$.

This representation, known as the *Mittag-Leffler integral representation*, allows to obtain many properties of the functions $E_{\alpha,\beta}$, specifically their asymptotic expansions.

Proposition 14.3 *Let $\alpha, \beta > 0$.*

1. *Assume $\alpha \in]0,2[$. Then, the following asymptotic expansions hold for all $\mu \in]\pi\alpha/2, \min(\pi,\pi\alpha)[$ and all $J \geqslant 2$:*

   (a) for $|t| \to +\infty$ and $|\arg(t)| \leqslant \mu$:

$$E_{\alpha,\beta}(t) = \frac{1}{\alpha}t^{(1-\beta)/\alpha}\exp\left(t^{1/\alpha}\right) - \sum_{j=1}^{J-1}\frac{1}{\Gamma(\beta-\alpha j)}\frac{1}{t^j} + O\left(\frac{1}{t^J}\right).$$

   (b) for $|t| \to +\infty$ and $\mu \leqslant |\arg(t)| \leqslant \pi$:

$$E_{\alpha,\beta}(t) = -\sum_{j=1}^{J-1}\frac{1}{\Gamma(\beta-\alpha j)}\frac{1}{t^j} + O\left(\frac{1}{t^J}\right).$$

2. *Assume $\alpha \geqslant 2$. Then, the following asymptotic expansion holds for all $J \geqslant 2$. For $|t| \to +\infty$ and $|\arg(t)| \leqslant \pi\alpha/2$:*

$$E_{\alpha,\beta}(t) = \frac{1}{\alpha} t^{(1-\beta)/\alpha} \sum_{\substack{j \geqslant 0 \\ |\arg(t) + 2\pi j| \leqslant \pi\alpha/2}} \exp\left(\frac{2i\pi j(1-\beta)}{\alpha}\right) \exp\left(t^{1/\alpha} \exp\left(\frac{2i\pi j}{\alpha}\right)\right)$$

$$- \sum_{j=1}^{J-1} \frac{1}{\Gamma(\beta - \alpha j)} \frac{1}{t^j} + O\left(\frac{1}{t^J}\right)$$

Note that the asymptotic expansion (1)-(b) implies in particular that the function $E_{\alpha,\beta}$ satisfy condition (2)-(c) of Definition 8.1 when $\alpha \in]0, 2[$.

A first proof of this result was given by M. G. Mittag-Leffler [86] in the case of the function E_α. This was then generalized by A. Wiman [132, 133] and, later, by R. P. Agarwal and P. Humbert [1, 2]. We refer to these various references for the detail of the proof of Proposition 14.3.

The functions $E_{\alpha,\beta}$ also admit many other important properties such as functional and recurrence relations, Mellin-Barnes type integral representations, associated functions, etc. For more details, we refer to [39] and the references therein.

Bibliography

1. Agarwal, R.P.: A propos d'une note de M. Pierre Humbert. C. R. Acad. Sci. Paris **236**(1), 2031–2032 (1953)
2. Agarwal, R.P., Humbert, P.: Sur la fonction de Mittag-Leffler et quelques-unes de ses généralisations. Bull. Sci. Math. **77**(2), 180–185 (1953)
3. Baleanu, D., Diethelm, K., Scalas, E., Trujillo, J.J.: Fractional Calculus. Models and Numerical methods, volume 5 of Series on Complexity, Nonlinearity and Chaos. World Scientific, Hackensack, NJ (2017)
4. Balser, W.: From Divergent Power Series to Analytic Functions, volume 1582 of Lecture Notes in Math. Springer (1994)
5. Balser, W.: Moment methods and formal power series. J. Math. Pures Appl. **76**, 289–305 (1997)
6. Balser, W.: Formal Power Series and Linear Systems of Meromorphic Ordinary Differential Equations. Universitext. Springer, New York (2000)
7. Balser, W.: Multisummability of formal power series solutions of partial differential equations with constant coefficients. J. Differential Equations **201**(1), 63–74 (2004)
8. Balser, W.: Power series and moment summability of finite order. Surikaisekikenkyusho Kokyuroku **1367**, 87–94 (2004)
9. Balser, W., Kostov, V.: Formally well-posed cauchy problems for linear partial differential equations with constant coefficients. In: Analyzable Functions and Applications, volume 373 of Contemp. Math., pp. 87–102. American Mathematical Society, Providence, RI (2005)
10. Balser, W., Loday-Richaud, M.: Summability of solutions of the heat equation with inhomogeneous thermal conductivity in two variables. Adv. Dyn. Syst. Appl. **4**(2), 159–177 (2009)
11. Balser, W., Malek, S.: Formal solutions of the complex heat equation in higher spatial dimensions. In: Global and Asymptotic Analysis of Differential Equations in the Complex Domain, volume 1367 of Kôkyûroku RIMS, pp. 87–94 (2004)
12. Balser, W., Miyake, M.: Summability of formal solutions of certain partial differential equations. Acta Sci. Math. (Szeged) **65**(3–4), 543–551 (1999)
13. Balser, W., Yoshino, M.: Gevrey order of formal power series solutions of inhomogeneous partial differential equations with constant coefficients. Funkcial. Ekvac. **53**, 411–434 (2010)
14. Borel, E.: Mémoire sur les séries divergentes. Ann. Sci. E.N.S. 3e sér **16**, 2–136 (1899)
15. Borel, E.: Leçon sur les séries divergentes, 3rd edn. Gauthier-Villars, Paris (1928)
16. Braaksma, B.L.J.: Multisummability of formal power series solutions of nonlinear meromorphic differential equations. Ann. Inst. Fourier (Grenoble) **42**(3), 517–540 (1992)
17. Braaksma, B.L.J., Faber, B.F.: Multisummability for some classes of difference equations. Ann. Inst. Fourier (Grenoble) **46**(1), 183–217 (1996)

P. Remy, *Asymptotic Expansions and Summability*, Lecture Notes in Mathematics 2351, https://doi.org/10.1007/978-3-031-59094-8

18. Braaksma, B.L.J., Faber, B.F., Immink, G.K.: Summation of formal solutions of a class of linear difference equations. Pacific J. Math. **195**(1), 35–65 (2000)
19. Canalis-Durand, M., Ramis, J.-P., Schäfke, R., Sibuya, Y.: Gevrey solutions of singularly perturbed differential equations. J. Reine Angew. Math. **518**, 95–129 (2000)
20. Carrillo, S.A.: Summability in a monomial for some classes of singularly perturbed partial differential equations. Publ. Mat. **65**(1), 83–127 (2021)
21. Carrillo, S.A., Hurtado, C.A.: Formal P-Gevrey series solutions of first order holomorphic PDEs. In: Formal and Analytic Solutions of Differential Equations, pp. 325–362. World Scientific Publishing Europe (2022)
22. Carrillo, S.A., Lastra, A.: Formal Gevrey solutions: in analytic germs – for higher order holomorphic PDEs. Math. Ann. **386**, 85 (2023)
23. Carrillo, S.A., Mozo-Fernández, J.: Tauberian properties for monomial summability with applications to Pfaffian systems. J. Differential Equations **261**(12), 7237–7255 (2016)
24. Carrillo, S.A., Mozo-Fernández, J.: An extension of Borel-Laplace methods and monomial summability. J. Math. Anal. Appl. **457**(1), 461–477 (2018)
25. Carrillo, S.A., Mozo-Fernández, J., Schäfke, R.: Tauberian theorems for k-summability with respect to an analytic germ. J. Math. Anal. Appl. **489**(2), 124174, 21 pp. (2020)
26. Costin, O., Park, H., Takei, Y.: Borel summability of the heat equation with variable coefficients. J. Differential Equations **252**(4), 3076–3092 (2012)
27. Daalhuis, A.B.O., Olver, F.W.J.: On the asymptotic and numerical solution of linear ordinary differential equations. SIAM Rev. **40**(3), 463–495 (1998)
28. Deeb, A., Hamdouni, A., Liberge, E., Razafindralandy, D.: Borel-Laplace summation method used as time integration scheme. In: Congrès SMAI 2013, volume 45 of ESAIM Proc. Surveys, pp. 318–327. EDP Sci., Les Ulis (2014)
29. Delabaere, E., Rasoamanana, J.-M.: Sommation effective d'une somme de borel par séries de factorielles. Ann. Inst. Fourier (Grenoble) **57**(2), 421–456 (2007)
30. Di Vizio, L., Zhang, C.: On q-summation and confluence. Ann. Inst. Fourier (Grenoble) **59**(1), 347–392 (2009)
31. Écalle, J.: Les fonctions résurgentes, tome I: les algèbres de fonctions résurgentes. Publ. Math. Orsay, 81-05 (1981)
32. Écalle, J.: Les fonctions résurgentes, tome II: les fonctions résurgentes appliquées à l'itération. Publ. Math. Orsay, 81-06 (1981)
33. Écalle, J.: Les fonctions résurgentes, tome III: l'équation du pont et la classification analytique des objets locaux. Publ. Math. Orsay, 85-05 (1985)
34. Gérard, R., Tahara, H.: Formal power series solutions of nonlinear first order partial differential equations. Funkcial. Ekvac. **41**, 133–166 (1998)
35. Gevrey, M.: Sur la nature analytique des solutions des équations aux dérivées partielles. Ann. Sci. Ecole Norm. Sup. **3**(25), 129–190 (1918)
36. Gomoyunov, M.I.: On representation formulas for solutions of linear differential equations with caputo fractional derivatives. Fract. Calc. Appl. Anal. **23**(4), 1141–1160 (2020)
37. Hadamard, J.: Sur les problèmes aux dérivées partielles et leur signification physique. Princeton Univ. Bull. **13**, 49–52 (1902)
38. Hashimoto, Y., Miyake, M.: Newton polygons and Gevrey indices for linear partial differential operators. Nagoya Math. J. **128**, 15–47 (1992)
39. Haubold, H.J., Mathai, A.M., Saxena, R.K.: Mittag-Leffler functions and their applications. J. Appl. Math. **2011**, Article 298628 (2011)
40. Hibino, M.: Borel summability of divergence solutions for singular first-order partial differential equations with variable coefficients. Part I. J. Differential Equations **227**(2), 499–533 (2006)
41. Hibino, M.: Borel summability of divergence solutions for singular first-order partial differential equations with variable coefficients. Part II. J. Differential Equations **227**(2), 534–563 (2006)
42. Hibino, M.: On the summability of divergent power series solutions for certain first-order linear PDEs. Opuscula Math. **35**(5), 595–624 (2015)

43. Hopf, E.: The partial differential equation $u_t + uu_x = \mu u_{xx}$. Comm. Pure Appl. Math. **3**, 201–230 (1950)
44. Ichinobe, K.: On k-summability of formal solutions for a class of partial differential operators with time dependent coefficients. J. Differential Equations **257**(8), 3048–3070 (2014)
45. Ichinobe, K., Miyake, M.: On k-summability of formal solutions for certain partial differential operators with polynomial coefficients. Opusc. Math. **35**, 625–653 (2015)
46. Immink, G.K.: On the summability of the formal solutions of a class of inhomogeneous linear difference equations. Funkcial. Ekvac. **39**, 469–490 (1996)
47. Jiménez-Garrido, J., Sanz, J.: Strongly regular sequences and proximate orders. J. Math. Anal. Appl. **438**(2), 920–945 (2016)
48. Jiménez-Garrido, J., Sanz, J., Schindl, G.: Injectivity and surjectivity of the asymptotic Borel map in Carleman ultraholomorphic classes. J. Math. Anal. Appl. **469**(1), 136–168 (2019)
49. Jiménez-Garrido, J., Sanz, J., Schindl, G.: Sectorial extensions, via Laplace transforms, in ultraholomorphic classes defined by weight functions. Results Math. **74**(1), 27, 44 pp. (2019)
50. Jiménez-Garrido, J., Sanz, J., Schindl, G.: Surjectivity of the asymptotic Borel map in Carleman-Roumieu ultraholomorphic classes defined by regular sequences. Rev. R. Acad. Cienc. Exactas Fís. Nat. Ser. A Math. RACSAM **115**(4), 181, 18 pp. (2021)
51. Jung, F., Naegelé, F., Thomann, J.: An algorithm of multisummation of formal power series solutions of linear ODEs. Math. Comput. Simul. **42**, 409–425 (1996)
52. Kilbas, A., Srivastava, H., Trujillo, J.: Theory and Applications of Fractional Differential Equations, volume 204 of North-Holland Math. Stud. Elsevier, Amsterdam (2006)
53. Kowalevskaya, S.: Zur Theorie der partiellen Differentialgleichungen, J. Reine Angew. Math. **80**, 1–32 (1875)
54. Laplace, P.-S.: Mémoire sur les approximations des formules qui sont fonctions de très grands nombres. In: Oeuvres complètes, volume 10 of Mémoires de l'Académie royale des sciences de Paris, pp. 209–291. Gauthier-Villars et fils, Paris (1894)
55. Lastra, A., Tahara, H.: Maillet type theorem for nonlinear totally characteristic partial differential equations. Math. Ann. **377**, 1603–1641 (2020)
56. Lastra, A., Malek, S., Sanz, J.: Summability in general Carleman ultraholomorphic classes. J. Math. Anal. Appl. **430**(2), 1175–1206 (2015)
57. Lastra, A., Michalik, S., Suwińska, M.: Estimates of formal solutions for some generalized moment partial differential equations. J. Math. Anal. Appl. **500**(1), 125094 (2021)
58. Lastra, A., Michalik, S., Suwińska, M.: Summability of formal solutions for a family of generalized moment integro-differential equations. Fract. Calc. Appl. Anal. **24**(5), 1445–1476 (2021)
59. Lastra, A., Michalik, S., Suwińska, M.: Summability of formal solutions for some generalized moment partial differential equations. Results Math. **76**(1), 22 (2021)
60. Lastra, A., Sanz, J., Sendra, J.R.: On the summability of a class of formal power series. Math. Inequal. Appl. **25**(4), 1101–1121 (2022)
61. Leroy, E.: Sur les séries divergentes et les fonctions définies par un développement de Taylor. Ann. Fac. Université de Toulouse, pp. 317–430 (1900)
62. Levin, B.Ya.: Lectures on Entire Functions, volume 150 of Transl. Math. Monogr. American Mathematical Society, Providence, RI (1996)
63. Loday-Richaud, M.: Stokes phenomenon, multisummability and differential Galois groups. Ann. Inst. Fourier (Grenoble) **44**(3), 849–906 (1994)
64. Loday-Richaud, M.: Rank reduction, normal forms and Stokes matrices. Expo. Math. **19**, 229–250 (2001)
65. Loday-Richaud, M.: Divergent Series, Summability and Resurgence II. Simple and Multiple Summability, volume 2154 of Lecture Notes in Math. Springer (2016)
66. Luo, Z., Chen, H., Zhang, C.: Exponential-type Nagumo norms and summability of formal solutions of singular partial differential equations. Ann. Inst. Fourier (Grenoble) **62**(2), 571–618 (2012)
67. Lutz, D.A., Miyake, M., Schäfke, R.: On the Borel summability of divergent solutions of the heat equation. Nagoya Math. J. **154**, 1–29 (1999)

68. Lysik, G.: Borel summable solutions of the Burgers equation. Ann. Polon. Math. **95**(2), 187–197 (2009)
69. Malek, S.: On the summability of formal solutions of linear partial differential equations. J. Dyn. Control Syst. **11**(3), 389–403 (2005)
70. Malek, S.: On Gevrey asymptotic for some nonlinear integro-differential equations. J. Dyn. Control Syst. **16**(3), 377–406 (2010)
71. Malgrange, B.: Sommation des séries divergentes. Expo. Math. **13**, 163–222 (1995)
72. Malgrange, B., Ramis, J.-P.: Fonctions multisommables. Ann. Inst. Fourier (Grenoble) **42**, 353–368 (1992)
73. Marotte, F., Zhang, C.: Multisommabilité des séries entières solutions formelles d'une équation aux q-différences linéaire analytique. Ann. Inst. Fourier (Grenoble) **50**(6), 1859–1890 (2000)
74. Martinet, J., Ramis, J.-P.: Théorie de Galois différentielle et resommation. In: Computer Algebra and Differential Equations, pp. 117–214. Academic Press (1989)
75. Martinet, J., Ramis, J.-P.: Elementary acceleration and multisummability. Ann. Inst. H. Poincaré Phys. Théor. **54**(4), 331–401 (1991)
76. Michalik, S.: Summability of divergent solutions of the n-dimensional heat equation. J. Differential Equations **229**, 353–366 (2006)
77. Michalik, S.: Summability of formal solutions to the n-dimensional inhomogeneous heat equation. J. Math. Anal. Appl. **347**, 323–332 (2008)
78. Michalik, S.: Summability and fractional linear partial differential equations. J. Dyn. Control Syst. **16**(4), 557–584 (2010)
79. Michalik, S.: Multisummability of formal solutions of inhomogeneous linear partial differential equations with constant coefficients. J. Dyn. Control Syst. **18**(1), 103–133 (2012)
80. Michalik, S.: Analytic solutions of moment partial differential equations with constant coefficients. Funkcial Ekvac. **56**, 19–50 (2013)
81. Michalik, S.: Analytic summable solutions of inhomogeneous moment partial differential equations. Funkcial Ekvac. **60**, 325–351 (2017)
82. Michalik, S.: Summable solutions of the Goursat problem for some partial differential equations with constant coefficients. J. Differential Equations **304**, 435–366 (2021)
83. Michalik, S., Suwińska, M.: Gevrey estimates for certain moment partial differential equations. In: Complex Differential and Difference Equations, pp. 391–408. De Gruyter Proc. Math. (2020)
84. Mittag-Leffler, M.G.: Sur la nouvelle fonction $E_\alpha(x)$. C. R. Acad. Sci. Paris **137**(15), 554–558 (1903)
85. Mittag-Leffler, M.G.: Une généralisation de l'intégrale de Laplace-Abel. C. R. Acad. Sci. Paris **136**(9), 537–539 (1903)
86. Mittag-Leffler, M.G.: Sur la représentation analytique d'une branche uniforme d'une fonction monogène (cinquième note). Acta Math. **29**, 101–181 (1905)
87. Miyake, M.: Newton polygons and formal Gevrey indices in the Cauchy-Goursat-Fuchs type equations. J. Math. Soc. Japan **43**(2), 305–330 (1991)
88. Miyake, M.: Borel summability of divergent solutions of the Cauchy problem to non-Kovaleskian equations. In: Partial Differential Equations and Their Applications (Wuhan, 1999), pp. 225–239. World Sci. Publ., River Edge, NJ (1999)
89. Miyake, M., Shirai, A.: Convergence of formal solutions of first order singular nonlinear partial differential equations in the complex domain. Ann. Polon. Math. **74**, 215–228 (2000)
90. Miyake, M., Shirai, A.: Structure of formal solutions of nonlinear first order singular partial differential equations in complex domain. Funkcial. Ekvac. **48**, 113–136 (2005)
91. Miyake, M., Shirai, A.: Two proofs for the convergence of formal solutions of singular first order nonlinear partial differential equations in complex domain. Surikaiseki Kenkyujo Kokyuroku Bessatsu, Kyoto Univ. **B37**, 137–151 (2013)
92. Mozo-Fernández, J., Schäfke, R.: Asymptotic expansions and summability with respect to an analytic germ. Publ. Mat. **63**(1), 3–79 (2019)

93. Naegelé, F., Thomann, J.: Algorithmic approach of the multisummation of formal power series solutions of linear ODE applied to the Stokes phenomena. In: The Stokes Phenomenon and Hilbert's 16th Problem, Groningen 1995, pp. 197–213. World Scientific Publishing, River Edge, NJ (1996)
94. Nagumo, M.: Über das Anfangswertproblem partieller Differentialgleichungen. Jap. J. Math. **18**, 41–47 (1942)
95. Nevanlinna, F.: Zur Theorie der Asymptotischen Potenzreihen. Ann. Acad. Sci. Fennicae Ser. A **XII**, 1–81 (1919)
96. Ouchi, S.: Multisummability of formal solutions of some linear partial differential equations. J. Differential Equations **185**(2), 513–549 (2002)
97. Ouchi, S.: Borel summability of formal solutions of some first order singular partial differential equations and normal forms of vector fields. J. Math. Soc. Japan **57**(2), 415–460 (2005)
98. Pliś, M.E., Ziemian, B.: Borel resummation of formal solutions to nonlinear Laplace equations in 2 variables. Ann. Polon. Math. **67**(1), 31–41 (1997)
99. Ramis, J.-P.: Dévissage Gevrey. Astérisque, Soc. Math. France, Paris **59–60**, 173–204 (1978)
100. Ramis, J.-P.: Les séries k-sommables et leurs applications. In: Complex Analysis, Microlocal Calculus and Relativistic Quantum Theory (Proc. Internat. Colloq., Centre Phys., Les Houches, 1979), volume 126 of Lecture Notes in Phys., pp. 178–199. Springer, Berlin (1980)
101. Ramis, J.-P.: Théorèmes d'indices Gevrey pour les équations différentielles ordinaires. Mem. Amer. Math. Soc. **48**, viii+95 (1984)
102. Ramis, J.-P., Sibuya, Y.: Hukuhara domains and fundamental existence and uniqueness theorems for asymptotic solutions of Gevrey type. Asymptotic Anal. **2**(1), 39–94 (1989)
103. Remy, P.: Gevrey order and summability of formal series solutions of some classes of inhomogeneous linear partial differential equations with variable coefficients. J. Dyn. Control Syst. **22**(4), 693–711 (2016)
104. Remy, P.: Gevrey order and summability of formal series solutions of certain classes of inhomogeneous linear integro-differential equations with variable coefficients. J. Dyn. Control Syst. **23**(4), 853–878 (2017)
105. Remy, P.: Gevrey properties and summability of formal power series solutions of some inhomogeneous linear Cauchy-Goursat problems. J. Dyn. Control Syst. **26**(1), 69–108 (2020)
106. Remy, P.: Gevrey regularity of the solutions of the inhomogeneous partial differential equations with a polynomial semilinearity. Rev. R. Acad. Cienc. Exactas Fís. Nat. Ser. A Math. RACSAM **115**(3), 145 (2021)
107. Remy, P.: On the summability of the solutions of the inhomogeneous heat equation with a power-law nonlinearity and variable coefficients. J. Math. Anal. Appl. **494**(2), 124656 (2021)
108. Remy, P.: Summability of the formal power series solutions of a certain class of inhomogeneous partial differential equations with a polynomial semilinearity and variable coefficients. Results Math. **76**(3), 118 (2021)
109. Remy, P.: Summability of the formal power series solutions of a certain class of inhomogeneous nonlinear partial differential equations with a single level. J. Differential Equations **313**, 450–502 (2022)
110. Remy, P.: Gevrey regularity and summability of the formal power series solutions of the inhomogeneous generalized Boussinesq equations. Asymptot. Anal. **131**(1), 1–32 (2023)
111. Remy, P.: Gevrey regularity of the solutions of some inhomogeneous nonlinear partial differential equations. Electron. J. Differential Equations **2023**(6), 1–28 (2023)
112. Sanz, J.: Flat functions in Carleman ultraholomorphic classes via proximate orders. J. Math. Anal. Appl. **415**(2), 623–643 (2014)
113. Sanz, J.: Asymptotic analysis and summability of formal power series. In: Analytic, Algebraic and Geometric Aspects of Differential Equations, Trends Math., pp. 199–262. Birkhäuser/Springer, Cham (2017)
114. Shirai, A.: Maillet type theorem for nonlinear partial differential equations and Newton polygons. J. Math. Soc. Japan **53**, 565–587 (2001)

115. Shirai, A.: Convergence of formal solutions of singular first order nonlinear partial differential equations of totally characteristic type. Funkcial. Ekvac. **45**, 187–208 (2002)
116. Shirai, A.: A Maillet type theorem for first order singular nonlinear partial differential equations. Publ. RIMS. Kyoto Univ. **39**, 275–296 (2003)
117. Shirai, A.: Maillet type theorem for singular first order nonlinear partial differential equations of totally characteristic type. Surikaiseki Kenkyujo Kokyuroku, Kyoto Univ. **1431**, 94–106 (2005)
118. Shirai, A.: Alternative proof for the convergence of formal solutions of singular first order nonlinear partial differential equations. J. School Educ. Sugiyama Jogakuen Univ. **1**, 91–102 (2008)
119. Shirai, A.: Gevrey order of formal solutions of singular first order nonlinear partial differential equations of totally characteristic type. J. School Educ. Sugiyama Jogakuen Univ. **6**, 159–172 (2013)
120. Shirai, A.: Maillet type theorem for singular first order nonlinear partial differential equations of totally characteristic type, part II. Opuscula Math. **35**(5), 689–712 (2015)
121. Slater, L.J.: Generalized Hypergeometric Functions. Cambridge University Press, Cambridge (1966)
122. Suwińka, M.: Gevrey estimates of formal solutions of certain moment partial differential equations with variable coefficients. J. Dyn. Control Syst. **27**, 355–370 (2021)
123. Suwińka, M.: Summability of formal solutions for a family of linear moment integro-differential equations. In: Recent Trends in Formal and Analytic Solutions of Differential Equations, Contemp. Math., pp. 167–192. American Mathematical Society, Providence, RI (2023)
124. Tahara, H.: Gevrey regularity in time of solutions to nonlinear partial differential equations. J. Math. Sci. Univ. Tokyo **18**, 67–137 (2011)
125. Tahara, H., Yamazawa, H.: Multisummability of formal solutions to the Cauchy problem for some linear partial differential equations. J. Differential Equations **255**(10), 3592–3637 (2013)
126. Thilliez, V.: Division by flat ultradifferentiable functions and sectorial extensions. Results Math. **44**(1–2), 169–188 (2003)
127. Thomann, J.: Resommation des séries formelles. Numer. Math. **58**(1), 503–535 (1990)
128. Thomann, J.: Procédés formels et numériques de sommation de séries solutions d'équations différentielles. Expo. Math. **13**, 223–246 (1995)
129. van der Hoeven, J.: Efficient accelero-summation of holomonic functions. J. Symbolic Comput. **42**(4), 389–428 (2007)
130. Walter, W.: An elementary proof of the Cauchy-Kowalevsky theorem. Amer. Math. Mon. **92**(2), 115–126 (1985)
131. Watson, G.N.: A theory of asymptotic series. Philos. Trans. R. Soc. Lond. Ser. A Math. Phys. Eng. Sci. **CCXI**, 279–313 (1911)
132. Wiman, A.: Über den fundamentalsatz in der theorie der funktionen $E_\alpha(x)$. Acta Math. **29**, 191–201 (1905)
133. Wiman, A.: Über die nullstellen der funktionen $E_\alpha(x)$. Acta Math. **29**, 217–234 (1905)
134. Yamazawa, H.: On multisummability of formal solutions with logarithm terms for some linear partial differential equations. Funkcial. Ekvac. **60**(3), 371–406 (2017)
135. Yonemura, A.: Newton polygons and formal Gevrey classes. Publ. Res. Inst. Math. Sci. **26**, 197–204 (1990)

Index

P. Remy, *Asymptotic Expansions and Summability*, Lecture Notes in Mathematics 2351, https://doi.org/10.1007/978-3-031-59094-8

Editorial Policy

1. Lecture Notes aim to report new developments in all areas of mathematics and their applications – quickly, informally and at a high level. Mathematical texts analysing new developments in modelling and numerical simulation are welcome.

   Manuscripts should be reasonably self-contained and rounded off. Thus they may, and often will, present not only results of the author but also related work by other people. They may be based on specialised lecture courses. Furthermore, the manuscripts should provide sufficient motivation, examples and applications. This clearly distinguishes Lecture Notes from journal articles or technical reports which normally are very concise. Articles intended for a journal but too long to be accepted by most journals, usually do not have this "lecture notes" character. For similar reasons it is unusual for doctoral theses to be accepted for the Lecture Notes series, though habilitation theses may be appropriate.

2. Besides monographs, multi-author manuscripts resulting from SUMMER SCHOOLS or similar INTENSIVE COURSES are welcome, provided their objective was held to present an active mathematical topic to an audience at the beginning or intermediate graduate level (a list of participants should be provided).

   The resulting manuscript should not be just a collection of course notes, but should require advance planning and coordination among the main lecturers. The subject matter should dictate the structure of the book. This structure should be motivated and explained in a scientific introduction, and the notation, references, index and formulation of results should be, if possible, unified by the editors. Each contribution should have an abstract and an introduction referring to the other contributions. In other words, more preparatory work must go into a multi-authored volume than simply assembling a disparate collection of papers, communicated at the event.

3. Manuscripts should be submitted either online at www.editorialmanager.com/lnm to Springer's mathematics editorial in Heidelberg, or electronically to one of the series editors. Authors should be aware that incomplete or insufficiently close-to-final manuscripts almost always result in longer refereeing times and nevertheless unclear referees' recommendations, making further refereeing of a final draft necessary. The strict minimum amount of material that will be considered should include a detailed outline describing the planned contents of each chapter, a bibliography and several sample chapters. Parallel submission of a manuscript to another publisher while under consideration for LNM is not acceptable and can lead to rejection.

4. In general, **monographs** will be sent out to at least 2 external referees for evaluation.

   A final decision to publish can be made only on the basis of the complete manuscript, however a refereeing process leading to a preliminary decision can be based on a pre-final or incomplete manuscript.

   Volume Editors of **multi-author works** are expected to arrange for the refereeing, to the usual scientific standards, of the individual contributions. If the resulting reports can be forwarded to the LNM Editorial Board, this is very helpful. If no reports are forwarded or if other questions remain unclear in respect of homogeneity etc, the series editors may wish to consult external referees for an overall evaluation of the volume.

5. Manuscripts should in general be submitted in English. Final manuscripts should contain at least 100 pages of mathematical text and should always include

   - a table of contents;
   - an informative introduction, with adequate motivation and perhaps some historical remarks: it should be accessible to a reader not intimately familiar with the topic treated;
   - a subject index: as a rule this is genuinely helpful for the reader.
   - For evaluation purposes, manuscripts should be submitted as pdf files.

6. Careful preparation of the manuscripts will help keep production time short besides ensuring satisfactory appearance of the finished book in print and online. After acceptance of the manuscript authors will be asked to prepare the final LaTeX source files (see LaTeX templates online: https://www.springer.com/gb/authors-editors/book-authors-editors/manuscriptpreparation/5636) plus the corresponding pdf- or zipped ps-file. The LaTeX source files are essential for producing the full-text online version of the book, see http://link.springer.com/bookseries/304 for the existing online volumes of LNM). The technical production of a Lecture Notes volume takes approximately 12 weeks. Additional instructions, if necessary, are available on request from lnm@springer.com.

7. Authors receive a total of 30 free copies of their volume and free access to their book on SpringerLink, but no royalties. They are entitled to a discount of 33.3 % on the price of Springer books purchased for their personal use, if ordering directly from Springer.

8. Commitment to publish is made by a *Publishing Agreement*; contributing authors of multiauthor books are requested to sign a *Consent to Publish form*. Springer-Verlag registers the copyright for each volume. Authors are free to reuse material contained in their LNM volumes in later publications: a brief written (or e-mail) request for formal permission is sufficient.

Addresses:

Professor Jean-Michel Morel, CMLA, École Normale Supérieure de Cachan, France
E-mail: moreljeanmichel@gmail.com

Professor Bernard Teissier, Equipe Géométrie et Dynamique,
Institut de Mathématiques de Jussieu – Paris Rive Gauche, Paris, France
E-mail: bernard.teissier@imj-prg.fr

Springer: Ute McCrory, Mathematics, Heidelberg, Germany,
E-mail: lnm@springer.com

Printed in the United States
by Baker & Taylor Publisher Services